D1287611

A GARDENER'S DICTIONARY OF PLANT NAMES

A. W. SMITH

A GARDENER'S DICTIONARY
OF
PLANT NAMES

A Handbook on the Origin and Meaning of
Some Plant Names

REVISED AND ENLARGED BY

WILLIAM T. STEARN

ISADORE LEIGHTON LUCE SMITH

ST. MARTIN'S PRESS · NEW YORK

For information, write:
St. Martin's Press, Inc.
175 Fifth Avenue, New York, N.Y. 10010

Printed in Great Britain

Library of Congress Catalogue Card Number:
72–79502

First published in this edition in the United
States of America in 1972

Affiliated Publishers:
Macmillan and Company, Limited, London
also at Bombay, Calcutta, Madras, and Melbourne
The Macmillan Company of Canada, Limited,
Toronto

Preface

My husband began work many years ago collecting the material which forms the basis for this present book to satisfy his own curiosity. Later, he hoped that it might eventually become a source of reference and pleasure for all gardeners who, however keen, find themselves wondering at the botanical names attached to their plants. What useful information lies there? Of what interest are the personal names involved? What, indeed, of the gardener's particular pleasure, noted in the seventeenth century by William Cole, 'to have the plants speaking Greek and Latin to him and putting him in mind of stories which otherwise he would never think of'. Familiar names of plants fade at even a short distance in space or time from their current bloom and gardeners may miss the just deserts of their labours by not communicating exactly with each other.

Archibald William Smith, 1899–1962, was a man whom Justice Oliver Wendell Holmes would have acknowledged to have lived 'at all', since he had so thoroughly 'shared the passion and action of his time'. He was born in Spain, spent his childhood in Russia, went home to school and passed from Shrewsbury into the Royal Military Academy at Sandhurst. Early in the first World War he arrived in the trenches of the Western Front with the Gloucestershire Regiment. Recovering from wounds towards the end of the war he trained cadets at Cambridge and went from there to serve on General Denikin's staff in South Russia. He rode with the Cossacks. Rejoining his regiment in India, he fought on the North-West Frontier. Tired, as he said, after six years of killing, he considered the Indian Forest Service but, instead, joined his uncle's firm and went to extract teak from the Burmese forests, using elephants. His first articles for *The Atlantic Monthly*, *Blackwood's*, and *The National Geographic Magazine* were about this experience. Later he wrote several novels, short stories, and his autobiography to the age of twenty-one, *A Captain Departed*.

From Burma he sent seeds to his uncle, Reginald Heber Macaulay, for his famous rock garden at Kirnan, and when on leave he met there many of the great collectors. It was then that the need for a book such as this for the average gardener was discussed and the filing cards began to accumulate.

Although throughout his life he had gardened wherever he could, his first permanent garden was in Ipswich, Massachusetts, where he 'settled' in 1934 on a little hill overlooking the marshes and the sea. By this time

Preface

his personal need for such a book as this *Gardener's Dictionary* was evident and he began to arrange the cards in book form. During the 1938 Munich crisis, however, he was notified that he might be recalled to the Army, and in the summer of 1939 he went to London and the General Staff at the War Office. During the second World War, while employed in M.I.5, he saw much of Africa and the South Pacific. At the end of the war he returned to the United States and helped to start the Conservation Foundation in New York. In Ipswich he worked for the Children's Medical Centre at Boston and began to get his plant name book ready for the publishers. Peter Green, then at the Arnold Arboretum of Harvard University and now at the Royal Botanic Gardens, Kew, encouraged him in his wish to make it useful on both sides of the Atlantic. He died soon after correcting the final proofs of the American edition of the book, which was published in 1963. The book had, however, already proved its worth to him as a gardener's companion.

My husband's intention was always to keep it a book for practising gardeners. I am all the more grateful, therefore, to Dr. W. T. Stearn, Dr. Averil Lysaght and all who have helped in the preparation for publication in Britain of this revised and enlarged edition, that they have striven to retain the personal flavour of the book and in many of their additions and emendations have provided just such touches as he would have enjoyed.

Isadore Leighton Luce Smith

Little Hill,
Ipswich,
Massachusetts.
February 1971

His helmet *now* shall make a hive for bees. . . .

GEORGE PEELE
in *Polyhymnia: A Farewell to Arms*

Contents

Editor's Introduction

IN order to pursue Colonel Smith's aim of providing gardeners with reliable information it has been necessary to go back to the original publication of names, and in this respect we must acknowledge the help and guidance provided by Backer's classic and unrivalled *Verklarend Woordenboek van wetenschappelijke Namen* (1936).

To avoid disappointment the user of this book, as indeed of the *Verklarend Woordenboek*, must bear in mind that the number of generic names is now so great that no book of reasonable size can include more than a small fraction of them; the selection of these must necessarily be arbitrary; the present edition includes most of the genera listed by Boom, Hylander, Sander and Zander as being cultivated in the British Isles, Germany, the Netherlands and Scandinavia and will certainly be helpful elsewhere. Colonel Smith's widow, Isadore Leighton Luce Smith, and Dr. Averil Lysaght, by painstakingly correlating his material with the works of these authors, provided a valuable basis for this present book. The need to keep it within bounds has made it impossible to include documentation, but in many cases of doubt or revision the original publication of names or authentic sources have been consulted. Thus Rumphius himself gave the origin of his name *Quis qualis* in his *Hortus Amboinense* 5:71 (1747). Schott published many names of genera of *Araceae* without explanation of their origin but he later provided etymologies for them in his *Genera Aroidearum* (1857). The derivation of *Anemone* is discussed by Toy in *Rhodora* 1: 40 (1899); unbeknown to him La Gardie had earlier reached the same conclusion as to its Semitic origin. The derivation of *Elaeagnus* (*elaiagnŏs*) is discussed by Thiselton-Dyer in *Journal of Philology* 34: 79 (1915). For biographic details, such as dates and forenames, Pritzel's *Thesaurus Literaturae botanicae*, 2nd ed. (1872), Britten & Boulger's *Biographical Index of British and Irish Botanists*, 2nd ed. (1931), Saccardo's *La Botanica in Italia* (1895–1901) and Barnhart's *Biographical Notes upon Botanists* (1965) have been particularly helpful, but other sources have also been used, for example Viviam's 1922 biography of Cesalpino. The information about Mitchell's dates comes from Dorman and Lewis's paper in the *Virginia Magazine of History* 76: 437–440 (1968). There exist indeed many such relevant articles scattered through learned journals. The publication of *A Gardener's Dictionary of Plant Names*, incorporating notes on plantnames I have gathered during the past forty years, has thus provided an opportunity of making some of their conclusions more readily accessible.

Specific epithets are even more difficult to cover adequately than generic names. The present edition includes, it is hoped, most of the commonly used descriptive epithets as well as some unusual ones but makes no claim whatever to anything approaching comprehensiveness.

It is a pleasure to record the help and interest of Mrs. Smith in the preparation of this monument to her late husband's enthusiasm, and to note that she herself, as Ann Leighton, is known also as the author of *Early English Gardens in New England* (1970). I am also indebted to Dr. Lysaght for some help in inserting accents and in other editorial matters, particularly those relating to Pacific exploration. Mr. Anthony Spalding, who kindly read the proofs of my *Botanical Latin*, has done a like service with like scholarship for this work.

William T. Stearn

Department of Botany,
British Museum (Natural History),
Cromwell Road,
London.
February 1971

BOTANICAL NAMES

An Introduction to Botanical Names

WILLIAM T. STEARN

MANY thousands of plants, among them some 250,000 species of flowering plants and 6,000 of ferns, as well as mosses, liverworts, fungi, lichens and algae beyond estimate, have been given botanical names of Latin form since the Swedish naturalist Carl Linnaeus (1707–1778) instituted modern botanical nomenclature in 1753. Of this huge number, most interest only botanists, who are content to use internationally intelligible botanical names for them. The inhabitants of their native lands have locally intelligible vernacular names only for those specially needing them. Hence relatively few possess or need English vernacular names. Indeed, a single English vernacular name may refer to a diversity of plants, its application depending on the region. The name 'bluebell' is applied to *Campanula rotundifolia* in Scotland, *Endymion non-scriptus* in England; on the American continent it may indicate a species of *Campanula*, *Polemonium*, *Mertensia* or *Penstemon*, all on a regional basis; in Australia it is used for *Wahlenbergia*. Local convenience and tradition determine the use of such names, which may be apt and pleasing and also historically and etymologically interesting, but are limited in utility by geographical and linguistic factors; they can never be other than local substitutes or alternatives for botanical names (see p. 340).

Botanical names such as *Campanula rotundifolia* cut across these local preferences. They are intended to be international, the common property of professional and amateur alike in all countries, and because at their first publication they have to be associated with descriptions based on specimens, they have usually a fixed and precise application. Internationally agreed rules and recommendations forming a periodically revised code control their formation, publication and acceptance.

In the eighteenth century, when the foundations were laid for the present system of naming plants, Latin was still the most widely used international language of science and scholarship. This was not, however, classical Latin but an expanded form of Latin derived from the Latin used for many purposes in the Middle Ages and the Renaissance. Botanical names in Latin form are a legacy from the eighteenth century, derived from much earlier usage. The Latin used by botanists today is very different indeed from that of the Romans because it has to deal with so many plants and with so many structures unknown in

3

classical times. Words such as *corolla* and *anthera* now possess special botanical meanings not recorded in dictionaries of classical Latin. A standard Latin dictionary may in fact be quite misleading when consulted for botanical information.

Classification and the subsequent naming of a plant begin with placing it in its plant **family** (formerly called its natural order). Such a family may contain only one genus, for example *Garryaceae, Ginkgoaceae* and *Eucryphiaceae*, only a few genera, for example *Berberidaceae, Cistaceae* and *Plantaginaceae*, or many genera, for example *Compositae, Labiatae, Leguminosae, Orchidaceae, Rosaceae* and *Umbelliferae*. Most family names consist of the name of one of the genera plus the ending *-aceae* which means 'belonging to, having the nature of'. The few exceptions with descriptive names, such as the *Cruciferae*, literally 'cross-bearers', *Gramineae*, literally 'grasses', and *Umbelliferae*, literally 'umbel-bearers', were the first families to be recognized and have been allowed to retain their time-honoured designations.

The name of a **species** does not include the family name but consists of a generic name, for example *Berberis*, followed by a specific epithet (formerly called a trivial name), for example *vulgaris*, the two words in combination forming a specific name, *Berberis vulgaris*. This suffices to distinguish the species by name from the many thousands of other known species, although for greater precision or to prevent confusion the name of its author is often indicated (for example L. for Linnaeus), by citing this after the name, thus *Berberis vulgaris* L.

Most botanical **generic names** are of Latin form and end in *-a*, *-um*, or *-us*, but only a limited number are genuine Latin names used by the Romans.

Thus, analysis of 200 alphabetically consecutive generic names in *Orchidaceae* showed 132 to be of Greek origin, 48 personal, 12 Latin, 4 obscure, 3 geographical, 1 vernacular.

According to their origin, several classes may be distinguished:

1. Latin plant names used by classical authors, such as Virgil, Columella, Pliny: examples are *Acer* (maple), *Arbutus* (strawberry tree), *Bellis* (daisy), *Cornus* (dogwood), *Hedera* (ivy), *Juniperus* (juniper), *Lilium* (lily), *Pinus* (pine), *Quercus* (oak), *Reseda* (mignonette), *Ruscus* (butcher's broom), *Ruta* (rue), *Taxus* (yew), *Ulex* (gorse).

2. Greek names taken into Latin by classical authors: examples are *Achillea, Acorus, Anemone, Bupleurum, Daphne* (spurge laurel), *Lychnis, Petroselinum, Phalaris, Scilla*. A number of Greek names were derived from Semitic names or from the lost languages of

people inhabiting Greece before the coming of the ancestors of the Greeks, for example *Anemone, Hyacinthus, Hyssopus, Terebinthus,* or even from Assyrian, for example *Mandragora.*

3. Names taken from classical mythology by later authors, but not employed as plant names in classical times: *Acacallis, Achlys, Andromeda, Calypso, Cassiope, Hebe, Nerine* are examples. Linnaeus derived much of his knowledge of classical mythology from the *Fabulae* of Hyginus.

4. Names of medieval origin: examples are *Alchemilla, Aquilegia, Borago, Calendula, Carlina, Doronicum, Filago, Filipendula, Linaria, Tanacetum.*

5. Modern names formed from Latin words: examples are *Arundinaria, Cimicifuga, Convallaria, Digitalis, Fissidens, Mobilabium, Pennisetum, Setaria.*

6. Modern names formed from Greek words: such as *Agrostemma, Chionodoxa, Chrysophyllum, Dimorphotheca, Echinochloa, Echinodorus, Enarthrocarpus, Eremurus, Galanthus, Leontodon, Meconopsis, Mesembryanthemum, Odontoglossum, Osmanthus, Pychnanthemum, Tetragonotheca, Trichocentrum.*

7. Modern names formed from personal names: examples are *Adansonia, Albizia* (after Albizzi), *Butea, Camellia, Carludovica* (after King Carlos and Queen Luisa), *Carmichaelia, Carnegiea, Cattleya, Damrongia, Dorotheanthus, Fuchsia, Kitaibela, Linnaea, Robinia, Sequoia, Tradescantia, Warscewiczella, Welwitschia, Yoania, Zaluzianskya.*

8. Modern names formed from vernacular names: for example *Ailanthus, Akebia, Amelanchier, Angraecum, Antiaris, Armeria, Bambusa, Butia, Camassia, Cantua, Caragana, Catalpa, Catha, Corokia, Cortaderia, Datura, Musa, Pandanus, Puya, Rorippa, Sasa, Sorghum, Swida, Tacca, Wattakaka.*

9. Modern names formed from geographical names: examples are *Arisanorchis, Araucaria, Austroamericium, Bogoria, Caribea, Cathaya, Cotopaxia, Domingoa, Hammarbya, Indopoa, Jacaima* (an anagram of Jamaica), *Librevillea, Lobivia* (an anagram of Bolivia), *Papuodendron, Patagonula, Salweenia, Tacazzea, Taiwania, Utahia.*

10. Anagrams made by altering the arrangement of letters in the names of a related genus: *Docynia* (from Cydonia), *Ivodea* (from Evodia), *Microtoena* (from Craniotome), *Nosema* (from Mesona), *Poacynum* (from Apocynum), *Tellima* (from Mitella), and *Gifola, Ifloga, Lifago, Logfia, Oglifa* (all from Filago). Such names excited the wrath of the learned Rev. R. T. Lowe who in his *A manual of the Flora of Madeira* (1868) earnestly invoked all botanists to unite in utterly repudiating and putting down such scandalously childish,

5

bald and witless trickery with names as the anagrammatic formations by Cassini . . . such unseemly barbarisms . . . such base name-coinage'—but all to no effect!

11. Modern names of no evident meaning, their authors having left them unexplained, perhaps wisely: *Aa, Asystasia, Ate, Iliamna, Ipsea, Liatris, Lonas, Ratibida, Venidium.*

For the naming of **intergeneric hybrids,** that is those with parents belonging to two or more different genera, special group names are formed by combining parts of the names of the parent genera into a single word not exceeding eight syllables, for example *Mahoberberis* from *Mahonia* and *Berberis,* or by adding the termination *-ara* to the name of a person connected with the group, for example *Burrageara* for orchid hybrids with *Cochlioda, Miltonia, Odontoglossum* and *Oncidium* in their parentage. Such names are regarded as condensed formulae or as substitutes for formulae. Some examples are:

> *Amelosorbus (Amelanchier × Sorbus), Brassocattleya (Brassia × Cattleya), Chionoscilla (Chionodoxa × Scilla), Cupressocyparis (Chamaecyparis × Cupressus), Crataemespilus (Crataegus × Mespilus), Fatshedera (Fatsia × Hedera), Heucherella (Heuchera × Tiarella), Osmarea (Osmanthus × Phillyrea), Philageria (Lapageria × Philesia), Ruttyruspolia (Ruspolia × Ruttya), Solidaster (Aster × Solidago), Sorbaronia (Aronia × Sorbus).*

Specific epithets display the same variety. Many descriptive epithets, such as *albus* (white), *bellus* (pretty), *calvus* (bald), *caudatus* (tailed), *centifolius* (hundred-leaved), *foetidus* (stinking), *glomeratus* (crowded), *glutinosus* (sticky), *horridus* (bristly), *laxus* (loose), *multiflorus* (many-flowered), *odoratus* (scented), *roseus* (rose), *sempervirens* (evergreen), and geographical epithets, such as *anglicus* (English), *gallicus* (French), *graecus* (Greek), *hispanicus* (Spanish), *italicus* (Italian) are genuine Latin words; others such as *biltmoreanus, canadensis, labradoricus, araucanus, missuricus, nantucketensis, nepalensis, japonicus, laurentianus* and *pensylvanicus* have been formed by analogy with them. As, however, the number of species known within certain genera has increased to such an extent that all appropriate Latin epithets have been used, so more and more epithets have been added from Greek, in which it is comparatively easy to combine words and form epithets. Thus in the genus *Rhododendron,* out of 747 alphabetically consecutive epithets, 259 were of Latin origin, 246 of Greek origin, 162 personal, and 80 from other sources, mostly geographical.

Descriptive specific epithets often convey a little information about the plant, its size (usually relative to other species of the genus), habit of growth, colour or habitat. Personal names usually refer to the collector or someone who has studied the group, or they may be purely complimentary; the epithets *armandii, davidii, kingdonii, wardii, georgei, forrestii, ernestii* and *wilsonii* are attached to many plants collected in eastern Asia by Armand David, F. Kingdon-Ward, George Forrest and Ernest H. Wilson, but the epithets *hemsleyi* and *franchetii* commemorate W. B. Hemsley and A. Franchet, who studied the plants of eastern Asia but never had an opportunity of going there.

The two-word system of naming, usually called **binomial nomenclature,** is a very old system in vernacular names of plants and animals all the world over. It was first consistently adopted for the scientific names of plants by Linnaeus in his *Species Plantarum* (1753), a two-volume work of 1,200 pages in which he set out to provide a concise guide to all the plants then known. In 1768 in his *Systema Naturae* he named all the known animals likewise. He used in these and later works two sets of names, a descriptive name which stated the main distinguishing features of the species, for example *Ranunculus seminibus aculeatis, foliis superioribus decompositis linearibus,* and a more convenient but less informative two-word name, for example *Ranunculus arvensis,* for everyday use. Many of his contemporaries were reluctant to adopt Linnaeus's binomials but by the end of his life they were accepted everywhere. A primary reason for this was the encyclopaedic nature of his works. They brought together so much information about the kinds of plants in such a very concise, methodical and practical manner that they became indispensable. Through their use people became accustomed to Linnaeus's binomial nomenclature and gradually realized its advantages over the cumbersome phrase-names. Linnaeus's system of nomenclature has persisted, but his temporary and artificial system of classifying the genera into large groups based on the number and relation of their stamens and styles or stigmas, the 'sexual system', has long been superseded by other classifications based on a greater range of structures.

When necessary, **varietal** or **cultivar epithets** are used to distinguish variants within a species or group of interspecific hybrids; a third epithet is added to the binomial. This is of Latin form when referring to a wild subspecies or variety (e.g. *Consolida regalis* subsp. *paniculata, Ranunculus marginatus* var. *trachycarpus*), but should nowadays be a vernacular or fancy name, for example 'Melody', 'Daydream', 'Pixie', 'Rosy Morn', or a personal or geographical one, such as 'Caerhays', 'Laurens Koster', 'Lucy', 'Hester', 'Princess Mary', when referring to a cultivar, that is, an assemblage of cultivated individual plants distinguished by characters significant in agriculture, forestry, or horticulture.

7

A Gardener's Dictionary of Plant Names

In many groups where there has been much hybridization and selection, it suffices merely to use a generic name such as *Narcissus* or a vernacular equivalent such as daffodil, and to follow this simply by the cultivar epithet, for example 'Fortune'.

The **correct pronunciation** of Latin has long been a matter of debate between scholars, among whom there is now general agreement, but their rules cannot consistently be applied to botanical names, so many of which are of non-Latin origin. Particularly difficult in this respect are names commemorating persons, such as *Dahlia* and *Dalea*, the first of which is often pronounced by English-speaking gardeners as if it were the second. Curiously enough *Clivia* (named for the noble family of Clive) is often pronounced as if 'Clive' rhymed with 'cleave' instead of 'hive'. In general people tend to pronounce botanical names as if these were words in their own language. The pronunciation of Latin as formerly taught in schools in the English-speaking world and as still used by most of its gardeners differs considerably from the reformed academic pronunciation accepted by classical scholars, and it is to the latter that the pronunciation of Continental European botanists more closely approximates. Hence some acquaintance with both systems may help towards international intelligibility. These may be summarized in the following table (taken from Stearn: *Botanical Latin*; 1966):

REFORMED ACADEMIC	TRADITIONAL ENGLISH
ā as in *fāther*	*fāte*
ă as in *ăpart*	*făt*
ae as *ai* in *aisle*	as *ea* in *meat*
au as *ou* in *house*	as *aw* in *bawl*
c always as in *cat*	before *a, o, u* as in *cat* / before *e, i, y* as in *centre*
ch (of Greek words) as *k* or *k-h* (if possible)	as *k* or *ch*
ē as in *thēy*	*mē*
ĕ as in *pĕt*	*pĕt*
ei as in *rein*	as in *height*
g always as in *go*	hard before *a, o, u* as in *gap, go* / soft before *e, i, y* as in *gem, giro*
ī as in *machīne*	*īce*
ĭ as in *pĭt*	*pĭt*
j (consonant i) as *y* in *yellow*	*j* in *jam*
ng as in *finger*	*finger*
ō as in *nōte*	*nōte*
ŏ as in *nŏt*	*nŏt*

REFORMED ACADEMIC	TRADITIONAL ENGLISH
oe as *oi* in *toil*	as *ee* in *bee*
ph as *p* or *p-h* if possible	like *f*
r always trilled	
s as in *sit, gas*	*sit, gas*
t as in *table, native*	*table* but *ti* within a word as in *nation*
ū as in *brūte*	*brūte*
ŭ as in *fŭll*	*tŭb*
ui as *oui* (French), *we*	*ruin*
v (consonant u) as *w*	as in *van*
ȳ as *u* in French *pur*	as in *cȳpher*
ÿ as in French *du*	as in *cÿnical*

Vowels such as those in cake, kite, eve, vote are described as 'long'; vowels such as those in cat, kit, egg, pot are 'short'. The stress or emphasis in a Latin word falls on the next to last (penultimate) syllable when this is 'long', for example *for*-mo-*sus*, *cru*-en-*tus*, but on the last but two (antepenultimate) syllable when the last but one is 'short', for example flo-*ri*-dus, *syl*-vat-*ti*-cus.

In the list of plant names which forms the bulk of this book, an accent is used to indicate the stressed syllable, *Ara'lia*, *alternifo'lius*, *rugo'sus*.

To determine accentuation, much use has been made of C. A. Backer: *Verklarend Woordenboek*; N. Hylander: *Våra Prydnadsväxters Namn på Svenska och Latin*; C. Wittstein: *Etymologisch-botanisches Handwörterbuch*; and R. Zander: *Handwörterbuch der Pflanzennamen und ihre Erklärungen*.

In transliteration of Greek words ŏ represents the Greek o (omicron) and ō the Greek ω (omega), ĕ the Greek ε (epsilon) and ē the Greek η (eta).

The scientific names of plants are treated as Latin regardless of their derivation and the **Latin grammatical rules of agreement in gender** are used when applicable. The name of a genus is either a noun or a word treated as such, and has accordingly masculine, feminine or neuter gender. In this book the gender is indicated by the letters 'm' (masculine), 'f' (feminine) or 'n' (neuter) placed after the generic name. There are no completely reliable easy rules for ascertaining the gender but the following notes may help:

1. Generic names ending in -*us* are mostly masculine, for example *Cistus, Hibiscus, Humulus, Lupinus, Narcissus*, unless they are names of trees, wherefor *Ailanthus, Alnus, Cedrus, Cornus, Fagus, Laurus, Pinus, Sorbus*, for example, are feminine.

Generic names ending in the Greek -*ōn* are masculine, for example *Cotyledon, Dendromecon, Platycodon, Platystemon*, but those ending in

A Gardener's Dictionary of Plant Names

the Greek -ŏn (which can also be rendered as -um) are neuter, as for example *Acantholimon, Clerodendrum, Rhododendron.*

2. Most generic names ending in -a are feminine, for example *Castanea, Diosma, Fuchsia, Hosta, Nandina, Rosa,* unless they are derived from neuter Greek words ending in -ma; such Greek words include *derma* (skin), *loma* (border), *nema* (thread), *paegma* (sport), *phragma* (partition), *sperma* (seed), *stema* (stamen), *stigma* (stigma); hence *Aglaeonema, Atherosperma, Ceratostigma, Duasperma, Pachyphragma, Phyteuma, Trichostema,* are neuter.

Generic names ending in -ago, -ix, -odes, -oides, are feminine, for example *Nymphoides, Omphalodes, Plumbago, Solidago, Salix, Tamarix,* as are most names ending in -is, for example *Bellis, Cannabis, Clematis, Nomocharis, Orchis, Oxalis, Pteris, Vitis,* or -es, for example *Cheilanthes, Menyanthes.*

3. Generic names ending in -dendron and -um are neuter, for example *Allium, Halimodendron, Epimedium, Lilium, Spartium, Trillium.*

Specific epithets may be either adjectives and participles or nouns. The adjectives agree in gender with the generic names they follow. For example *Philadelphus* is of masculine gender and the epithets *coronarius, grandiflorus, incanus, mexicanus, microphyllus* etc. applied to species of *Philadelphus* are likewise masculine. *Rosa* is of feminine gender and the associated epithets *alba, floribunda, japonica, rosea,* etc., are likewise feminine. *Lilium* is of neuter gender and the associated epithets *auratum, canadense, candidum, pumilum, sulphureum,* etc., are likewise neuter.

The adjectives and participles form three main groups:

1. Those in which the masculine, feminine and neuter forms each have a different ending:

MASCULINE -US	FEMININE -A	NEUTER -UM
albus (white)	*alba*	*album*
florifer (flowery)	*florifera*	*floriferum*
glaber (glabrous)	*glabra*	*glabrum*
graecus (Greek)	*graeca*	*graecum*
hirsutus (hairy)	*hirsuta*	*hirsutum*
triphyllus (three-leaved)	*triphylla*	*triphyllum*

2. Those in which the masculine and feminine forms have the same ending but the neuter a different one:

MASCULINE	FEMININE	NEUTER
acaulis (stemless)	*acaulis*	*acaule*
acaulos (stemless)	*acaulos*	*acaulon*
brevis (short)	*brevis*	*breve*

MASCULINE	FEMININE	NEUTER
campestris (of plains)	*campestris*	*campestre*
canadensis (Canadian)	*canadensis*	*canadense*
communis (common)	*communis*	*commune*
laevis (smooth)	*laevis*	*laeve*
tenuis (thin)	*tenuis*	*tenue*
triphyllos (three-leaved)	*triphyllos*	*triphyllon*

3. Those in which the masculine, feminine and neuter forms all have the same ending:

MASCULINE	FEMININE	NEUTER
bicolor (two-coloured)	*bicolor*	*bicolor*
bryoides (moss-like)	*bryoides*	*bryoides*
macrobotrys (large-clustered)	*macrobotrys*	*macrobotrys*
praecox (early)	*praecox*	*praecox*
pubescens (pubescent)	*pubescens*	*pubescens*
repens (creeping)	*repens*	*repens*
reptans (creeping)	*reptans*	*reptans*
simplex (unbranched)	*simplex*	*simplex*

Nouns used as specific epithets are not affected by the gender of the generic name. They remain the same whatever the gender. Such nouns may be obsolete names as in *Hydrocharis morsus-ranae, Liriodendron tulipifera, Sedum rosea, Teucrium chamaedrys,* or current generic names, as in *Corokia cotoneaster, Diervilla lonicera, Rosa rubus, Rubus rosa,* or vernacular names, as in *Codonopsis tangshen, Gentiana kurroo, Ligustrum ibota, Pinus mugo, Podocarpus totara, Prunus mume, Sabal palmetto, Schinus molle, Silene schafta, Taraxacum kok-saghyz.** They may also be nouns in the singular or genitive plural, for example *deserti* (of the desert), *desertorum* (of the deserts), *clivorum* (of the hills), *tectorum* (of roofs), *tinctorum* (of the dyers), or personal names in the singular or genitive plural, for example *baileyi* (of Bailey), *baileyorum* (of the Baileys), *elisabethae* (of Elizabeth), *helenae* (of Helen), *heleni* (of Helenus), *clusii* (of Clusius), *farreri* (of Farrer), *nuttallii* (of Nuttall), *Willmottiae* (of Miss Willmott).

Generic names are always written with a capital letter, specific and varietal epithets with a small letter, cultivar epithets with a capital letter. For example *Viburnum farreri, V.* × *bodnantense* 'Dawn', *V. plicatum* var. *tomentosum, Bellis perennis* 'Dresden China'.

* The cryptic epithet *mitejea* in *Kalanchoe mitejea* of Alice Leblanc and Raymond Hamet is not an obscure African vernacular name, as might be supposed, but a Gallic linguistic prank, being an anagram of 'je t'aime'! The lady involved is commemorated in *Kalanchoe aliciae, Kalanchoe leblanciae* and *Sedum leblanciae,* but is otherwise unknown to botany.

A Gardener's Dictionary of Plant Names

The **technical legality of botanical names** is, unfortunately, little understood by many gardeners. Botanical names have no standing, that is have no claim to use and acceptance, unless they have been published and are otherwise in accordance with the *International Code of Botanical Nomenclature*. The rules and recommendations in this code have been the subject of debate and consideration at international congresses over a long period. Before 1866 there were no such internationally agreed rules, but long before then botanists had come to accept certain guiding principles and to follow procedures made standard by the example of leading botanists, notably Augustin Pyramus de Candolle (1778–1841) of Geneva. De Candolle, who was the author of numerous monographic works, set out in his *Théorie élémentaire de la Botanique* (Paris; 1813) the main considerations, of which the basic ones were that the names should be (1) either Latin or of Latin form, because Latin was then the common language of civilized people, and (2) formed according to the rules of grammar, avoiding bilingual hybrids, such as a word part Latin, part Greek like *aculeaticarpa*, and that (3) the first to discover or to distinguish a plant had the right to name it, hence names should not be changed without good cause. He accordingly emphasized priority as a means of fixing names.

By 1866 the Candollean regulations and recommendations had been found not to cover all the problems of nomenclature made evident in monographic studies. At an international botanical congress in London a German professor, Karl Koch, proposed that such congresses should deal with matters of nomenclature. Alphonse de Candolle (1806–1893; son of A. P. de Candolle), who had a legal training, accordingly drew up his *Lois de la Nomenclature botanique* (1867), which the Paris botanical congress of 1867 adopted. These *Laws* were replaced in 1905 by a new set of *International Rules of Botanical Nomenclature* adopted by the Vienna botanical congress. They were further modified at the Cambridge botanical congress of 1930 and at the subsequent congresses of 1935 (Amsterdam), 1950 (Stockholm), 1954 (Paris), 1959 (Montreal), 1964 (Edinburgh) and 1969 (Seattle), the number of significant changes becoming less and less at each successive congress; since 1952 they have formed the *International Code of Botanical Nomenclature*, a detailed and complicated document which must be carefully studied by anyone concerned with establishing the correct name of a plant.

The naming of garden and other cultivated plants is controlled by the *International Code of Nomenclature for Cultivated Plants*, first published in 1953, with revised editions in 1958, 1961 and 1969. It deals with the naming of agricultural, silvicultural and horticultural cultivars which are normally given fancy names such as apple 'Cox's Orange Pippin', barley 'Balder', *Petunia* 'Rosy Morn', etc.

Changes in names which have been long accepted are a source of annoyance to all concerned but are often unavoidable. Some result from changes in classification. Thus the Madagascar periwinkle has long been known as *Vinca rosea* but it differs in so many details, including its chemical products, from the true periwinkles (*Vinca*), that it is now put in a separate genus (*Catharanthus*) and its correct name is now *Catharanthus roseus*. For similar reasons the genus *Chaenomeles*, containing the Japanese quince and its Chinese allies, has been separated from *Cydonia*, the true quince. The reverse process takes place when groups kept separate by some botanists are put together by other botanists because the differences once used to separate them appear to be inadequate: thus *Chamaenerium* has been returned to *Epilobium*, and *Siphonosmanthus* to *Osmanthus*. The same applies to species. If, for example, *Portulaca pilosa* is considered to be a widespread and extremely variable species, with the colour and the size of the seeds not important enough to merit breaking up the group into minor species, then the well-known *P. grandiflora* becomes *P. pilosa* subsp. *grandiflora*. Such changes are called taxonomic changes.

Other changes result from application of the principle of priority. Thus the name *Hypericum inodorum* published by Miller in 1768 has priority over *Hypericum elatum* published by Aiton in 1789 for the same plant. Similarly the name *Scilla mischtschenkoana* published by Grossheim in 1927 has priority over *Scilla tubergeniana* published by Hoog in 1936 and accepted by Stearn in 1950. Moreover, a specific name once validly published for a given plant cannot later be used for a different species in the same genus. Similar names applied to different plants are called homonyms. The rejection of later homonyms has caused a number of unavoidable but regrettable name changes. Thus in 1833 the Russian botanist Bunge named *Viburnum fragrans* a species cultivated at Peking, which Farrer and Purdom found growing in a wild state in 1914 on hillsides of Kansu, Western China, and introduced into European gardens. Bunge was unaware that in 1824 a French botanist, Loiseleur, had used the name *Viburnum fragrans* for a different species. Bunge's *Viburnum fragrans*, the popular garden plant, was accordingly renamed *Viburnum farreri* in 1966.

Some changes are due merely to correction of past mis-identifications or mis-applications of names. For example, the genera *Tritonia*, *Montbretia* and *Crocosmia* are considered to be distinct genera of *Iridaceae*. The common montbretia of gardens lacks the special features of *Montbretia* proper and *Tritonia* and is a hybrid between two species of *Crocosmia*, *C. aurea* and *C. pottsii* (also known as *Tritonia* or *Montbretia pottsii*); its correct name is accordingly not *Montbretia* or *Tritonia crocosmiiflora* but *Crocosmia × crocosmiiflora* because it must be put in the same genus as that in which both its parents are now placed. Under the name *Nepeta mussinii* two distinct kinds of catmint are known in gardens. The less ornamental one, which sets seed freely, is the Caucasian species to which the name *N. mussinii* was originally given and for which it must be retained. The other, which is sterile, more ornamental and more widely grown, had accordingly to be distinguished from it by receiving a new name, *N. × faassenii*; unknown in a wild state, it is a hybrid between *N. mussinii* and *N. nepetella*.

In general, however, the effect of applying the internationally accepted rules of nomenclature has been to stabilize the names of plants.

W. T. S.

February 1971

Some Sources of Further Information

André, J. 1956. *Lexique des Termes de Botanique en Latin*. Paris (Klincksieck).

Backer, C. A. 1936. *Verklarend Woordenboek van wetenschappelijke Namen*. Groningen & Batavia.

Beckmann, J. 1801. *Lexicon botanicum exhibens Etymologiam Orthographiam et Prosodiam Nominum botanicorum*. Göttingen.

Brown, R. W. 1956. *Composition of Scientific Works*. Washington D.C.

Carnoy, A. 1959. *Dictionnaire étymologique des Noms grecs des Plantes*. Louvain (Publications Universitaires).

Drewitt, F. D. 1927. *Latin Names of common Plants*. London.

Du Molin, J. B. 1856. *Flore poétique ancienne ou Etudes sur les Plantes des Poètes anciens, grecs et latins*. Paris.

Else, G. F. 1967. The pronunciation of classical names and words in English. *Classical Journal* 62: 210–14.

Featherly, H. I. 1954. *Taxonomic Terminology of the Higher Plants*. Ames, Iowa (Iowa State College Press).

Gilbert-Carter, H. 1964. *Glossary of the British Flora*. 3rd edition. Cambridge (Cambridge University Press).

Green, M. L. 1927. History of plant nomenclature. *Kew Bulletin* 1927: 403–415.

Gunawardena, D. C. 1968. *Genera et Species Plantarum Zeylaniae. An etymological and historical Account of the Flowering Plants of Ceylon*. Colombo (Lake House Investments).

Jackson, B. D. 1928. *A Glossary of botanic Terms*. 4th edition (reprinted 1960). London (Duckworth); New York (Hafner).

Some Sources of Further Information

Kobuski, C. E. 1955. 'A revised glossary of the more common botanical and horticultural terms.' *Arnoldia* 15: 25–44.

Lemprière, J. 1949. *Lemprière's Classical Dictionary of Proper Names mentioned in Ancient Authors*. New edition by F. A. Wright. London (Routledge & Kegan Paul).

Marafioti, R. L. 1970. The meaning of generic names of important economic plants. *Economic Botany* 24: 189–207.

McClintock, D. 1969. *A Guide to the Naming of Plants, with special reference to Heathers*. Horley, Surrey (Heather Society).

Murr, J. 1890. *Die Pflanzenwelt in der griechische Mythologie*. Innsbruck.

Rose, H. J. 1958. *A Handbook of Greek Mythology including its Extension to Rome*. 6th edition (reprinted 1965). London (Methuen).

Schultes, R. E. & Pease, A. S. 1963. *Generic Names of Orchids, their Origin and Meaning*. New York and London (Academic Press).

Sprague, T. A. 1957. English plant names. *Proc. Cotteswold Naturalists' Field Club* 32: 22–27.

Stearn, W. T. 1966. *Botanical Latin*. London (Nelson); New York (Hafner).

Thiselton-Dyer, W. T. 1914–15. On some ancient plant names. *Journal of Philology* 33: 195–207 (1914), 34: 78–96 (1915), 290–312 (1915).

Thompson, R. Campbell. 1924. The migration of Assyrian plant-names into the West. *Classical Review* 38:148–149.

Wittstein, G. C. 1852. *Etymologisch-botanisches Wörterbuch*. Ansbach.

A

a- Prefix in compound words of Greek origin signifying a negative, lacking, or contrary to. Thus *apetalus*, lacking petals; *Alyssum*, against madness.

abbrevia'tus, -a, -um Shortened; abbreviated.

Abe'lia f. Ornamental shrubs named for Dr. Clarke Abel (1780–1826), who, at the suggestion of Sir Joseph Banks, accompanied Lord Amherst on his embassy to Peking (1816–1817) as botanist. Much of his collection was lost by shipwreck on the way home to Kew. Except for Alexander von Bunge who accompanied a Russian ecclesiastical mission in 1830, no European naturalist was to visit China for nearly thirty years thereafter, Robert Fortune (see *Fortunella*) being among the first to follow. Abel died in India while serving as personal physician to Lord Amherst, who was by that time Governor-General.

Abeliophyl'lum n. *Abelia*; Gr. *phyllon*, a leaf; from its resemblance in foliage to an Abelia.

Abelmo'schus m. Apparently from Arabic *abu-l-mosk*, father of musk; in allusion to the smell of the seeds; often included in *Hibiscus*, but distinguished by the split calyx.

aberconwa'yi In honour of Henry Duncan McLaren (1879–1953), 2nd Baron Aberconway, owner of a fine garden at Bodnant, Denbighshire, North Wales, President of the Royal Horticultural Society 1931–1953.

aber'rans Deviating from the normal.

A'bies f. Fir. The classical Latin name.

abieti'nus, -a, -um Resembling the fir tree.

Abob'ra f. Latinized form of the Brazilian name for this vine of the cucumber family.

aborti'vus, -a, -um With parts missing; imperfect; producing abortion.

Abro'ma n. Gr. *a*, not; *broma*, food. These evergreen trees are mildly poisonous, in contrast to *Theobroma*.

Abro'nia f. Sand verbena, wild lantana. Gr. *abros*, delicate; in allusion to the appearance of the bracts beneath the flower of these herbs.

Abrophyl'lum n. Gr. *abros*, delicate; *phyllon*, a leaf. The leaves of these Australian shrubs have a delicate appearance.

abrotanifo'lius, -a, -um Having leaves resembling southernwood.

abrotanoi'des Resembling southernwood.

abro'tanum Ancient Latin name and now the specific epithet of southernwood (*Artemisia abrotanum*, originally spelled *abrotonum*).

abrup'tus, -a, -um Ending suddenly; abrupt.

Ab'rus m. Wild licorice. Possibly from Gr. *abros*, delicate; more probably an Arabic name. The root has the property of licorice of poor quality. The poisonous seeds, bright scarlet with black spots, are used in India as weights and are strung as beads.

abscha'sicus, -a, -um Referring to Abkhasia, a region of the Caucasus.

abscis'sus, -a, -um Ending abruptly; cut off.

absin'thium Latin and pre-Linnaean name for wormwood, the botanical name for which is now *Artemisia absinthium*. It is used to flavour absinthe. In biblical days it was a symbol of calamity and sorrow.

abundiflo'rus, -a, -um Profusely flowered.

Abu'tilon n. Flowering maple. From the Arabic name for a mallowlike plant.

abyssin'icus, -a, -um Abyssinian; native to Ethiopia (Abyssinia).

Acacal'lis f. Gr. *Akakallis*, name of a nymph in Gr. mythology.

Aca'cia f. The Greek name for the tree *Acacia arabica*. Derived from Gr. *akis*, a sharp point.

acaciifo'lius, -a, -um Leaved like an Acacia.

Acae'na f. New Zealand bur, biddy-bid. Gr. *akaina*, a thorn. The name derives from the spines on the calyx of these low-growing shrubs or herbs.

Acaly'pha f. Copper leaf. Variant of *akalēphē*, ancient Greek name for nettle but applied by Linnaeus to this genus of shrubs or herbs because of the nettlelike appearance of the leaves.

Acam'pe f. Gr. *akampes*, inflexible or brittle. The flowers of these epiphytic orchids are brittle and break off easily.

acanth- In compound words of Greek origin signifying spiny, spiky, or thorny, from Gr. *akantha*, thorn, prickle.

acanthifo'lius, -a, -um With leaves like *Acanthus*.

acanthocar'pus With spiny fruit.

Acanthoce'reus m. Trailing or climbing cactus. Gr. *akantha*, a thorn; *Cereus*, cactus.

acanthoco'mus, -a, -um Having spiny hairs.

Acantholi'mon n. Prickly thrift. Gr. *akantha,* a thorn; *limonium,* sea-lavender, to which it is related.

Acanthopa'nax m. Gr. *akantha,* a thorn; *Panax,* ginseng, which this genus of spiny trees and shrubs resembles.

Acanthophoe'nix f. Spine-areca. Gr. *akantha,* a thorn; *Phoenix,* which is both the Greek and the modern generic name for the date palm, which *Acanthophoenix* resembles.

Acanthorrhi'za f. Gr. *akantha,* a thorn; *rhiza,* a root. The rootlets of these palms are spiny.

Acanthosta'chys f. Gr. *akantha,* a thorn; *stachys,* a spike. In allusion to the spiny bracts of the flowerhead.

Acan'thus m. Latin form of Gr. *akanthos,* from *akantha,* thorn, prickle. In Europe it is sometimes called 'bear's breech' from the size and appearance of the leaf which in some species is big, broad, and hairy. The acanthus leaf was a favourite decoration in classical sculpture, as in the capital of the Corinthian column.

acau'lis, -e Stemless or with only very short stems.

-aceae A plural adjectival ending used in the names of plant families denoting 'belonging to'. Thus *Rosaceae* means belonging to the rose family. The eight exceptions such as *Compositae* (daisies), *Gramineae* (grasses), *Leguminosae* (legumes), refer to families which were first recognized as natural groups and given descriptive names.

ace'phalus, -a, -um Without a head.

A'cer n. Latin name for the maple tree. The word also means sharp and is in reference to the hardness of the wood, which the Romans used for spear hafts.

a'cer, a'cris, a'cre Sharp; pungent. Used both in relation to taste and touch.

A'ceras n. Gr. *a,* without, lacking; *kĕras,* horn, spur; referring to the spurless flowers.

acer'bus, -a, -um Bitter; sour. Also rough to touch.

acerifo'lius, -a, -um With leaves resembling maple.

aceroi'des Resembling maple.

acero'sus, -a, -um Needlelike.

aceto'sa, acetosel'la Pre-Linnaean names for common sorrel and other plants with acid leaves. From L. *acētum,* vinegar.

Achille'a f. Yarrow or sneezewort. The Greek name honours Achilles, the heroic warrior of the Trojan wars. As a youth, he was taught the properties of this plant in healing wounds by his tutor Chiron the Centaur, who was half horse and half man. This useful piece of knowledge was highly regarded and was regularly applied in medicine until relatively recent times. Achilles himself had the good fortune to be almost entirely invulnerable—his mother, Thetis, having dipped him as a baby into the River Styx. As a result he was vulnerable only in the heel where his mother's forefinger and thumb had held him during the process of immersion. In the end, the god Apollo directed the arrow of Paris to this one spot and it proved fatal.

'Sneezewort' derives from the fact that since ancient times one common species *Achillea ptarmica* (Greek for sneeze-making) has been used as a kind of snuff.

achilleifo′lius, -a, -um With leaves resembling *Achillea millefolium* or milfoil.

Achime′nes f. Name obscure. Central American herbaceous plant.

Ach′lys f. Deer-foot. Vanilla leaf. Named for a minor Greek goddess of hidden places—an allusion to the woodland habitat of this genus of perennial herbs.

Ach′ras f. A tropical evergreen tree with edible fruit. Name of Greek derivation for a kind of wild pear which Linnaeus transferred to the sapodilla or naseberry (*Manilkara zapota, Achras zapota*) and the sapote, mammee sapote or marmalade plum (*Pouteria sapota, Calocarpum sapota*).

acicula′ris, -is, -e Shaped like a needle; needlelike.

acicu′lifer, aciculi′fera, -um Needle-bearing.

Acidanthe′ra f. Gr. *akis*, a point; *anthera*, anther.

acidis′simus, -a, -um Very sour indeed.

acido′sus, -a, -um Acid; full of sourness.

a′cidus, -a, -um Acid; sour.

acina′ceus, -a, -um Shaped like a curved sword or scimitar.

acinacifo′lius, -a, -um With leaves shaped like a curved sword or scimitar.

acinacifor′mis, -is, -e Like a curved sword or scimitar.

Acine′ta f. Gr. *akinētos*, without movement, in reference to the immobile lip of the flower of these epiphytic orchids.

Aci′nos f. Gr. *akinos*, name of an aromatic herb mentioned by Pliny.

Acka′ma f. From the Maori name *maka-maka* for these New Zealand trees.

Acokanthe′ra f. Gr. *akis*, a spike; *anthera*, anther. The anthers are pointed. Certain species of these shrubs and small trees are the source of deadly arrow poisons used by some African tribes.

aconitifo′lius, -a, -um With leaves like aconite or monkshood.

Aconi′tum n. The ancient Latin name, from Gr. *akŏnitŏn,* for these poisonous herbs. The English names monkshood and wolf's-bane are derived from the shape of the flower and the fact that the plant was used for poisoning wolves. The leaves and roots of all species are the source of a strong alkaloid poison which has uses in pharmacy. Care must be taken on all occasions when handling the roots, which have sometimes been mistaken for horse-radish with disastrous results.

A′corus m. Sweet flag. The Latin name, from Gr. *akŏrŏn,* originally for the yellow flag (*Iris pseudacorus*), later applied to the sweet flag. The root stocks yield the sweet-scented calamus root once much used in making cosmetics.

Acroco′mia f. Gr. *akros,* highest, topmost, terminal; *kome,* a tuft of hair. The leaves of this tropical American feather palm are at the top of the stem.

Acrony′chia f. Gr. *akros,* terminal; *onux,* a claw. The points of the petals of these trees and shrubs are curved and look like claws.

Acrophyl′lum n. Gr. *akros,* terminal; *phryllon,* leaf. In allusion to the terminal clusters of leaves.

Acro′stichum n. Gr. *akrŏs,* terminal; *stichos,* a row, a tropical genus of ferns.

Actae′a f. Baneberry. Cohosh. Latin name adopted by Linnaeus from Pliny, apparently from Gr. *aktĕa* or *aktē,* elder. The fruit is poisonous; hence the English name.

actinacan′thus, -a, -um With radiating spines.

Actini′dia f. From Gr. *aktis,* a ray, in allusion to the styles of these climbing shrubs which radiate like the spokes of a wheel.

Actino′meris f. From Gr. *aktis,* ray; *mĕris,* a part.

Actinostro′bus m. From Gr. *aktis,* a ray; *strŏbos,* a cone; with reference to the structure of the cone scales on these shrubs.

Actino′tus m. Flannel flower. Gr. *aktinōtus,* furnished with rays. Referring to the showy spreading bracts which make these plants of *Umbelliferae* look like *Compositae.*

acu- In compound words signifying sharply pointed. Thus:

aculeatiss′imus	very prickly
aculea′tus	prickly
aculeola′tus	with small prickles
acumina′tus	tapering into a long narrow point
acuminatifo′lius	with leaves tapering quickly into long narrow points

acutan'gulus	with sharp angles
acutifo'lius	with sharply pointed leaves
acuti'lobus	with sharply pointed lobes
acutipe'talus	with sharply pointed petals
acu'tus	with a sharp but not a tapering point
acuti'ssimus	very acutely pointed

Adanso'nia f. Baobab or monkey-bread tree of Africa. Named in honour of a celebrated French botanist, Michel Adanson (1727-1806), who resided in Senegal, west tropical Africa from 1748 to 1754 and later published *Familles des Plantes* (1763).

As a tree, the baobab is frequently a remarkable sight. Although seldom as much as forty feet in height it may be thirty feet in diameter at the bottom. With this enormous girth, so disproportionate to the height, it ranks as one of the largest and most grotesque trees in the world. Its appearance is often that of a huge bottle with few branches and sparse leaves stuck into the top.

aden- Prefix which in compound Greek words indicates the presence of glands, the minute bodies common on several parts of many plants, sometimes making them very sticky indeed.

Adenan'dra f. From Gr. *adēn* gland; *anēr*, *andros*, man, hence male organ, stamen.

Adenanthe'ra f. Red sandalwood tree of India. From Gr. *adēn*, gland; *anthera*, anther. The anthers are tipped with a gland. The red, lens-shaped seeds are used in necklaces.

Adenocar'pus m. From Gr. *adēn*, gland; *karpos*, fruit. The pods of these shrubs are sticky.

adenogy'nus, -a, -um with glandular or sticky ovary.

Adeno'phora f. Gland bellflower. From Gr. *adēn*, gland; *phoreo*, to bear. A sticky nectary surrounds the base of the style in the flowers of these perennial herbs.

adeno'phorus, -a, -um Gland-bearing, generally in reference to a nectary.

adenophyl'lus, -a, -um Having sticky leaves.

adeno'podus, -a, -um Having sticky pedicels.

Adenosto'ma n. Gr. *adēn*, gland; *stoma*, mouth; in allusion to the glands at the mouth of the calyx of these evergreen shrubs.

Adhato'da f. Latinized version of the native Tamil and Sinhalese name for these shrubs. From *ada*, goat; *thodai*, not touch; with reference to the bitter leaves.

adiantifo'lius, -a, -um Having leaves like *Adiantum* or maidenhair fern.

adianto'ides Resembling *Adiantum* or maidenhair fern.

Adian'tum n. Maidenhair fern. Gr. *adiantos*, dry, unwetted. So named because the leaflets repel water in a remarkable way—if plunged into water the fronds remain dry.

ad'lamii Species named for R. W. Adlam of Johannesburg who sent South African plants to the Royal Botanic Gardens, Kew in the 1890's.

Adlu'mia f. Climbing fumitory, Allegheny vine. Named for John Adlum (1759–1836), a native of Pennsylvania and a well-known grape breeder. In 1819 he put Catawba, the first great American grape, on the market from his Georgetown D.C. nursery. It is still a leading American grape for both wine and table.

admira'bilis, -is, -e Noteworthy.

adna'tus, -a, -um Adnate; joined together. Usually used in connection with two unlike structures growing together naturally but in a way which appears to be abnormal.

adonidifo'lius, -a, -um With leaves like *Adonis* or pheasant's-eye.

Ado'nis f. Pheasant's-eye. The Greek name.
The flower is supposed to have sprung from the blood of Adonis (Semitic Adon, 'lord', Naaman or Tamnuz) who was gored to death by a wild boar. He was beloved by Aphrodite and by some accounts was unsuccessfully wooed by her. Adonis was regarded by the Greeks as the god of plants. It was believed that he disappeared into the earth in autumn and winter only to reappear in spring and summer. To celebrate his return, the Greeks adopted the Semitic custom of making Adonis gardens, consisting of clay pots of quickly growing seeds. See below under *Anemone*.

Ado'xa f. Gr. *a*, without; *dŏxa*, repute, glory; referring to the inconspicuous greenish flowers.

adpres'sus, -a, -um Adpressed or appressed; signifying the manner in which scales are pressed against a cone or certain leaves against a stem.

Adromis'chus m. From Gr. *hadrŏs*, thick, stout, thickset; *mischos*, stalk. In allusion to the short pedicels of these South African succulents.

adscen'dens Ascending, mounting; generally implies a somewhat gradual rise or upward curve from a nearly prostrate base.

adsur'gens Rising erect; pushing straight upwards.

adun'cus, -a, -um Hooked. The uncus was the hook used by Roman executioners to drag away the bodies of the executed.

adus'tus, -a, -um Sunburnt; swarthy.

ad'venus, -a, -um Adventive; newly arrived; not native.

Aech'mea f. Gr. *aichme*, a point; in reference to the stiff points of the sepals of these epiphytic herbs.

aegeus, -a, -um From the Aegean Sea (Aegaeum Mare).

Ae'gilops f. Gr. *aigilōps*, a kind of bearded grass.

Ae'gle f. Bael fruit of India. Named for one of the naiads, those female divinities of Greek mythology who presided over springs, rivers, and lakes. The bael tree, sacred to the god Siva, is widely cultivated in India for its fruit which, besides being edible, is regarded as a specific against dysentery.

Aegopo'dium n. Goutweed, bishopsweed. Gr. *aix*, a goat; *podion*, a little foot. It was once thought to cure gout.

aegypti'acus, -a, -um; aegyp'ticus, -a, -um Egyptian.

ae'mulus, -a, -um Rivalling; imitating.

ae'neus, -a, -um Of a bronze colour.

Aeo'nium, n. Latin name for one of this genus of succulents sometimes included in *Sempervivum*.

aequa'lis, -is, -e Equal.

aequi'lobus, -a, -um With equal lobes.

aequinoctia'lis, -is, -e Belonging to the equinoctial zone; from the equatorial regions.

aequitrilo'bus, -a, -um With three equal lobes.

Aeran'gis f. Gr. *aer*, air; *aggos* (*angos*) or *aggeion* (*angeion*), a vessel. These orchids are epiphytes, which are sometimes called air plants or tree perchers. They draw their sustenance from the air and some epiphytes of the family *Bromeliaceae* are quite capable of growing on such unlikely hosts as telephone wires. They are parasites only to the extent that they steal a growth site from their hosts. They are common in the moist tropics.

Aeran'thes m. From Gr. *aer*, air; *anthos*, flower.

Aer'ides n. From Gr. *aer*, air. These orchids are epiphytes.

aer'ius, -a, -um Airy, aerial, high.

aeruginosus, -a, -um Rust-coloured.

Aer'va f. From the Arabic name of the plant. A herbaceous species from the East Indies.

Aeschynan'thus m. From Gr. *aischune*, shame; *anthos*, flower; referring to the red flowers.

aesculifo'lius, -a, -um With leaves like horse-chestnut.

Aes′culus f. Horse-chestnut, buckeye. The Latin name for a kind of oak bearing edible acorns but applied by Linnaeus to this genus.

aestiva′lis, -is, -e Pertaining to summer.

aesti′vus, -a, -um Ripening or developing in summer.

Aethione′ma n. Stone cress; candy mustard. Origin uncertain, possibly from Gr. *aëthes*, unusual, or *aitho*, to scorch; *nema*, a thread.

aethio′picus, -a, -um African, usually South African.

aetnen′sis, -is, -e Referring to Mount Etna, the volcanic mountain of Sicily.

a′fer, af′ra, af′rum African, often referring to the North African coast (Algeria, Tunis, etc.).

affi′nis, -is, -e Related or similar to.

Aframo′mum n. From *Africa* and *Amōmum*, a related genus.

africa′nus, -a, -um African.

Agalmy′la f. From Gr. *agalma*, a glory, pleasing gift for the gods. In allusion to the showy flowers, plus probably *hylē*, woodland; in allusion to its habitat.

Agani′sia f. From Gr. *aganos*, gentle; referring to the neat appearance of these orchids.

agan′niphus, -a, -um Much snowed on.

Aganos′ma f. From Gr. *aganŏs*, mild, gentle; *ŏsmē*, fragrance.

Agapan′thus m. African lily. From Gr. *agape*, love; *anthos*, a flower.

Agape′tes f. Gr. *agapētos*, beloved, desirable, lovable.

Aga′ricus m. Old generic name for field and other mushrooms, many of which are now reclassified in other genera.

Aga′stache f. Gr. *agan*, very much; *stachys*, an ear of wheat; in reference to the many flower spikes of these perennial herbs.

agas′tus, -a, -um Charming.

Ag′athis f. Dammar or kauri pine. Gr. *agathis*, a ball of thread; from the appearance of the catkin on the female trees. The dammar pine is the source of gum dammar or copal used in the manufacture of varnish.

Agathos′ma f. From Gr. *agathos*, good; *ŏsmē*, fragrance. Some species of these evergreen shrubs are aromatic.

Aga′ve f. Century plant, sisal. Gr. *agauos*, admirable, in allusion to the splendid appearance of the plants in flower. These plants are used for rope and fibre. They are also the source of both tequila and pulque, the distilled and fermented liquors much appreciated in Mexico. The substance causing cloudiness in the latter is an important source of vitamins.

agavoi'des Resembling *Agave*.

ageratifo'lius, -a, -um With leaves like *Ageratum*.

ageratoi'des Like *Ageratum*.

Age'ratum n. Greek name *agērator*, presumably from the Gr. *a*, not; *geras*, old age; presumably because the flowers of these annuals or perennials retain their clear colour for a long time.

agglutina'tus, -a, -um Stuck together.

aggrega'tus, -a, -um Aggregate; clustered in a dense mass. The raspberry and strawberry are aggregate fruits, being the massed product of several ovaries.

Aglai'a f. Evergreen trees and shrubs named for one of the three Graces because of the beauty and sweet scent of the flowers.

Aglaone'ma n. Gr. *aglaos*, bright, clear, manifest; *nēma*, a thread; referring to the stamens, according to Schott, author of the name.

Ago'nis f. Gr. *agōn*, a gathering or an assembly, in allusion to the number of the seeds on these Australian evergreen trees and shrubs.

agra'rius, -a, -um Growing in the fields.

agres'tis, -is, -e Growing in the fields.

Agrifo'lium n. A widely used medieval Latin name for holly; a variant of *Aquifolium*.

agrifo'lius, -a, -um With rough or scabby leaves.

Agrimo'nia f. Agrimony. Apparently a misrendering of *argemonia*, from the Greek name *Argemone*. It yields a yellow dye. This herb is used in medicine as an astringent and tonic.

Agropy'rum n. Gr. *agros*, field; *pyros*, wheat.

Agrostem'ma n. Gr. *agros*, field; *stĕmma*, crown, garland.

Agros'tis f. Gr. *agrōstis*, a kind of grass.

Aichry'son n. A classical Greek name for *Aeonium arboreum*, which is related to these plants from the Canary Islands resembling house leeks.

Ailan'thus f. Tree-of-heaven. The Latinized version of the native Moluccan name *ailanto*, meaning 'sky tree', for a species of this genus.

aiolosal'pinx Specific name meaning the trumpet of Aeolus, the god of the winds.

Ai'phanes f. Gr. *aiphnes*, ragged or jagged. The leaves of these palms have jagged tips.

Aipyan'thus m. Gr. *aipys*, high and steep; *anthos* flower. In allusion to mountain habitat of this plant often put in *Arnebia* or *Macrotomia*.

26

Ai'ra f. Hair grass. From the classical Greek name for another plant.

aitchi'sonii In honour of Dr. James Edward Tierney Aitchison (1836–1898), British physician and botanist who served with the delimitation commission in Afghanistan (1884–1885) and made important collections in India and Afghanistan.

aizoid'es Like *Aizoon*, a genus related to the ice plants.

aja'cis In honour of Ajax, the Greek hero who at the siege of Troy committed suicide in a fit of pique because the armour of Achilles was awarded to Odysseus. This specific epithet was applied to *Delphinium ajacis* because the markings on the flower were thought to resemble the Greek letters *AIAI*.

ajanen'sis, -is, -e From Ajan on the coast of Siberia.

Aju'ga Bugle. Origin obscure.

ajugifo'lius, -a, -um With leaves like an *Ajuga*.

Ake'bia f. Latinized version of the Japanese name *akebi* for these twining shrubs.

alabamen'sis, -is, -e From the state of Alabama, U.S.A.

alacriporta'nus, -a, -um From Porto Alegre, Brazil.

Alan'gium n. From the Malabar (Tamil) name *alangi*.

alata'vicus, -a, -um From the Ala Tau mountains of Turkistan, U.S.S.R.

ala'tus, -a, -um Winged, e.g. the seed of the maple.

alb-, albi-, albo- In compound words signifying white. Thus:

albes'cens	whitish
al'bicans	off-white, becoming white
albicau'lis	white-stemmed
al'bidus	whitish
albiflo'rus	white-flowered
al'bifrons	with white fronds
albiple'nus	with double white flowers
albispi'nus	with white thorns
albocin'ctus	with a white crown or girdle
albomacula'tus	spotted with white
albopic'tus	painted with white
albopilo'sus	with white hairs; shaggy white
albople'nus	with double white flowers
albovariega'tus	variegated with white

albanen'sis, -is, -e From St. Albans in Hertfordshire, England.

alber'ti In honour of Albert von Regel (1845–1908), explorer in Turkistan, central Asia, eldest son of Eduard von Regel.

Albi'zia f. Pink siris, nemu tree. In honour of F. del Albizzi, a Florentine nobleman who in 1749 introduced *A. julibrissin* into cultivation.

albrech'tii In honour of M. Albrecht, 19th-century Russian naval surgeon.

al'bus, -a, -um White.

Al'cea f. L. name, from Gr. *alkaia*, of a kind of mallow.

Alchemil'la f. Lady's mantle. Origin uncertain, perhaps the Latinized version of an Arabic name.

alcicor'nis, -is, -e Palmated like the antlers of the European elk (*Alces alces*), called moose in America.

alcockia'nus, -a, -um In honour of Sir Rutherford Alcock (1809–1897), British consular official in China.

aldenhamen'sis, -is, -e Originated at Aldenham, Elstree, environs of London, where the Hon. Vicary Gibbs (1853–1932) had a celebrated garden.

alep'picus, -a, -um Of Aleppo (Haleb) in Syria.

A'letris f. Gr. *alētris*, the female slave who ground the meal. The name refers to the powdered appearance of these herbs.

Aleuri'tes f. Gr. *aleuron*, floury. The young growth of some species appears to be dusted with flour. The genus includes the Tung Oil or China Oil tree, the seed of which produces an oil valuable in making varnish and high-quality paint. A more recent use, at least as far as Western medicine is concerned, is the use of Tung Oil in treating leprosy.

aleu'ticus, -a, -um Of the Aleutian Islands off the coast of Alaska.

alexan'drae Of Queen Alexandra (1844–1925), wife of Edward VII of England.

alexandri'nus, -a, um Of Alexandria in Egypt.

algarven'sis, -is, -e From the province of Algarve, southern Portugal.

algerien'sis, -is, -e Algerian.

algi'dus, -a, -um Cold; originating in high mountains.

alie'nus, -a, -um Foreign, not related.

Alis'ma f. Water plantain. The classical Greek name for this plant.

Alkan'na f. From Arabic *al-hinna*, henna.

alkeken'gi Bladder cherry. From Arabic *al kakendi*, derived from Gr. *halikakabos*. Epithet for a species *Physalis*.

Allaman'da f. In honour of Frederik Allamand (b. about 1735, d. after

1776), Swiss botanist, who collected in Surinam about 1770, and sent seeds of this climbing plant to Linnaeus.

alleghanien'sis, -is, -e From the Allegheny mountains, eastern North America.

allia'ceus, -a, -um Like onion or garlic in flavour or appearance.

alliariifo'lius, -a, -um With leaves like *Alliaria*, a plant named by Leonhart Fuchs for its garlic smell and of no gardening interest.

Allio'nia f. A North American herb named in honour of Carlo Allioni (1705–1804), Italian botanist, professor at Turin.

Al'lium n. Onion, chive, garlic and related plants. From the classical Latin name for garlic. The whole group was prized by the ancients as possessing medicinal and aphrodisiac qualities as well as flavour.

Alloplec'tus m. Gr. *allos*, diverse; *pleco*, to plait. The allusion is to the overlapping of the calyx segments.

Alloso'rus m. From Gr. *allos*, different; *sŏrŏs*, heap, hence sorus; referring to the variously shaped sori of these ferns.

alnifo'lius, -a, -um With leaves like the alder.

Al'nus f. Alder. The classical Latin name.

Aloca'sia f. From Gr. *a*, without; *Colocasia*, the name of a closely allied genus, from which it was separated. Both of these foliage plants are very closely related to the genus *Caladium*.

A'loe f. From the Arabic name of these perennial succulents.

aloi'des (sometimes **alooi'des**) Aloe-like.

aloifo'lius, -a, -um With leaves like an aloe.

Aloino'psis f. From *Aloe* and *opsis*, likeness.

Alon'soa f. Mask-flower. Andean plants named in honour of Alonzo Zanoni, Secretary of State of Colombia when it was a Spanish colony in the 18th century.

Alopecu'rus m. Foxtail grass. Gr. *alōpĕkŏurŏs*, grass like a fox's tail.

Aloy'sia f. In honour of Maria Louisa (d. 1819), princess of Parma.

alpes'tris, -is, -e Of the lower mountains, with the implication of coming from below timber line, although not invariably.

Alphito'nia f. From Gr. *alphiton*, barley meal, in reference to the dry, mealy quality of the fruit pulp or mesocarp of these tall trees.

alpi'cola A dweller in high mountains.

alpi'genus, -a, -um Originating in the mountains.

Alpi'nia f. Perennial tropical herbs, named in honour of Prospero Alpino (1553–1616), Italian botanist, professor of botany at Padua, who wrote on the plants of Egypt.

alpi′nus, -a, -um Alpine; from high mountains above timber line.

Alseuos′mia f. Gr. *alsos*, a grove; *euosmos*, fragrant. The flowers of these shrubs are strongly fragrant.

Alsi′ne f. Gr. *alsinē*, a plant-name of obscure application.

Also′phila f. Gr. *alsos*, a grove; *philos*, loving. They are shade-loving tree ferns.

Alsto′nia f. Evergreen shrubs or trees with milky juice, named in honour of Dr. Charles Alston (1685–1760), professor of botany, Edinburgh University, 1716–1760.

Alstroeme′ria f. Peruvian lily. In honour of Baron Claus Alstroemer (1736–1794), a friend of Linnaeus.

altacleren′sis, -is, -e Raised at Highclere, Hampshire, England.

alta′icus, -a, -um From the Altai Mountains in Central Asia.

alter′nans Alternating.

Alternanthe′ra f. Joy-weed. L. *alternans*, alternating; *anthera*, anther. Alternate anthers in this genus of coloured-leaved perennials are barren.

alternifo′lius, -a, -um Having alternate leaves; the leaves on each side of a stem not opposite to each other.

alter′nus, -a, -um Alternate; not opposite.

Althae′a f. Hollyhock. Gr. *althaia*, a cure, something that heals; in allusion to the use of some species in medicine.

althaeoi′des Resembling hollyhock.

alti′cola Dwelling in the heights.

al′tifrons Having tall fronds or foliage.

altis′simus, -a, -um Very tall; tallest.

al′tus, -a, -um Tall.

alula′tus, -a, -um Narrow-winged.

aluta′ceus, -a, -um Like leather in colour or texture.

Alys′sum n. Madwort. From Gr. *a*, not or against; *lyssa*, rage or madness. This herb was regarded as a specific against madness and the bites of mad dogs. The well-known *A. saxatile* is placed by some botanists in another genus as *Aurinia saxatilis*.

Alyxi′a f. Gr. *halusis*, chain; apparently referring to the fruits.

ama′bilis, -is, -e Lovely.

amagia′nus From Mount Amagi, Japan.

ama′nus, -a, -um From the Amanus mountains of Southern Turkey.

Amara′cus m. From Gr. *amarakos*, marjoram.

amaranthoi'des Resembling amaranth.

Amaran'thus m. Love-lies-bleeding. From Gr. *amarantos*, unfading. The flowers of some species of these annuals retain their colour for a long time like everlastings.

amaranti'color Colour of amaranth or purple.

Amarcri'num n. From a contraction of *Amaryllis* and *Crinum*, this being a hybrid of these two. The earlier and more correct name is *Crindonna*.

amarel'lus, -a, -um Somewhat bitter.

amaricau'lis, -is, -e Having a bitter-tasting stem.

ama'rus, -a, -um Bitter.

Amaryl'lis f. Showy, bulbous plants, named after a beautiful shepherdess Amaryllis in classical poetry and equally irresistible to the English pastoral poets of the 16th and 17th centuries.

Amaso'nia f. Tropical American sub-shrubs, named in honour of Thomas Amason, a traveller in America in colonial days.

amazo'nicus, -a, -um, amazo'num From the region of the Amazon River.

ambi'guus, -a, -um Uncertain; doubtful.

amboinen'sis, -is, -e From the island of Amboina (Ambon), Indonesia.

Ambro'sia f. Gr. *ambrŏsia*. In ancient mythology, ambrosia was the substance which, with nectar, provided the food and drink of the gods, making immortal those who partook of them. Now the generic name of some weedy herbs.

ambrosi'acus, -a, -um Ambrosial; sweet.

ambrosioi'des Resembling *Ambrosia*.

Amelan'chier f. From *amelancier*, the French Provençal name of *A. ovalis* (*A. vulgaris*). It is also called 'sarvis' or 'servis berry' from its resemblance to the service, a forgotten English fruit rather like a pear which was eaten overripe like a medlar.

amelloi'des Resembling *Amellus*, a genus of no gardening interest.

amel'lus Species name of an Italian aster.

america'nus, -a, -um From America, North or South.

amesia'nus, -a, -um In honour of two members of the Ames family of Boston. Frederick Lothrop Ames (1835–1893) was a well-known horticulturist and amateur grower of orchids. Oakes Ames (1874–1950) was Supervisor of the Arnold Arboretum and Professor of Botany at Harvard. His teaching of economic botany was deeply interesting to generations of students. On his death he left his orchid herbarium of 64,000 specimens to Harvard University. *Rhododendron amesiae* commemorates Mrs. Mary S. Ames.

amethys'tinus, -a, -um Violet-coloured.

Amhers'tia f. A handsome flowering tree, native to India, named after Lady Sarah Amherst (d. 1838), wife of the Earl who took Dr. Clarke Abel (for whom *Abelia* was named) on his mission to Peking (1816) and who later became Governor-General of India. An amateur botanist, she collected plants on her travels.

Ami'cia f. In honour of Giovanni Battista Amici (1786–1863), professor of astronomy and microscopy at Florence.

amico'rum Of the Friendly or Tonga Islands in the South Pacific.

Am'mi n. Gr. and L. *ammi*, name of an umbelliferous plant.

Ammo'bium n. Winged everlasting. Gr. *ammos*, sand; *bio*, to live; in reference to the native habitat of these Australian annuals and perennials.

Ammo'charis f. Gr. *ammos*, sand; *charis*, grace, beauty; an allusion both to the habitat and to the quality of these bulbous plants.

Ammo'phila f. Marram or beach grass. Gr. *ammos*, sand; *philos*, loving.

ammo'philus, -a, -um Sand-loving.

amoe'nus, -a, -um Pleasant; delightful.

Amo'mum n. Gr. *amōmŏn*, an Indian spice-plant. *A. cardamon* is not true cardamon but provides a cheap and fairly acceptable substitute.

Amor'pha f. False indigo. Gr. *amorphos*, shapeless or deformed; in allusion to the corolla of these shrubs lacking wings and keel.

Amorphophal'lus m. Gr. *amorphos*, shapeless or deformed; plus *phallus*; in allusion to the form and shape of the tubers of these herbs.

Ampelopra'sum n. A kind of wild leek found in Greek vineyards, from Gr. *ampelos*, vine; *prason*, leek.

Ampelo'psis f. Pepper vine. Gr. *ampelos*, a vine; *opsis*, likeness. It looks like and is closely related to the grape vine.

amphi'bius, -a, -um Growing both in water and on land; amphibious.

Amphi'come f. Gr. *amphi*, both; *kŏmē*, a head of hair. The seeds of these herbaceous plants have a hairy tuft at each end.

amplexicau'lis, -is, -e Stem-clasping.

amplexifo'lius, -a, -um Leaf-clasping.

amplia'tus, -a, -um Enlarged.

ampliss'imus, -a, -um Very large.

amp'lus, -a, -um Large.

ampulla'ceus, -a, -um Flask-like.

Amso'nia f. Blue-flowered herbaceous perennials named in honour of Dr. Charles Amson, 18th-century Virginia physician.

amuren'sis, -is, -e From the area of the Amur River which forms the boundary between Manchuria and Soviet Russia in eastern Asia.

amygdalifor'mis, -is, -e Shaped like an almond.

amygda'linus, -a, -um Almond-like.

amygdaloi'des Resembling almond.

Amyg'dalus f. Greek name for almond, but now generally classified under *Prunus*.

Anacam'pseros f. Gr. *anakampserōs*, a herb which brought back love when touched, possibly a stonecrop.

Anacamp'tis f. From Gr. *anakampto*, bend back; referring to the spur of the flower of these orchids.

anacan'thus, -a, -um Without thorns.

anacardioi'des Resembling the cashew or *Anacardium*.

Anacar'dium n. Cashew nut. A name first used by 16th-century pharmacists for the heart-shaped fruit of the marking-nut tree of India, *Semecarpus anacardium*, from the Gr. *kardia*, heart, but adopted by Linnaeus as the generic name for the cashew, called *Ajacoo* by Tournefort.

Ana'charis f. Gr. *ana*, without; *charis*, beauty. A genus of North American water plants usually included in *Elodea*, whose principal importance is to act as oxygenators in aquaria and to provide shelter for young fish.

Anacy'clus m. Shortened form of *Ananthocyclus*; from Gr. *an*, without; *anthos*, flower; *kuklos*, ring.

Anagal'lis f. Greek plant name. The genus of low-growing herbs includes scarlet pimpernel or poor man's weather glass, so called because it opens only in fine weather.

Anagy'ris f. Gr. name for *A. foetida*.

anagyro'ides Resembling *Anagyris*. Small ornamental trees.

A'nanas f. Pineapple. From the South American (Tupi) Indian name. Many languages use this or some related form of this word.

Ana'phalis f. Pearly everlasting. From the classical Greek name for another of the everlastings.

Anasta'tica f. Rose of Jericho, resurrection plant. From Gr. *anastasis*, resurrection, in allusion to the fact that, no matter how dry this Near East plant (*A. hierochuntica*) may have become, the plant recovers its shape on being placed in water.

anato'licus, -a, -um From Anatolia, Turkey in Asia Minor.

an'ceps Two-sided, hence applied to flattened two-edged stems; in classical L. *anceps* sometimes meant wavering, doubtful, uncertain, and *Arundinaria anceps*, with round stems, was so named because of doubts as to its country of origin.

Anchu'sa·f. Bugloss, alkanet. From Gr. *ankousa*, alkanet. Some species of these plants can be used to make rouge.

ancistr- In Greek compounds means hooked or barbed, e.g. *Ancistrochilus* (hooked lip).

ancyren'sis, -is, -e From Ankara, Turkey.

andegaven'sis, -is, -e From Angers, France.

andi'cola Native of the Andes.

andi'nus, -a, -um Belonging to the Andes.

Andi'ra f. Cabbage tree. The Latinized form of the Brazilian name for these ornamental evergreen trees.

Andrach'ne f. Gr. *andrachne*, ancient plant-name applied both to purslane and strawberry-tree.

andro'gynus, -a, -um Having male and female flowers separate but on the same spike; hermaphrodite.

Androm'eda f. Bog rosemary. Named by Linnaeus after the mythological maiden who was chained to a rock as an offering to the sea-monster and rescued by Perseus; see Gourlie, *The Prince of Botanists*, 62 (1953), for Linnaeus's comical sketch of the type-species (*Andromeda polifolia*) as found by him in Lapland and the trivial incident which suggested the application of the name *Andrŏmĕda* to it. The name has been used for a confusing variety of plants. Today among the plants wrongly so-called in some nursery catalogues and by gardeners are *Pieris japonica*, *Pieris floribunda*, *Enkianthus*, *Gaultheria*, and *Leucothoe*.

Andropo'gon m. Beard grass. From Gr. *aner*, a man; *pogon*, a beard; in reference to the hairs on the spikelets of some species of these grasses.

Andro'sace f. Rock jasmine. From Gr. *aner*, a man; *sakos*, a shield; a name used by Dioscorides for some other plant.

androsaemifo'lius, -a, -um With leaves like *Androsaemum*, an old generic name of *Hypericum*.

Androsae'mum m. Gr. *androsaimon*, plant-name used by Dioscorides for plant with blood-like sap; from *aner, andros*, man; *haima*, blood.

androsa'ceus, -a, -um Resembling *Androsace*.

Androste'phium n. Gr. *aner, andros* man; *stephanos*, a crown. The stamens form a corona or crown within the flower of these bulbous plants.

Anemarrhe'na f. From Gr. *a*, without; *nēma*, thread, filament; *arrēn*, male; in allusion to the sessile anthers.

Ane'mia f. Gr. *aneimōn*, naked, unclad; from *heima*, clothing.

Anemo'ne f. Often said to be derived from Gr. *anemos*, wind, with which there is no evident connexion, but more likely a corrupted Greek loan word of Semitic origin referring to the lament for slain Adonis or Naaman, whose scattered blood produced the blood-red *Anemone coronaria* or *Adonis*; see above under *Adonis*.

Anemonel'la f. Diminutive of *Anemone*.

anemoniflo'rus, -a, -um *Anemone*-flowered.

anemonoi'des Resembling *Anemone*.

Anemono'psis f. From Gr. *Anemone*; *opsis*, likeness; because of this perennial's resemblance to *Anemone*.

Anemopaeg'ma n. From Gr. *anemos*, wind; *paigma*, sport. These are rampant tropical climbing shrubs.

Anemo'psis f. From Gr. *anemone*; *opsis*, likeness; from a resemblance of the flower to *Anemone* in this aquatic herb.

anethifo'lius, -a, -um With leaves like dill.

Ane'thum n. Greek name for dill. Best known in pickles, dill seeds are also used in making gin and in pharmacy.

anfractuo'sus, -a, -um Bent; twisted.

Ange'lica f. So named on account of its supposed angelic qualities in medicine. This showy herb was regarded as an aphrodisiac, a specific against plague, an antidote to poisons, and the root, chewed, protected against witches. Deep green in colour, the candied stalks of *A. archangelica* have a delectable and unusual flavour.

Angelo'nia f. Latinized version of the South American vernacular name of one of the species of these perennial herbs and sub-shrubs.

Angiop'teris f. Gr. *aggeion*, a vessel; *pteros*, a wing; from the form of the receptacle in which the spores of these ferns are produced.

ang'licus, -a, -um From England (Anglia), English.

Ango'phora f. Gum myrtle. From Gr. *aggeion*, a vessel; *phoreo*, to carry; in allusion to the form of the fruit of these Australian trees and shrubs.

Angrae'cum n. Latinized version of the Malayan name (*angurek*) for epiphytic plants. Now the name of a large genus of orchids.

anguici'dus, -a, -um Snake-killing.

angui'nus, -a, um Serpentine.

angula'ris, -is, -e; angula'tus, -a, -um Angular.

angu'lidens Having hooked teeth.

angu'liger, anguli'gera, -um Having hooks.

Angulo'a f. Terrestrial orchids named in honour of Don Francisco de Angulo, Spanish botanist of the latter part of the 18th century.

angulo'sus, -a, -um Full of corners; many angled; with marked angles.

angu'ria A specific epithet derived from the Greek for a cucumber.

angusta'tus, -a, -um Narrowed.

angustifo'lius, -a, -um Having narrow leaves.

A'nia f. Gr. *ania*, trouble; a name possibly suggested by difficulty in placing this genus of orchids.

Anigozan'thos m. Kangaroo's paw, Australian sword lily. Gr. *anoigo*, to open; *anthos*, a flower. The petals of these perennials flare open almost to the base.

anisa'tus, -a, -um Like anise.

anisodo'rus, -a, -um Anise-scented.

anisophyl'lus, -a, -um With unequal leaves, generally implying one of a pair of opposite leaves being much larger than the other.

Aniso'tome f. Gr. *anisos*, unequal; *tŏmĕ*, cut; in allusion to the manner in which the edges of the leaves are cut on these New Zealand herbaceous plants.

ani'sum The old name as well as the specific epithet for anise (*Pimpinella anisum*), much used for flavouring in various condiments, sweets, medicines, and liqueurs, also providing a distinctive scent for fox and other hounds to follow in drag-hunting. The leaves are said to remove freckles.

annamen'sis, -is, -e Of Annam in south-eastern Asia.

Anno'na f. Latinized version of the American Indian (Taino) vernacular name for the Cherimoya, soursop or custard-apple.

annula'ris, -is, -e Ring-shaped.

annula'tus, -a, -um Furnished with rings.

an'nuus, -a, -um Annual.

Ano'da f. Sinhalese (Ceylon) name for a species of *Abutilon*.

Anoigan'thus m. From Gr. *anoigo*, to open; *anthos*, flower.

ano'malus, -a, -um Deviating from the normal.

anope'talus, -a, -um With erect or ascending petals.

Ano'pteris f. From Gr. *ana*, upward; *pteron*, a wing; in allusion to the appearance of the flap, or indusium, over the sori or spore-producing part of this fern.

anos'mus, -a, -um Lacking scent.

Ansel'lia f. Epiphytic orchids named in honour of John Ansell, (d. 1847), English gardener who accompanied Vogel on an expedition (1841–1842) to the river Niger in west tropical Africa.

anseri'nus, -a, -um Relating to geese.

antarc'ticus, -a, -um Of the South Polar region, used in botany for the region south of 45° S., including Tierra del Fuego.

Antenna'ria f. Everlasting. L. *antenna*, literally, the yard of a sailing ship. The pappus of the flower of these silvery-leaved plants is supposed to look like the antennae of a butterfly.

An'themis f. Chamomile, also spelled camomile. *Anthĕmis* is the Greek name for this highly scented plant which has a long history as a flavouring herb; in medicine it was valued for its bitter and tonic properties, and is well known as the old household remedy of chamomile tea. The Elizabethans used it in lawns since it stands mowing and makes a strong turf with a characteristic odour, but this use is rare today.

anthemoi'des Resembling chamomile.

Anthe'ricum n. St. Bernard's lily. From Gr. *antherikon*, name for asphodel.

anthocre'ne Flower fountain, from Gr. *anthos*, flower; *krēnē*, spring.

Antholy'za f. Gr. *anthos*, flower; *lyssa*, rage. The opened flower of these plants looks like the mouth of an angry animal.

anthopo'gon Flower beard; with reference to the hairs on the tube of the corolla.

Anthoxan'thum n. Sweet vernal grass. From Gr. *anthos*, a flower; *xanthos*, yellow; in allusion to the yellow of the spikelets when ripe.

Anthris'cus m. Salad chervil. From the Latin name for another unidentified plant.

Anthu'rium f. Flamingo flower. From Gr. *anthos*, a flower; *oura*, a tail; in allusion to the tail-like spadix of these foliage plants.

anthyllidifo'lius, -a, -um With leaves like woundwort or ladies' fingers.

Anthyl'lis f. Kidney vetch, woundwort, ladies' fingers. The ancient Greek name. This herb once had a place in medicine as a vulnerary.

anti- Gr. *anti*, over against, opposite, instead, in place of, like, in return for; used in forming compounds with other Greek words.

Anti'aris f. Upas tree of the East Indies. From the Javanese word *anjar*, the poisonous juice of the tree which is used as an arrow poison. There is a legend that it is deadly merely to sleep in the shade of the upas tree.

Antides'ma n. From Gr. *anti*, against; *desma*, actually a head-band, but used by Burman to mean 'poison'; the name thus being intended to refer to the plant's supposed efficacy against snake-bite.

Anti'gonon n. Coral vine. From Gr. *anti*, in place of; *polygŏnŏn*, knotweed; indicating its affinity to *Polygonum*.

antilla'rum Of the Antilles, the two groups of islands (Greater and Lesser) which comprise the West Indies.

antipo'dus, -a, -um Of the Antipodes.

antiquo'rum Of the ancients.

anti'quus, -a, -um Ancient, antique.

antirrhiniflo'rus, -a, -um With flowers resembling snapdragon.

antirrhinoi'des Resembling snapdragon.

Antirrhi'num n. Snapdragon. From Gr. *anti*, like; *rhis*, a nose or snout. This is an allusion to the appearance of the flower, which looks like the long snout of a dragon. The mouth can be made to open very satisfactorily by gently pinching the sides of the corolla.

Anu'bias f. From Anubis, the jackal-headed ancient Egyptian god.

Ao'tus m. From Gr. *a*, without; *ous*, *ŏtŏs*, ear; in allusion to the lack of two appendages on the calyx characteristic of the related genus *Pultenaea*.

Apari'ne f. Greek name for cleavers (*Galium aparine*).

aperan'thus, -a, -um Boundless; of unlimited growth.

ape'rtus, -a, -um Exposed; bare; open.

apet'alus, -a, -um Without petals.

Aphano'stephus m. Gr. *aphanes*, inconspicuous; *stephos*, a garland; from the small size of the flower heads of these herbs.

Apheland'ra f. Gr. *apheles*, simple; *aner*, male; an allusion to the fact that the anthers of these shrubs and herbs are one-celled.

Aphyllan'thes f. From Gr. *a*, without; *phyllon*, leaf; *anthos*, flower; the flowers appearing on the leafless rush-like stems.

aphyl'lus, -a, -um Without leaves or apparently so.

apia'tus, -a, -um Dotted, spotted.

apicula'tus, -a, -um Terminating abruptly in a short and often sharp point.

a'pifer, api'fera, -um Literally, bee-bearing. Specific epithet of *Ophrys apifera*, the bee orchid; the labellum of the flower so resembles a bee that drones attempt copulation.

apiifo'lius, -a, -um Having leaves like celery (*Apium*).

A'pios f. Potato-bean, groundnut. Gr. *apios*, a pear; from the shape of the tubers of *Apios americana* which were eaten sometimes by North American Indians.

A'pium n. Celery. The classical Latin name for parsley and celery, possibly from *apis*, bee.

Aplect'rum n. Adam and Eve, puttyroot. Gr. *a*, without; *plēktron*, a spur. The flower of this small North American terrestrial orchid has no spur.

Apo'cynum n. Dogbane. The classical Greek name for this or a related plant. From *apo*, away; *kyon*, a dog. Supposed to have been poisonous to dogs. The root of one species, *A. cannabinum* (Choctaw root or Indian hemp), furnishes an emetic and cathartic.

apodan'thus, -a, -um With stalkless flowers.

apodec'tus, -a, -um Gr. *apŏdĕktŏs*, welcome, acceptable.

apo'dus, -a, -um Footless, sessile, without a stalk.

Apo'gon m. From Gr. *a*, not; *pōgōn*, a beard; applied to the group of beardless irises.

Aponoge'ton m. Cape pondweed. From the Latin name of the healing springs at Aquae Aponi (now Bagni d'Abano), Italy; plus *geiton*, a neighbour; originally applied to a water-plant (now *Zannichellia*) found here, later transferred to the African and Indian genus.

Aporocac'tus m. Rat-tail cactus. From the Greek, meaning impenetrable cactus (Gr. *apŏrŏs*, having no way in, out or through).

appendicula'tus, -a, -um Having appendages, such as a crown, crest, or hairs.

appeni'nus, -a, -um Of the Apennine Mountains in central Italy.

applana'tus, -a, -um Flattened out.

Apple All domestic apples are considered to be cultivars of the species *Malus pumila*, a crab with very wide distribution, growing wild from Norway to the Himalaya, and from Asia to Spain. It is true that there may have been some minor intervention of other species (such as *M. prunifolia* or *M. sylvestris*) but for practical purposes apples have been developed over centuries of cultivation and selection from the one species. Over the years innumerable varieties have been brought into cultivation. The majority have been discarded as better and more productive sorts have come on the scene. But around the world in both temperate zones apple varieties are grown in great numbers. While this is in part the result of local taste and custom, relatively small differences in climate, soil, and elevation materially affect both quality and fruit production.

Existing varieties must be propagated by vegetative means—nursery stock generally by budding or by grafting. The odds are enormously against any seedling resembling the parent tree. In fact, the chances are against its being of any real value at all. Thus, while new varieties are generally the result of careful selection in a deliberate and planned attempt to combine the good qualities of both parents, the grower is almost inevitably faced with the frustrations of having to discard possibly hundreds of seedling trees before one can be found which might be worth a further try.

A great many of today's best apples are of known parentage and are the result of careful selection of mutations. There still remain, however, quite a large number of admirable varieties which were discovered first as more or less chance seedlings and which have stood the test of time. Among the oldest of apples in America is the Sweet Rhode Island Greening. The first named variety in America, it was introduced by the Reverend William Blaxton around 1640 as Blaxton's Yellow Sweeting. Blaxton originally planted an orchard in Boston around 1625, approximately at what is now the corner of Beacon and Charles Streets. One of the best gardeners of his day, he also trained a bull to the saddle and used it to ride around Boston on his daily business.

Among other American apples, Rhode Island Greening appeared around 1748 at a Rhode Island village now known as Green's End. The tree was killed by being stripped of scions for grafting purposes. The Baldwin arrived as a chance seedling soon after 1740 on the farm of John Ball of Wilmington, Massachusetts, where it was originally known as Woodpecker or Butters. It was brought to the notice of Colonel Loammi Baldwin who introduced it widely about 1784. Similarly, McIntosh was found as a wild seedling on the Ontario farm of John McIntosh in 1796 and rapidly spread from there.

Green Newtown, which originated along the Hudson, was sent to England by Benjamin Franklin in 1759. It is seldom seen today in eastern American markets which are dominated by colour and size at the expense of flavour and other qualities. It is exported in considerable quantity to England where people seem to be prepared to sacrifice appearance to quality. In fact, the distinctions, so carefully drawn in England, of dessert, cooking, and cider apples are hardly known in America and certainly the terms are quite unfamiliar. This may account in part for the insipid flavour (save the heresy) of most American apple pies. A dessert apple has more sugar, a cooking apple more acid, a cider apple more tannin than the other two groups. In the West Country of England there are cider-apple varieties which are known to have been in cultivation for six or seven centuries. Of the cookers, Bramley's Seedling has been estimated to account for about 75 per cent of the apples in that category. It seems to have been raised from one of several pips sown by a Miss Brailsford in a flower pot and later planted in the garden of Matthew Bramley, a Nottinghamshire innkeeper and butcher, about 1850.

Ribston Pippin appears to be the sole survivor of a planting of apple pips brought from Normandy to England about 1688. The original tree is known to have lived for possibly 240 years. It was finally blown down in 1928.

That most delectable of all English eating apples, Cox's Orange Pippin, originated in 1830, probably as a pip from Ribston Pippin. Richard Cox was a brewer of Bermondsey who retired to Colnbrook where the apple was grown.

The word 'seedling' in the name of an apple is understood generally to mean that the apple appeared more or less by chance—either volunteered or as one from an unknown source. 'Pippin' and 'Pearmain' merely mean a kind of apple and their inclusion in any varietal name is largely decoration.

appres'sus, -a, -um Pressed close to; lying flat against, as some leaves against a stem, or the scales of a fir cone.

ap'ricus, -a, -um Sun-loving; open to the sun.

Apte'nia f. From Gr. *aptēn*, unfledged, unwinged; in allusion to the absence of wings on the valves of the capsule of these plants formerly included in *Mesembryanthemum*.

ap'terus, -a, -um Wingless.

a'pulus, -a, -um From Apulia (Puglia), southern Italy.

a'pus Footless, sessile, without a stalk.

aqua'ticus, -a, -um; aquat'ilis, -is, -e Growing in or near water.

Aquifo'lium n. Classical name for holly (*Ilex*).

Aquile'gia f. Columbine. From L. *aquila*, an eagle, from the form of the petals. The English name is also in allusion to the form of the flower which has the appearance of doves (L. *columba*) drinking—especially the short-spurred varieties of these perennial herbs.

aquilegiifo'lius, -a, -um With leaves like columbine.

aquili'nus, -a, -um Eaglelike; aquiline.

ara'bicus, -a, -um; a'rabus, -a, -um Of Arabia.

Arabido'psis f. From *Arabis* and Gr. *opsis*, appearance.

A'rabis f. Rock-cress. Origin obscure.

Ara'chis f. Peanut, groundnut, monkeynut, earthnut. Old Greek plant name transferred to this plant. One of the most valuable oil seeds. After flowering, the stalks bend down and push the fruits into the soil where they develop and ripen.

Arachnan'the f. Gr. *arachnis*, a spider; *anthos*, a flower; from the shape of the flower of these orchids.

arachnoi'des, arachnoi'deus, -a, -um Covered with entangled hairs giving the appearance of a spider's web.

aragoa'nus, -a, -um In honour of J. E. V. Arago (1790–1855), French botanist who accompanied Admiral de Freycinet in his explorations, 1817–1820.

aragonen'sis, -is, -e From the province of Aragon in north-eastern Spain.

araiophyl'lus, -a, -um With thin or slender leaves.

Ara'lia f. Latinization of the old French-Canadian name *aralie*. The genus includes wild sarsaparilla.

araliifo'lius, -a, -um With leaves like *Aralia*.

araneo'sus, -a, -um Cobwebby.

arauca'nus, -a, -um From the district Arauco, Chile, where the monkey-puzzle tree was discovered.

Arauca'ria f. Monkey-puzzle tree. A genus of conifers; now confined to Australasia and South America, taking both the generic name and the specific epithet, *araucana*, of the type species from the Araucani Indians of central Chile to whose territory it is native. The Norfolk Island species (*A. heterophylla*) was discovered on Captain Cook's second voyage (1772–1775) and introduced into cultivation in 1793.

Arauj'ia f. From the Brazilian vernacular name of these vines.

arbores'cens; arbor'eus, -a, -um Tending to be woody; growing in treelike form.

arbus'culus, -a, -um Resembling a small tree.

arbutifo'lius, -a, -um Having leaves like *Arbutus*.

Ar'butus f. Strawberry tree. The Latin name. Mayflower (*Epigaea repens*) is often popularly called trailing arbutus.

Archange'lica f. Referring to the Archangel Raphael, who, according to medieval legend, revealed its virtues.

Archontophoe'nix f. Gr. *archon, archontos*, a chieftain; *phoenix*, the date palm. In allusion to the majestic appearance of these palms.

arc'ticus, -a, -um From the polar regions; arctic.

Arc'tium n. Gr. *arction*, name of a plant, perhaps a mullein, also called *arcturus* by Pliny.

Arctosta'phylos f. Bearberry. Gr. *arktos*, a bear; *staphylē*, bunch of grapes.

Arcto'tis f. African daisy. Gr. *arktos*, a bear; *ous, otos*, an ear. The scales of the pappus are supposed to look like the ears of a bear.

Arctu'rus, n. Name of plant also called *arction* or *arctium*.

arcua'tus, -a, -um Bent like a bow; arched.

ar'dens Glowing.

Ardis'ia f. Spearflower, marlberry. From Gr. *ardis*, a point; in allusion to the pointed anthers of these flowering trees and shrubs.

Ardu'ina f. In honour of Pietro Arduino (1728–1805), Italian botanist and horticulturist, professor at Padua.

Are'ca f. Betelnut. From the vernacular name of the palm used by the natives of Malabar on the south-west coast of India. The sliced nut, wrapped in a paan leaf smeared with lime, is chewed by people throughout tropical Asia. It gives the blood-red colour to their frequent and copious expectorations. It stains the teeth black but is considered to aid digestion.

Arecas'trum n. Queen palm. From *Areca* plus *-astrum* indicating resemblance.

Arege'lia f. Ornamental epiphytic plants named in honour of E. Albert von Regel (1815–1892), Russian botanist of German origin, Director of the Imperial Botanical Gardens at what was then called St. Petersburg, now Leningrad. The correct name is *Neoregelia*.

Arena'ria f. Sandwort. From L. *arena*, sand; an allusion to the fact that most of the species of these herbs prefer sandy places.

arena'rius, -a, -um Relating to sand, hence growing in sandy places.

Aren'ga f. From the Malayan name, *areng*, for this palm from which palm-sugar or jaggery is made as well as palm wine.

areno'sus, -a, -um Full of sand, sandy.

areola'tus, -a, -um Marked out in small areas, areolate.

Arethu'sa f. Bog-orchid. Named in honour of Arethusa, a wood nymph, who, pursued by Alpheus, a river god, was changed by Artemis (L. Diana) into a stream, which, running under the sea, came up in Sicily as a fountain.

Arga'nia f. Latinized version of the local Moroccan name for this tree whose kernels provide a substitute for olive oil.

Argemo'ne f. Prickly or crested poppy. Greek name of a poppy-like plant. Once reputed to cure cataract of the eye. The word *argĕmŏn* is Greek for cataract; hence *argĕmōnē*.

argen'teo- In compound words signifying silver. Thus:
argenteogutta'tus   spotted with silver
argenteomargina'tus   with silver edges
argenteovariega'tus   silver-variegated

argen'teus, -a, -um Silvery.

argenti'nus, -a, -um Of Argentina.

argilla'ceus, -a, -um Whitish. Literally, pertaining to white or potter's clay.

argophyl'lus, -a, -um Whitish-leaved.

argu'tus, -a, -um Sharply toothed or notched.

argyrae'us, -a, -um Silvery.

Argyre'ia f. Silver morning-glory. Gr. *argyreios*, silvery. The undersides of the leaves are silvery.

argyro- In compound words signifying silver. Thus:
argyroco'mus   silver-haired
argyroneu'rus with silver veins
argyrophyl'lus silver-leaved

Argyroder'ma, n. From Gr. *argyrŏs*, silver; *derma*, skin. In allusion to the whitish leaves of these South African succulent plants.

a'ria An old name for *Sorbus aria*, whitebeam.

a'ridus, -a, -um Dry, growing in dry places.

arieti'nus, -a, -um Like a ram's head; like the horns of a ram.

arifo'lius, -a, -um With leaves like *Arum*.

ariifo'lius, -a, -um With leaves like *Sorbus aria*, whitebeam.

Ariocar'pus m. Gr. *aria*, the whitebeam; *karpos*, fruit. The fruits of these cacti resemble those of whitebeam.

Arisae'ma n. Jack-in-the-pulpit, Indian turnip, dragon arum. Gr. *aris* or *aron*, arum; *haima*, blood, used in the sense of 'relationship', i.e. akin to arum.

Ari'sarum n. Greek name, *arisaron*, for *Arisarum vulgare*. Tuberous herb mentioned by Dioscorides.

arista'tus, -a, -um Bearded; furnished with an awn, e.g. oats.

Aris'tea f. Apparently from Gr. *aristos*, best, used by Solander, its author, in the sense of 'pleasing'.

Aristolo'chia f. Dutchman's pipe, snakeroot. Gr. *aristos*, best; *lochia*, childbirth, the curved form of the flower with base and top together recalling the human foetus in the right position prior to birth; hence sometimes called birthwort; it was supposed to ease parturition. The form of the root suggested it as a remedy for snake bite. In modern pharmacy it is the source of an astringent bitter.

aristo'sus, -a, -um Bearded; awned.

Aristote'lia f. A genus of trees and shrubs named in honour of Aristotle (384–322 B.C.), Greek philosopher in whom is said to have reposed all the knowledge of his time, which he much extended.

arize'lus, -a, -um Gr. *arizĕlŏs*, conspicuous, notable.

arizon'icus, -a, -um Of Arizona, U.S.A.

arman'dii In honour of the French missionary Armand David (1826–1900); see *Davidia*.

arma'tus, -a, -um Armed with thorns, spines, and other such features.

armenia'cus, -a, -um Of Armenia, Western Asia.

arme'nus, -a, -um Armenian.

Arme'ria f. Statice, thrift, sea-pink. Latinized from the old French name *armoires* for a cluster-headed dianthus.

ar'miger, armi'gera, -um Arm-bearing, hence spiny or prickly.

armilla'ris, -is, -e Encircled, as with a bracelet or collar.

armilla'tus, -a, -um Like a bracelet or collar.

Armora'cia f. Horse-radish. The classical Latin name of a related plant.

Arne'bia f. Prophet flower. Arabian primrose. Latinized version of the Arabic name *sagaret el arneb* for the Arabian primrose.

Ar'nica f. The classical name for this plant. Derivation obscure. Tincture of arnica is obtained from the roots of *A. montana*, called mountain tobacco, and is used in the treatment of bruises and sprains.

arnoldia'nus, -a, -um Originated at the Arnold Arboretum, near Boston, Massachusetts, thus commemorating its benefactor by bequest, James Arnold (1781–1863).

arnol'dii Of Joseph Arnold, M.D. (1782–1818), naturalist under Sir Stamford Raffles (see *Rafflesia*).

aroma'ticus, -a, -um Fragrant; aromatic.

Aro'nia f. Chokeberry, from *aria*, the Greek name for whitebeam (a species of *sorbus*), the fruits of which resemble chokeberry.

Arpophyl'lum n. Gr. *arpe*, a sickle; *phyllon*, a leaf; in reference to the shape of the leaves of these epiphytic orchids.

Arrabidae'a f. In honour of Antonio da Arrabida, bishop in Brazil in first half of 19th century.

Arraca'cia f. From the Spanish name of these South American perennial herbs.

arrec'tus, -a, -um Upright; erect.

Arrhena'therum Gr. *arren*, male; *anther*, a bristle. The male or staminate flowers of this genus of perennial grasses are furnished with awns or bristles.

Arta'botrys m. Gr. *artao*, to support; *botrys*, a bunch of grapes. The bunched fruits are suspended by a tendril. In the East Indies, notably Java, the leaves of these woody climbers are regarded as a preventive of cholera.

Artemi'sia f. Wormwood; lad's love; old man. Named in honour of Artĕmis, the Greek goddess of chastity. Of considerable medicinal value, plants of the genus yield vermifuges, stimulants, and a vulnerary, as well as the active principle of absinthe.

artemisioi'des Resembling *Artemisia*.

Arthroce'reus m. Gr. *arthron*, joint, plus *Cereus*.

Arthro'lobium n. From Gr. *arthron*, joint; *lobos*, pod.

Arthropo'dium n. Gr. *arthron*, joint; *podion*, little foot. With reference to the pedicels being jointed at the middle.

articula'tus, -a, -um Jointed, e.g. some bamboos and grasses.

Artocar'pus m. Breadfruit. From Gr. *artos*, bread; *karpos*, fruit. The breadfruit when baked and warm has some resemblance to fine white bread. The 18th-century explorers thought it delicious as, indeed, it must have seemed after several months voyaging and a prolonged diet of salt meat and hard weevilly biscuit. It was the idea of Sir Joseph Banks, who became familiar with it on Tahiti when accompanying Captain James Cook on his first voyage round the world, to introduce it into Jamaica and the West Indies generally where food for the slaves was always short. He organized the expedition of Captain Bligh in the *Bounty* to bring plants from Tahiti to Jamaica.

A'rum n. Lords-and-ladies, cuckoo-pint. From *aron*, the Greek name for these poisonous plants which are closely related to the American jack-in-the-pulpit (*Arisaema*).

Arun'cus m. Goat's beard. The classical name for these herbs.

arundina'ceus, -a, -um Resembling a reed.

Arundina'ria f. Southern cane of the cane brakes of the south-east United States. From L. *arundo*, a reed.

Arun'do f. Genus of ornamental grasses. From L. *arundo*, a reed.

46

arva'lis, -is, -e; arva'ticus From near Arvas (Arbas) in the Cantabrian Mountains, northern Spain, between Oviedo and León.

arven'sis, -is, -e Growing in or pertaining to cultivated fields.

asarifo'lius, -a, -um With leaves resembling *Asarum* or wild ginger.

Asari'na f. Spanish vernacular name for an *Antirrhinum*.

Asa'rum n. Wild ginger. Latin and Greek name for these low-growing, nearly stemless herbs.

ascalo'nicus, -a, -um From the neighbourhood of Ashkelon, seaport city in Palestine, from which the English name 'shallot' is derived; the perennial forms of the common onion (*A. cepa*) to which the name 'shallot' is now applied have, however, no connexion with Ashkelon.

ascen'dens Rising upwards in a sloping fashion.

asclepia'deus, -a, -um Resembling milkweed or *Asclepias*.

Ascle'pias f. Milkweed; butterfly-weed. The Greek name for these plants, in honour of Asklēpiŏs, god of medicine, who in Latin was called Aesculāpius. Some of the species were used in medicine. Asklepios was generally shown with Hygieia, goddess of wise living, on one side and on the other, Panakeia, goddess of cure-alls. In hygiene and panaceas or wonder drugs, the cult of both is still widely observed.

Asclepiodo'ra f. From *Asklepios* and *dorea*, gift, i.e. the gift of Asklepios. These plants closely resemble *Asclepias*.

Ascy'rum n. Gr. *askyrŏn*, the name of a kind of *Hypericum*.

asia'ticus, -a, -um Asian.

Asi'mina f. Paw-paw. The Latinized version of what is said to be the French form of the Indian name for this North American genus of shrubs and small trees.

asparagi'nus, -a, -um; asparagoi'des Resembling asparagus.

Aspa'ragus m. Classical name for the plant which was well known to the ancients. Old English gardeners called it 'Sperage' and regarded it as a mild aphrodisiac. Nowadays it is often referred to vulgarly as sparrow-grass.

Aspa'sia f. Probably commemorating Aspasia, the beautiful mistress of the Athenian statesman Pericles (d. 429 B.C.) from *aspasies*, delightful.

as'per, -a, -um Rough.

aspera'tus, -a, -um Roughened.

aspericaulis, -is, -e Rough-stemmed.

asperifo'lius, -a, -um Rough-leaved.

asper'rimus, -a, -um Very rough.

Asperu'go f. From *asper*, rough.

Aspe'rula f. Woodruff. Diminutive of *asper*; an allusion to the rough stems. *A. odorata* is very fragrant and is used by Germans to flavour May wine.

Asphodeli'ne f. Jacob's rod. Yellow asphodel. Named from its relationship to *Asphodelus*.

asphodeloi'des Resembling asphodel (*asphŏdēlos*).

Aspho'delus m. Greek name for the true asphodel, *Asphodelus ramosus* and allied species.

Aspidis'tra f. Gr. *aspidion*, a small round shield; in reference to the shape of the stigma. This foliage plant is often used as decoration in bars, cinemas, and the front windows of boarding houses. Because of its ability to withstand unlimited ill-treatment it is called the cast-iron plant and, sometimes, the beerplant for the reason that bartenders often water it with beer dregs as the most readily available liquid. The flowers appear at ground level.

Aspi'dium n. From Gr. *aspidion*, a shield; referring to the indusium of these ferns.

asplenifo'lius, -a, -um With leaves like spleenwort which are fine, feathery, and fernlike.

Asple'nium n. Spleenwort. Gr. *a*, not; *splēn*, the spleen; a reference to this fern's traditional virtues in afflictions of the spleen and liver. It could not, however, be given to women because it was supposed to cause barrenness.

As'pris f. Classical Greek name for these ornamental grasses.

assi'milis, -is, -e Similar; like unto.

assur'gens Ascending.

assurgentiflo'rus, -a, -um With flowers in ascending clusters.

A'ster m. Starwort, Michaelmas daisy. L. *aster*, a star; an allusion to the form of the flower. The termination -*aster* in many plant names such as *Cotoneaster* has, however, nothing to do with *aster* 'star' but indicates incomplete resemblance or inferiority.

Asteranthe'ra f. From L. *aster*; star, *anthera*, anther; with reference to the anthers joined in a star-like manner.

asteroi'des Resembling aster.

Astil'be f. False goatbeard. Gr. *a*, without; *stilbe*, brightness; in allusion to the dullness of the leaves of the type species.

astilboi'des Resembling *Astilbe*.

astraca'nicus, -a, -um Of the Astrakhan region, European U.S.S.R., near the Caspian Sea.

Astra'galus m. Milk vetch. Greek word for some leguminous plants but applicable to this enormous genus of plants of the pea family.

Astran'tia f. Masterwort. Derivation obscure, a medieval plant-name possibly corrupted from *Magistrantia* and so derived from L. *magister*, master, or from *aster*, star, plus the ending *-antria*.

Astro'loba f. Gr. *astron*, star; *lobos*, lobe; in allusion to the stellately spreading lobes of the perigon.

Astro'phytum n. Gr. *astron*, a star; *phyton*, a plant; in allusion to the flattened form of these Mexican cacti.

astu'ricus, -a, -um; asturien'sis, -is, -e From the province of Asturia in Spain.

Atalan'tia f. Woody plants, useful as stocks for citrus fruits. Named in honour of Atalanta, daughter of King Schoeneus of Scyros. The fruit of the plant is golden-coloured. The allusion is to the story that Atalanta consented to accept in marriage any suitor who could outrun her. Hippomenes did so, but only by a ruse. He threw down three of Aphrodite's golden apples in front of her. In stopping to gather them up she lost the race.

a'ter, a'tra, a'trum Dead-black.

Athaman'ta f. Candy carrot. After Mount Athamas in Sicily where some species of these herbaceous plants are found.

Atherosper'ma n. Gr. *athēr*, *atheros*, barb, spine; *sperma*, seed; in allusion to the long-pointed or aristate fruits.

Athrota'xis f. Gr. *athroos*, crowded together; *taxis*, arrangement; referring to the leaves of these Tasmanian evergreens.

Athy'rium n. Lady fern, swamp spleenwort. Derivation unknown, possibly from Gr. *athuros*, spiritless.

atlan'ticus, -a, -um From the shores of the Atlantic Ocean, or, alternatively, from the Atlas Mountains in North Africa.

atra'tus, -a, -um Blackened.

At'riplex f. The Greek name for orach, a species of this genus of herbs and shrubs which can be used like spinach but which is generally regarded as a weed. The genus also includes salt-bush which is much appreciated by Australian sheep.

atriplicifo'lius, -a, -um With leaves like salt-bush.

atro- In compound words signifying dark. Thus:

atropurpu'reus	dark purple
atro'rubens	dark red
atrosangui'neus	dark blood-red
atroviola'ceus	dark violet
atro'virens	dark green

atrocar'pus, -a, -um With black or very dark fruit.

atrocau'lis, -is, -e With black or very dark stems.

At'ropa f. Deadly nightshade, belladonna. Named for Atrŏpŏs, one of the three Fates whose particular business it was to snip the thread of life. This genus of herbs, while very poisonous, has important uses in medicine. The specific name belladonna, meaning pretty lady, arises from the fact that it was used in cosmetics to make the eyes appear larger and brighter.

Atta'lea f. L. *attalus*, splendid or magnificent; in allusion to the beauty of these palms.

attenua'tus, -a, -um Narrowing to a point; attenuated.

at'ticus, -a, -um From Attica, the classical name for that part of Greece of which Athens was the chief city.

Aubrie'ta f. Purple rock-cress. Perennial trailers named in honour of Claude Aubriet (1668–1743), French botanical artist.

aubrietioi'des Resembling *Aubrieta*. (Sometimes called *Aubretia*.)

Aucu'ba f. Latinized version of the Japanese name of these ornamental evergreen shrubs.

aucupa'rius, -a, -um Specific epithet for the rowan tree, *Sorbus aucuparia*. From L. *aucupor*, go bird-catching, from *auceps* or *aviceps*, a bird-catcher; *avis*, bird; *capere*, catch.

Audiber'tia f. Herbs and small shrubs named for Urbain Audibert, well-known nurseryman of Tarascon, France, in the 19th century.

Audoui'nia f. Small shrubs from South Africa named for J. V. Audouin (d. 1841), Professor of Natural History at the Sorbonne in Paris.

augusti'nii Commemorating Dr. Augustine Henry (1857–1930).

augustis'simus, -a, -um Very notable or majestic.

augus'tus, -a, -um Majestic; notable.

aulacosper'mus, -a, -um From Gr. *aulax*, furrow; *sperma*, seed; with furrowed seeds.

auranti'acus, -a, -um; auran'tius, -a, -um Orange-coloured.

aurantifo'lius, -a, -um With leaves like the orange tree (*Citrus aurantium*).

aura'tus, -a, -um Ornamented with gold.

aurelianen'sis, -is, -e Of Orleans (formerly Aurelia and Aurelianum), France.

aureo- In compound words signifying golden. Thus:
aureomac'ulatus golden-spotted
aureomar'ginatus edged with gold

aureoregi′na   golden queen
aureoreticula′tus   veined with gold
aureovariega′tus   gold-variegated

au′reus, -a, -um Golden.

auri′comus, -a, -um Having golden hair.

auri′cula Pre-Linnaean name for plants now included in the genus *Primula* but still commonly used in England where auriculas are favourites both for show and for Alpine gardens.

auricula′tus, -a, -um Eared; with an ear-shaped appendage or basal lobe.

Auri′nia f. From *aureus*, golden, in allusion to the flowers of *A. saxatilis* (*Alyssum saxatile*).

auri′tus, -a, -um Having long or large ears.

au′rum Gold (the metal), represented by the symbol *Au*. It was once believed that gold was of botanic origin, the reason being that the metal chiefly occurs embedded in igneous rocks in a crystalline, wiry, branchlike form which gives it the appearance of a plant in the rocks. This pleasant misconception persisted until the 16th century when Agricola, the Swiss metallurgist, refuted it.

australien′sis, -is, -e Australian.

austra′lis, -is, -e Southern.

austri′acus, -a, -um Austrian.

austri′nus, -a, -um Southern.

autumna′lis, -is, -e Pertaining to autumn.

avella′nus, -a, -um Of Avella Vecchia near Naples, southern Italy.

Ave′na f. Oats. The classical name.

avena′ceus, -a, -um Oat-like.

Averrho′a f. Ornamental trees, named after a Moorish physician, Averroes (1126–1198) who lived in Cordoba, Spain. He translated Aristotle into Arabic.

avicula′ris, -is, -e Relating to small birds.

a′vium Of the birds. Appropriately, the common sweet cherry is *Prunus avium*.

axilla′ris, -is, -e Borne in the axil; axillary.

Aza′lea f. Gr. *azaleos*, dry; in allusion to the occurrence of the original species, *Azalea procumbens* of Linnaeus (now *Loiseleuria procumbens*), in dry places in Lapland. The name was later transferred to the deciduous rhododendrons, for which it is not appropriate. While these azaleas are technically classified as rhododendrons, they are generally kept apart by gardeners and often by nurserymen.

azaleoi'des Resembling Azalea.

Aza'ra f. Evergreen shrubs with fragrant flowers, named in honour of J. N. Azara, a Spanish patron of science of the early 19th century.

Azaro'lus f. Latinized form of Italian vernacular *azzerulo*.

aze'darach Contracted form of Persian vernacular name *azaddhirakt*, noble tree, for *Melia azedarach*.

Azol'la f. Fairy moss. Genus of minute but quickly spreading aquatics. Possibly from Gr. *azo*, to dry; *olluo*, to kill; the plants die when they become dry.

azo'ricus, -a, -um From the Azores Islands in the eastern Atlantic.

azu'reus, -a, -um Sky-blue; azure.

B

Babia'na f. Latinized version of an Afrikaans word, babiaan (baboon). Baboons are said to eat this bulbous plant which is sometimes called baboon-root.

babingto'nii Commemorating Charles Cardale Babington (1808–1895), Professor of Botany at Cambridge.

babylo'nicus, -a, -um Babylonian. Specific name of the weeping willow (*Salix babylonicus*) which was thought by Linnaeus to be a native of south-west Asia. It belongs, in fact, to the Far East. The biblical 'willows' of the waters of Babylon are generally thought to be *Populus euphratica*.

bac'cans Berrying; becoming berry-like.

bacca'tus, -a, -um Berry-like; having fruits with a pulpy texture.

Bac'charis f. Deciduous and evergreen trees, shrubs, and herbs named in honour of Bacchus, the god of wine. It is suggested that the roots were sometimes used to spice wine. Groundsel tree is *B. halimifolia*.

bac'cifer, bacci'fera, -um Berry-bearing.

bacilla'ris, -is, -e Stick- or staff-like.

Backhou'sia f. In honour of James Backhouse (1794–1869), English nurseryman who travelled as a Quaker missionary in Australia.

Baco'pa f. From a South-American Indian name.

Bac'tris f. From Gr. *bactron*, a walking staff; suggesting the use to which the young stems of these palms are often put.

Baeo'thryon n. From Gr. *baios*, small; *thryon*, rush.

Bae'ria Herbs, mostly annuals, named in honour of Karl Ernst von Baer (1792–1876), celebrated Estonian anatomist and embryologist of German origin, from 1819 to 1834 professor at Königsberg, then academician in St. Petersburg, best known for his discovery of the mammalian egg.

bahia'nus, -a, -um Of Bahia (Salvador), Brazil.

baicalen'sis, -is, -e From Lake Baikal in eastern Siberia.

bai'leyi Of Bailey, honouring one of the following:

(*a*) Frederick Manson Bailey (1827–1915), Australian botanist, author of the *Queensland Flora*.

(*b*) Lieut.-Colonel Frederick Marshman Bailey (1882–1967), Indian Army, explorer of western China, western and south-eastern Tibet, author of *No Passport to Tibet*. His most celebrated botanical contribution was *Meconopsis betonicifolia*, for a time also known as *M. baileyi*. Few men have had a more exciting career as secret agent and explorer.

(*c*) Major Vernon Bailey, U.S. Army, who collected cacti from 1900 on.

(*d*) Liberty Hyde Bailey (1858–1954), from 1888 to 1903 Professor of Horticulture at Cornell University, founder of the Bailey Hortorium and author of many authoritative books (*Standard Cyclopedia of Horticulture, Hortus, Manual of Cultivated Plants*, etc.).

Baillo'nia f. Deciduous shrubs named in honour of the French botanist Henri Baillon (1827–1895).

ba'keri Usually in honour of John Gilbert Baker (1834–1920), British botanist, for many years a leading authority on petaloid monocotyledons (lilies, etc.), Keeper of the Kew Herbarium 1890–1899.

Ba'laka f. The vernacular name for these feather palms from the Fiji Islands.

balan'sae In honour of Benedict Balansa (1825–1891), French botanical collector in Asia Minor, North Africa and Indo-China.

balca'nicus, -a, -um From the Balkan Peninsula.

Baldel'lia f. In honour of a 19th-century Italian nobleman Bartolomeo Bartolini-Baldelli.

balden'sis, -is, -e Of Monte Baldo, northern Italy, east of Lake Garda.

baldschua'nicus Of Baljuan, Turkistan, central Asia, U.S.S.R.

balea'ricus, -a, -um From the Balearic Islands off the Mediterranean coast of Spain.

balfouria'nus, -a, -um; balfou'rii In honour of Sir Isaac Bayley Balfour (1853–1922), professor of botany and director (Regius Keeper) of Royal Botanic Garden, Edinburgh.

Ballo'ta f. Greek name for black horehound, one of the species of this genus of sub-shrubs or perennial herbs.

balsa'meus, -a, -um Similar to balsam; balsamic.

balsami'fera, -um Balsam-bearing.

bal'ticus, -a, -um Of the area of the Baltic Sea.

Bambu'sa f. Bamboo. Latinized version of the Malayan vernacular name.

bambusoi'des Resembling bamboo.

bana'ticus, -a, -um From the former Austria crownland of Banat, Roumania.

Banksia f. Australian honeysuckle. Named for Sir Joseph Banks (1743–1820), President of the Royal Society and virtual Director of Royal Botanic Gardens, Kew. Banks was a distinguished botanist in his own right as well as a wealthy and generous patron of science. A man of boundless interests, he made great contributions, both scholarly and material, to research. When he died, he left immense natural-history collections for the public good, now incorporated in those of the British Museum and the British Museum (Natural History), London. When Captain James Cook went to the Pacific in 1768 to observe the transit of Venus and investigate the possible existence of a great southern continent in the Pacific Ocean, Banks, as a young man, provided the scientific equipment of the expedition and engaged other trained naturalists and artists to accompany him on the voyage. He returned three years later with a rich harvest of plants, seeds, and general biological collections.

He made other expeditions, also at his own expense, to Labrador, Newfoundland, and Iceland. Later (1787) he promoted a scheme for bringing the breadfruit tree from Tahiti to be established in the West Indian islands where food for the slaves had become a very difficult problem. Captain William Bligh (see *Blighia*) was selected to undertake the mission in the *Bounty*. It ended in the mutiny of the crew and the loss of the ship. A later expedition, also commanded by Bligh, success-fully transported the breadfruit and thus made a contribution to West Indian food supplies.

banksia'nus, -a, -um; bank'sii In honour of Sir Joseph Banks (see *Banksia*).

Ba'phia f. Gr. *baphe*, a dye. The tree provides the bar or cam wood of commerce which yields a red dye. Violin bows are made from the wood.

Bapti'sia f. Wild indigo of North America. From Gr. *bapto*, to dye. It has sometimes been used as a substitute for true indigo.

barbaden'sis, -is, -e Of Barbados, West Indies.

Barba'rea f. Yellow rocket, winter or upland cress. Once generally known as herb of St. Barbara (*Herba Sanctae Barbarae*), patron saint of artillerymen and miners and protectress in thunderstorms.

bar'barus, -a, -um Foreign.

barba'tulus, -a, -um Somewhat bearded.

barba'tus, -a, -um Bearded; furnished with long weak hairs.

bar'bigera, -um Bearing barbs or beards.

barbiner'vis, -is, -e With veins barbed or bearded.

barbino'dis, -is, -e With beards at the nodes or joints.

barbula'tus, -a, -um Somewhat bearded or with a short beard.

barcinonen'sis, -is, -e From the neighbourhood of Barcelona, Spain.

Barcla'ya f. In honour of Robert Barclay (1757–1830), English botanist and horticulturist.

Barkhau'sia f. In honour of Gottlieb Barkhausen, 18th-century German botanist.

Barklya f. A large tree, a native of Australia, named in honour of Sir Henry Barkly (1815–1898), a British colonial governor.

Barle'ria Evergreen shrubs named in honour of Jacques Barrelier (1606–1673), Dominican monk, French botanist.

bar'ometz Specific epithet of *Cibotium barometz*, woolly fern or Scythian lamb. Barometz is from a Tartar word meaning lamb and refers to the woolly rootstocks. Cotton was for centuries supposed to be derived from this fern which was commonly considered to be half animal and half plant.

Baros'ma f. Gr. *barys*, heavy; *ŏsmē*, odour; in reference to the powerful scent of the leaves of these evergreen shrubs.

Barringto'nia f. Evergreen trees and shrubs, named after the Hon. Daines Barrington (1727–1800), English jurist, antiquary and botanist, correspondent of Gilbert White, and close friend of Joseph Banks.

Basel'la f. Malabar nightshade. The Latinized version of the vernacular name. The plant is edible and is cultivated in India and elsewhere in the tropics as a potherb.

baselloi'des Resembling *Basella*.

basila'ris, -is, -e Pertaining to the base or bottom.

basi'licus, -a, -um Princely, royal. The classical and now the specific epithet of the herb sweet basil in allusion to its reputed healing qualities.

bata'tas Specific epithet for the sweet potato, *Ipomoea batatas*. The word is the vernacular Carib (Haitian) Indian name for sweet potato. From it derives the English word potato.

Bateman'nia f. In honour of James Bateman (1811–1897), English orchid-grower, whose immense book *Orchidaceae of Mexico* (1837–1841) has some amusing illustrations by George Cruikshank.

batesia'nus, -a, -um; batesii West African plants so-named commemorate George Latimer Bates (1863–1940), American ornithologist and botanist (formerly a school-teacher), from 1895 to 1928 in west tropical Africa. South African succulents commemorate John Thomas Bates (1882–1966), British Army sergeant in first World War, then London bus conductor, who became a very successful grower and keen student of succulent plants. His unrivalled collection went to the University of California from Hounslow in 1960.

bathyphyl'lus, -a, -um Thickly leaved.

batrachoi'des Resembling water-buttercup (*Ranunculus* sect. *Batrachium*).

battandi'eri In honour of Jules Aimé Battandier (1848–1922), French botanist, authority on Algerian plants.

Baue'ra f. Small evergreen shrubs named by Sir Joseph Banks after Franz (1758–1840) and Ferdinand (1760–1826) Bauer, Austrian brothers, both botanical artists, never excelled for their beautiful and accurate illustrations of plants.

bauera'nus, -a, -um In honour of Ferdinand Bauer who was botanical artist on Captain Flinders' expedition to Australia.

Bauhi'nia f. Evergreen shrubs named after two illustrious Swiss botanists, the brothers Johann (Jean) and Caspar (Gaspard) Bauhin. Johann (1541–1613) was responsible for, among other works, the great *Historia Plantarum*, published nearly forty years after his death. Caspar (1560–1624) produced a valuable index of plant names and synonyms, *Pinax Theatri botanici* (1620). The two lobes of the leaf exemplify the two brothers.

bava'ricus, -a, -um Bavarian.

beania'nus, -a, -um; beanni In honour of William Jackson Bean (1863–1947), curator of the Royal Botanic Gardens, Kew, which he served for 46 years, celebrated above all as a dendrologist.

Beaufor'tia f. Flowering, heathlike shrubs named after Mary Somerset (*c.* 1630–1714), Duchess of Beaufort, patroness of botany early in the 18th century.

Beaumon'tia f. Nepal trumpet-flower, herald's-trumpet. Trees, shrubs, and climbers with showy white flowers, named in honour of Lady Diana Beaumont (d. 1831) of Bretton Hall, Yorkshire.

beauverdia'nus In honour of Gustave Beauverd (1867–1940), Swiss botanist, for many years curator of the Barbey-Boissier herbarium at Geneva, and an accomplished botanical artist.

beesia'nus, -a, -um Commemorating the seed and nursery firm of Bees, founded by A. K. Bulley of Neston, Cheshire and based on a nickname for Bulley and his sister, the 'busy Bs'. Bulley sent George Forrest to China for the introduction of new plants.

Befaria See *Bejaria*.

Bego'nia f. A great group of cultivated ornamental plants named for Michel Bégon (1638–1710), Governor of French Canada and a patron of botany.

begonifo'lius, -a, -um Begonia-leaved, i.e. with markedly asymmetric leaves like a begonia.

Beja'ria f. Evergreen shrubs named in honour of José Bejar, 17th-century Spanish botanist. The name was erroneously published by Linnaeus as *Befaria*.

Belamcan'da f. Blackberry lily, leopard flower. Latinized version of the East Asiatic vernacular name, Malayalam *balamtandam*, Sanskrit *mātākanda*, for this tuberous-rooted, herbaceous perennial.

bel'gicus, -a, -um Of the Netherlands as a whole or of Belgium.

be'lladonna Italian word meaning beautiful lady. Specific epithet of *Atropa* and *Amaryllis*. Ladies used it to give brilliancy to the eyes—a property of the juice being to dilate the pupil. That vision was affected was probably not considered important.

Belleva'lia f. Small-flowered bulbous plants commemorating Pierre Riche de Belleval (1564–1632), founder in 1593 of the Montpellier botanic garden.

bellidifo'lius, -a, -um With leaves like *Bellis*.

bellidifor'mis, -is, -e Daisy-like.

bellidioi'des Resembling *Bellium*.

Bel'lis f. Daisy. L. *bellus*, pretty.

Bell'ium n. From *Bellis*, daisy. Only for technical reasons is it not included in *Bellis* which it much resembles. The English name 'Little Mary' is a pun on the Latin name *Bellium minutum* perpetrated by Clarence Elliott.

belloi'des Resembling *Bellis* or daisy.

bel'lus, -a, -um Beautiful, handsome.

Belope'rone f. Gr. *bĕlŏs*, an arrow; *pĕrŏnē*, buckle, rivet; in allusion to the arrow-shaped connective, the part of the filament of the stamen connecting the anther lobes.

belophyl'lus With arrow-shaped leaves.

benedic'tus, -a, -um Blessed; well spoken of.

bengalen'sis, -is, -e (sometimes **benghalensis**) From Bengal, India.

Beninca'sa f. Annual running squash-like herbs with edible fruits named for Count Giuseppe Benincasa (d. 1596), Italian botanist who founded the Botanic Garden in Pisa.

benthamia'nus, -a, -um; bentha'mii In honour of the celebrated English botanist George Bentham (1800–1884), author of numerous scholarly and fundamental works including the *Flora Australiensis* and most of the *Genera Plantarum* of Bentham and Hooker.

Benthami'dia f. In honour of George Bentham (see above).

ben'zoin From an Arabic vernacular word meaning aromatic gum. See *Lindera*.

Berberido'psis f. From *Berberis*, barberry, and Gr. *opsis*, similar to.

Ber'beris f. Barberry. The Latinized form of the Arabian name for the fruit. This large genus of shrubs comprises several hundred species.

Berche'mia f. Deciduous shrubs named for M. Berchem, French botanist of the 17th century.

Berge'nia f. Decorative perennials named for Karl August von Bergen (1704–1760), professor at Frankfurt an der Oder.

Bergerocac'tus m. Named for Alwyn Berger (1871–1931), German horticulturist and botanist with a special interest in succulents, for some years Superintendent of the Hanbury Garden, La Mortola, Italy.

Berkhe'ya f. In honour of Jan Le Francq van Berkhey (1729–1812), Dutch botanist.

Berlandie'ra f. Perennial herbs and sub-shrubs named for J. L. Berlandier (d. 1851), Belgian botanist, who explored in Texas and New Mexico.

bermudia'nus, -a, -um From Bermuda.

berolinen'sis, -is, -e Of Berlin, Germany.

Bertero'a f. Dwarf herbs named for Carlo Giuseppe Bertero (1789–1831), Italian physician who botanized in Guadeloupe and other West Indian islands and later practised in Chile. He was drowned at sea when returning from Tahiti to Chile.

Bertholle'tia f. Brazil-nut. Named for Claude-Louis Berthollet

(1748–1822), French chemist, who discovered the bleaching qualities of chlorine. The Brazil-nut tree (*B. excelsa*) grows to great size. The globular fruits consist of a hard, woody casing packed with about twenty of the nuts of commerce, each in its own shell.

Bertolo'nia f. Foliage plants named for Antonio Bertoloni (1775–1869), Italian botanist, professor at Bologna.

bessarab'icus, -a, -um From Bessarabia in eastern Roumania.

Bes'sera f. In honour of W. S. J. Th. von Besser (1784–1842), Austrian botanist, professor at Kremenets.

Be'ta f. Latin name. Cultivated from remote times, the genus includes the familiar red and the white sugar beet, both varieties of *B. vulgaris*, and Swiss chard (var. *cicla*). About half of the world's sugar supplies derive from beet.

beta'ceus, -a, -um Beetlike.

Beto'nica f. Variant of *Vettonica*, name of a plant which grew in Spain.

betonicifo'lius, -a, -um With leaves like betony.

Be'tula f. Birch. The Latin name. The tree has special emotional appeal for the Northern peoples and is surrounded with legend and symbolism. Oil of birch gives to Russian leather its particular fragrance.

betulifo'lius, -a, -um Having leaves like birch.

betuli'nus, -a, -um; betuloi'des Resembling birch.

bi- Two.

bi'color Of two colours.

bicor'nis, -is, -e; bicornu'tus, -a, -um Two-horned; having two horns or hornlike spurs.

Bi'dens f. Tickseed, beggar's ticks. L. *bis*, twice; *dens*, a tooth; in reference to the two teeth on what is commonly called the seed but which, correctly, is the cypsela or dry one-seeded fruit. Other names are stick-tight and bur-marigold.

bidenta'tus, -a, -um Having two teeth.

bien'nis, -is, -e Biennial.

bi'fidus, -a, -um Cleft into two parts.

biflo'rus, -a, -um Twin-flowered.

bifo'lius, -a, -um Twin-leaved.

Bifrena'ria f. L. *bis*, twice; *frenum*, a bridle; in allusion to the two caudicles by which the pollen masses are connected with their gland in these epiphytic orchids.

bifurca'tus, -a, -um Bifurcate; forked into two generally almost equal stems or branches.

Bigelovia

Bigelo'via f. Plumed golden rod. A North American shrubby and herbaceous perennial named in honour of Dr. John M. Bigelow (1787–1879), Boston physician who assisted Engelmann (see *engelmannii*) on the U.S.–Mexican boundary mission. He collected regularly while visiting his patients in the Boston area on horseback. One plant collected and described by him, *Rubus sempervirens*, does not seem to have been noted since. Professor Liberty Hyde Bailey, who was reclassifying the genus *Rubus* at the close of his life, had Bigelow's herbarium specimen and would encourage anyone he knew to be going to the Boston area to hunt for the plant—even offering to pay for a telegram if it were found. It is not, however, a very noticeable plant and it would take the eye of a Bigelow or a Bailey to detect it.

Bigno'nia f. Trumpet-creeper or cross vine. Named in honour of Abbé Jean Paul Bignon (1662–1743), Librarian to Louis XIV.

bignonioi'des Resembling *Bignonia*.

bi'hai Tropical American vernacular name for the wild plantain.

biju'gus, -a, -um Double-yoked; two pairs joined.

Billardie'ra f. Apple berry. Evergreen climbers named in honour of Jacques Julien de La Billardière (1755–1834), French explorer and botanist.

Billber'gia f. Evergreen plants named in honour of J. G. Billberg (1772–1844), Swedish botanist.

bipinna'tus, -a, -um Two-pinnated. A compound leaf is pinnate when the leaflets are arranged feather fashion. It is bipinnate when the leaflets themselves have leaflets arranged feather fashion.

birma'nicus, -a, -um From Burma.

Bischo'fia f. An ornamental tropical tree named in honour of Gottlieb Wilhelm Bischoff (1797–1854), professor of botany at Heidelberg, author of important works on botanical terminology.

Biscutel'la f. L. *bis*, twice; *scutella*, a small flat dish; in allusion to the form of the fruits of these annual or perennial herbs.

bisec'tus, -a, -um Divided into two equal parts.

biserra'tus, -a, -um Double-toothed, i.e. with the teeth on the leaves being themselves toothed.

Bismar'ckia f. A palm, named in honour of Prince Otto von Bismarck (1815–1898), first Chancellor of the German Empire.

Bistor'ta f. Medieval plant name from *bis*, twice; *tortus*, twisted; referring to the twisted root.

biterna'tus, -a, -um Twice ternate. When leaflets are borne in threes, as in clover, they are described as ternate. A plant is biternate when the three divisions each bear three leaflets.

bitumino′sus, -a, -um Tarry; in some way resembling bitumen.

bival′vis, -is, -e Having two valves. The two halves of a pea-pod are the valves.

Bi′xa f. Latinized version of the South American vernacular name for this evergreen flowering tree, the source of annatta dye.

Blacksto′nia f. In honour of John Blackstone (d. 1753), English apothecary and botanist.

Blandfor′dia f. In honour of George Spencer-Churchill (1766–1840), Marquis of Blandford, then 5th Duke of Marlborough, who had a celebrated garden at Whiteknights, Reading.

blan′dus, -a, -um Mild; not strong or bitter; pleasing, charming.

Blech′num n. From the classical Greek name *blēchnon* for a fern, probably not this one.

ble′o Brazilian vernacular name for *Pereskia bleo*.

blepharoca′lyx With fringed calyx.

blepharophyl′lus, -a, -um With leaves fringed like eyelashes, from Gr. *blĕpharon*, eye-lid; *phyllon*, leaf.

Bletil′la f. Diminutive of *Bletia*, a genus of terrestrial orchids of no horticultural interest which it closely resembles. Named in honour of Luis Blet, a Spanish apothecary who had a botanic garden in Algeciras towards the end of the 18th century.

Bli′ghia f. Akee tree. A tropical tree with edible fruit named in honour of William Bligh, master of the *Resolution* on Cook's third voyage, and captain of the *Bounty*. Bligh is notorious for the mutiny of the *Bounty* and less well known for his extraordinary feat of navigating 4,000 miles across the Pacific to the island of Timor in an open boat with the loyal remnant of his crew, virtually without loss of a man. His mission on the *Bounty* was to bring the breadfruit tree from Tahiti to the West Indies, as planned by Sir Joseph Banks (see *Banksia*). Bligh succeeded in a later voyage and the breadfruit is now established in the West Indies. Later he became Governor of New South Wales where, in 1808, he was again involved in a mutiny.

blitoi′des Resembling *Blitum* (an old name for strawberry-blite), a coarse weed with a red fruit.

Bloome′ria f. Bulbous plants named in honour of Dr. H. G. Bloomer (1821–1874), pioneer Californian botanist.

Blumenba′chia f. Plants covered with stinging hairs. Named in honour of Johann Friederich Blumenbach (1752–1840), Professor of Medicine at Göttingen, and the virtual founder of modern anthropology.

Bly′xa f. Apparently from Gr. *blyzo*, gush forth.

Bocco′nia f. Plume poppy. Named in honour of Paolo Boccone (1633–1703), Italian monk and physician, who spent much of his life in Sicily.

bodini′eri In honour of Emile Marie Bodinieri (1842–1901), French missionary in Kweichow, China, who made big plant collections.

Boehme′ria f. China-grass. Named in honour of George Rudolf Boehmer (1723–1803), professor of botany and anatomy at Wittenberg, Germany. One of the species, *B. nivea*, is the source of ramie, a very fine fibre, taken from the inner bark. Ramie is one of the strongest fibres known but, owing to certain technical drawbacks, it is not widely used commercially. Its principal use is in the manufacture of gas mantles.

Boenninghause′nia f. After Clemens M. F. von Boenninghausen (1785–1864), doctor at Münster, Germany, author of a local flora.

boeo′ticus, -a, -um Of Boeotia, the district around Thebes northwest of Athens, Greece.

boissi′eri In honour of Edmond Boissier (1810–1885), celebrated Swiss botanist, describer of many species from Spain and the Near and Middle East, author of *Flora Orientalis*, etc. The great tragedy of Boissier's life was the death of his young wife Lucille (*née* Butini, 1822–1849) when accompanying him on a botanical expedition in Spain; *Chionodoxa luciliae* and *Omphalodes luciliae* commemorate her.

bolan′deri In honour of Henry Nicholson Bolander (1831–1897), Californian botanist.

Bolbi′tis f. From Gr. *bolbos*, bulb. With reference to the bulbously thickened veinlets of these ferns.

bol′dus Latinized form of a Chilean vernacular name.

bolivia′nus, -a, -um From Bolivia, South America.

Bol′lea f. In honour of Carl Bolle (1821–1909) of Berlin, German dendrologist and ornithologist who collected in the Canary and Cape Verde islands.

Bolto′nia f. False chamomile. Herbaceous perennials and annuals, named in honour of James Bolton (d. 1799), British botanist.

Boma′rea f. Handsome flowering climbers, named after Jacques Christophe Valmont de Bomare (1731–1807), French patron of science.

Bom′bax n. From Gr. *bombyx*, silk; in allusion to the fluffy, silky hairs which fill the seed capsule. This is the cotton-tree of the East Indies.

bomby′cinus, -a, -um Silky.

bo′na-nox Good-night. Specific epithet for a night-flowering morning glory.

bonarien'sis, -is, -e Of Buenos Aires, Argentina.

bon'duc Arabic name for the hazel nut. Specific epithet of the leguminous *Guilandina bonduc*, with nut-like seeds.

Bongar'dia f. In honour of August Gustav Heinrich Bongard (1786–1839), German botanist who spent most of his life in Russia.

bononien'sis, -is, -e Of Bologna (formerly Bononia), Italy.

bonus-henri'cus Literally, good Henry. Specific epithet of a *Chenopodium* called good-King-Henry and sometimes mercury. It is a hardy, rank vegetable which may be eaten like spinach.

boo'thii In honour of T. J. Booth (b. 1829), nephew of Thomas Nuttall, who collected in Assam in 1850.

Bora'go f. The herb borage (pronounced burrij) possibly from L. *burra*, a hairy garment; in allusion to the hairy leaves. The leaves have a fragrance like cucumber and are used in such mixtures as claret-cup, giving a delicate but distinctive flavour and coolness.

Borassus f. From Gr. *bŏrassŏs*, immature spandix of date palm.

Borbo'nia f. Evergreen shrubs named for Gaston de Bourbon (1608–1660), Duke of Orleans and son of Henry IV of France, a patron of botany.

borbo'nicus, -a, -um
(a)   From the island of Réunion in the Indian Ocean, once called Île Bourbon.
(b)   In honour of the Bourbon kings of France.

borea'lis, -is, -e Northern.

borinque'nus, -a, -um From the island of Puerto Rico, called Borinquen by the Spanish.

borneen'sis, -is, -e From Borneo.

bornmuel'leri In honour of Joseph Bornmüller (1862–1948), traveller and botanical collector especially in Balkan Peninsula and Asia Minor, for many years curator of the Haussknecht Herbarium at Weimar.

Boro'nia f. Fragrant-flowered evergreen shrubs named by J. E. Smith for Francesco Borone (d. 1794 at age of 25), an Italian lad from Milan, engaged by J. E. Smith on his European tour in 1787, who accompanied Afzelius as an assistant in Sierra Leone and John Sibthorp in Greece. He was killed by an accidental fall from a window in Athens. Smith wrote: 'I think he had more acuteness in finding out specific differences of plants than anybody I ever knew.'

Borzicac'tus m. A genus of *Cactaceae* named in honour of Antonino Borzi (1832–1921), Italian botanist.

Bossiae'a f. Evergreen shrubs, usually with yellow flowers, named in honour of de Boissieu La Martinière, doctor of medicine and botanist, on the voyage of La Pérouse, the noted French navigator, who was lost at Vanikoro in the Santa Cruz Islands in 1788.

Boswel'lia f. Evergreen trees, source of the biblical resin, frankincense. Named in honour of James Boswell (1740–1795), friend and biographer of Dr. Samuel Johnson.

Botry'chium n. Moonwort. Grape fern. From Gr. *botrys*, a bunch of grapes; a reference to the bunchlike formation of the spore-bearing organs of these deciduous ferns.

botryoi'des Resembling a bunch of grapes.

Bougainvill'ea f. Perhaps the handsomest and most widely planted tropical vine. Named in honour of Louis Antoine de Bougainville (1729–1811), sailor and explorer, who sailed round the world 1767–1769, his name being commemorated in several South Pacific place names. He was a noted mathematician, scientist, lawyer, soldier and author, besides being a Fellow of the Royal Society of London.

bourgeaua'nus, -a, -um; bourgea'ui Named in honour of Eugène Bourgeau (1813–1877), French botanical traveller, who collected in France, Spain, Corsica, North Africa, Asia Minor and North America.

Boussingaul'tia f. Madeira-vine. Named for Jean-Baptiste Joseph Boussingault (1802–1887), noted French agricultural chemist.

Bouvar'dia f. Flowering evergreen shrubs named for Dr. Charles Bouvard (1572–1658), Superintendent of the Jardin du Roi, Paris.

Bowie'a f. In honour of James Bowie (c. 1789–1869), Kew gardener who collected in Brazil and South Africa.

Bowke'ria f. Evergreen shrubs named for James Henry Bowker (d. 1900) and his sister Mrs. Mary Elizabeth Barber (d. 1899), South African botanists. *Aloe barberae* was named after the latter.

bowlesia'nus, -a, -um In honour of Edward Augustus Bowles (1865–1954), distinguished English amateur horticulturist and botanist, an authority especially on *Crocus* and *Narcissus*.

Boyki'nia f. Perennial herbs named in honour of an American field botanist, Dr. Samuel Boykin (1786–1846) of Milledgeville, Georgia.

Brabe'ium n. From Gr. *brabĕiŏn*, a prize.

brachia'tus, -a, -um Branched at right angles; armlike.

brachy- In compound words signifying short. Thus:

brachyan'thus	with short flowers
brachy'botrys	short-clustered
brachycar'pus	with short fruits
brachy'cerus	short-horned

brachype′talus short-petalled
brachyphyl′lus with short leaves
brachysi′phon with short tube

Brachy′chiton m. Bottle tree. From Gr. *brachys*, short; *chitōn*, a tunic; a reference to the overlapping scales.

Brachy′come f. Swan River daisy. From Gr. *brachys*, short; *kŏmē*, hair; in allusion to the short bristles of the pappus.

bractea′tus, -a, -um Having bracts. A bract is a leaflike and often a brightly coloured organ generally just below the flower cluster. The red 'flower' of the *Poinsettia* is, in fact, composed of bracts, the real flower being somewhat insignificant and generally yellow. Other showy bracts are the 'flowers' of dogwood and *Bougainvillea*.

bracteo′sus, -a, -um With well-developed or conspicuous bracts.

Bra′hea f. Fan-leaved palms, named in honour of Tycho Brahe (1546–1601), noted Danish astronomer, who was responsible for some of the great scientific advances of his day.

brandegea′nus, -a, -um In honour of Townsend Stith Brandegee (1843–1925), American civil engineer, noted for his studies of plants of California and Mexico.

Brase′nia f. Water-shield. The origin of the botanical name for this aquatic plant is obscure.

brasilien′sis, -is, -e Brazilian.

Brassa′vola f. A genus of epiphytic orchids named after Antonio Musa Brassavola (1500–1555), Italian botanist and physician, professor at Ferrara.

Bras′sia f. A genus of epiphytic orchids named after William Brass (d. 1783), English botanist, who collected in West Africa for Sir Joseph Banks (see *Banksia*) and others in 1782–1783.

Bras′sica f. The cabbage tribe. The classical Latin name for cabbage.

brassicifo′lius, -a, -um Cabbage-leaved.

brau′nii In honour of Alexander Carl Heinrich Braun (1805–1877), professor of botany at Berlin.

Bravo′a f. In honour of Leonardo and Miguel Bravo, obscure 19th-century Mexican naturalists.

bre′vis, -is, -e Short. Also:
brevifol′lius, -a, -um with short leaves
brevipeduncula′tus, -a, -um with a short flower stalk
brevisca′pus, -a, -um with a short scape
brevisty′lis, -is, -e with a short style

Brevoor'tia f. Floral firecracker. Named for J. C. Brevoort (1811–1887) of New York State University. The single species, *B. ida-maia*, is named for Ida May Burke, daughter of a California stage-coach driver, who brought it to the attention of Alphonso Wood (1810–1881), American botanist.

bre'weri In honour of William Henry Brewer (1828–1910), American botanist, Professor of Agriculture at Yale, pioneer botanical explorer in California.

Brey'nia f. A genus of shrubs and small trees. Named in honour of Jacob Breyne (1637–1697), merchant in Danzig, and his son Johann Philipp Breyne (1680–1764), physician in Danzig, authors of works on rare and little-known plants.

Brickell'ia f. Herbs or small shrubs named in honour of Dr. John Brickell (*c.* 1749–1809), Irish physician who settled in Georgia, U.S.A.

britan'nicus Of Great Britain.

Brittonas'trum n. The name of a labiate. In honour of Nathaniel Lord Britton (1859–1934), for many years director of the New York Botanical Garden and author of important floras of the eastern United States and various West Indian islands.

Bri'za f. Quaking grass. The Greek name for one of the food grains, possibly rye.

brizifor'mis, -is, -e; brizoi'des Resembling quaking grass or *Briza*.

Brodiae'a f. Cormous plants named for James Brodie (1744–1824) of Brodie, Elgin, Scottish botanist, primarily interested in *Algae*.

Brome'lia f. Perennial herbs with stiff, pineapple-like leaves, named in honour of Olof Bromel (1629–1705), Swedish botanist.

Bromhea'dia f. In honour of Sir Edward Thomas ffrench Bromhead (1789–1855), High Steward of Lincoln.

bromoi'des Resembling brome grass or *Bromus*.

Bro'mus m. Brome grass. From Gr. *brŏmŏs*, oats.

bronchia'lis, -is, -e Useful in treating bronchitis.

Bro'simum n. From Gr. *brōsimŏs*, edible. The genus includes the West Indian bread-nut which also provides the beautifully marked snake-wood. Another species of the genus is the cow-tree, the *palo de vaca*, of South America, which has milky sap reputed to be as rich and wholesome as cow's milk.

Broughto'nia f. Epiphytic orchids named for Arthur Broughton (d. 1796), English botanist and physician from Bristol who emigrated to Jamaica in 1783 and made drawings of its plants.

Broussone'tia f. Named for Pierre Marie Auguste Broussonet (1761–1807), professor of botany at Montpellier, France. One species is the paper mulberry, the bark of which yields a kind of paper. It is a good street tree, and a few specimens may be seen in New York.

Browal'lia f. Shrubs or herbs with blue or white flowers named in honour of Johan Browall (1707–1755), Bishop of Abo, a Swedish botanist, who in 1739 defended the sexual system of classification proposed by Linnaeus.

Brow'nea f. Evergreen flowering trees and shrubs named for Patrick Browne (1720–1790), Irish physician and author of *A Civil and Natural History of Jamaica*.

Bruckentha'lia f. Spike heath. Named in honour of Samuel von Bruckenthal (1721–1803) and Michael von Bruckenthal (*fl.* 18th century), Austrian noblemen resident in Transylvania.

Brugman'sia f. Named for Sebald Justin Brugmans (1763–1819), professor of natural history at Leiden. This genus of South American tropical shrubs is often included in *Datura*.

bruma'lis, -is, -e Winter-flowering. Literally, pertaining to the winter solstice.

Brunel'la f. Alternative spelling of *Prunella*, first used by Brunfels.

Brunfel'sia f. Evergreen, free-flowering shrubs named for Otto Brunfels (1489–1534), Carthusian monk and one of the earliest German botanists, who in 1530 published the first good figures of plants.

brunneifo'lius, -a, -um Brown-leaved.

Brun'nera f. Perennial herbs named for Samuel Brunner (1790–1844), Swiss botanist who collected in Italy, the Crimea and west tropical Africa.

bru'nneus, -a, -um Brown.

Bruno'nia f. In honour of Robert Brown (1773–1858), Scottish botanist, who circumnavigated Australia on Flinders' voyage in 1801–1803 and later became keeper of botany in the British Museum, London. Among his many important discoveries was the Brownian movement of minute particles. The epithets *brownii, brunonianus* and *brunonis* commemorate him and also this Australian herbaceous perennial.

Brunsdon'na f. Hybrid between *Amaryllis* (*Belladonna*) and *Brunsvigia*.

Brunsvi'gia f. In honour of Carl Wilhelm Ferdinand (1713–1780), Duke of Brunswick-Lüneburg, an enlightened German prince who encouraged the arts and sciences in his domains. This is Josephine's lily of South Africa which differs only technically from *Amaryllis*.

Bry'a f. In honour of J. T. de Bry (1564–1617), engraver.

bryoi'des Resembling moss.

Bryo'nia f. Bryony. Gr. name *bryōnia* used by Dioscorides.

Bryophyl'lum n. From Gr. *bryo*, to sprout; *phyllon*, a leaf; in allusion to the vegetative buds on the edges of the leaves of these succulents which need only to be laid on damp sand to start new plants. Their sturdy constitution and ease of propagation make them popular house plants, with such English names as air plant, life-plant, and floppers.

buccinator'ius, -a, -um; buccina'tus, -a, -um Shaped like a crooked trumpet or horn.

bucha'ricus, -a, -um From the vicinity of Bokhara, Turkistan, U.S.S.R.

Buck'leya f. A parasitic shrub. After Samuel Botsford Buckley (1809–1884), American botanist.

Buddle'ja f. Butterfly bush. Ornamental small trees and shrubs. Named in honour of the Reverend Adam Buddle (1660–1715), English botanist and vicar of Farmbridge in Essex. Often spelled *Buddleia* but spelled *Buddleja* by Linnaeus.

buddleoi'des Resembling *Buddleja*.

bufo'nius, -a, -um Pertaining to toads; growing in damp places.

bul'bifer, bulbi'fera, -um Bulb-bearing; often used of plants with flowers replaced by bulbils.

bulbifor'mis, -is, -e Shaped like a bulb.

Bulbi'ne f. Herbaceous plants of the Lily family. Gr. *bŏlbinē*, name of a bulbous plant, from Gr. *bŏlbŏs*, a bulb.

Bulbinel'la f. Diminutive of *Bulbine*, now used as the name of a particular herbaceous tuberous-rooted perennial.

Bulboco'dium n. Gr. *bŏlbŏs*, a bulb; *kodion*, wool.

Bulbophyl'lum n. From Gr. *bŏlbŏs*, a bulb; *phyllon*, a leaf. The leaves of these epiphytic orchids grow from a pseudobulb.

bulbo'sus, -a -um Bulbous; may be applied to any swollen underground stem, not necessarily a bulb.

bulga'ricus, -a, -um Bulgarian.

bulla'tus, -a, -um Bullate; blistered or puckered; usually applying to foliage.

Bumal'da f. After Ovidio Montalbani (1601–1671), of Bologna, Latinized name Bumaldus. Epithet for a *Staphylea*.

Bume'lia f. From Gr. *boumĕlios*, Greek name for the ash tree, but its application to these small American trees and shrubs is obscure.

bungea'nus, -a, -um; bun'gei In honour of Alexander von Bunge

(1803–1890), Russian botanist, professor of botany at Dorpat, monographer of *Astragalus*.

Bu'nium n. Gr. *bounĭŏn*, name of the earth-nut.

Buphthal'mum n. From Gr. *bous*, an ox; *ophthalmos*, an eye; an allusion to the appearance of the flower.

Bupleu'rum n. Hare's ear. Herbaceous plants or shrubs. From the Gr. *boupleuros*, meaning ox-rib, a name for another plant.

Burchel'lia f. Evergreen shrubs. In honour of W. J. Burchell (1781–1863), English traveller in Africa and Brazil.

burea'vii In honour of Edouard Bureau (1830–1918), French botanist, authority on *Bignoniaceae*.

burma'nicus, -a, -um From Burma.

bur'sa-pasto'ris Old generic name, now the specific epithet for the weed shepherd's purse, which is the literal translation.

Bursa'ria f. L. *bursa*, a purse; in allusion to the shape of the seed pod of these shrubs or small trees.

Burse'ra f. Trees named after Joachim Burser (1583–1649), German physician and botanist, maker of a celebrated herbarium studied by Linnaeus and now at Uppsala, Sweden.

Bu'tea f. Named for John Stuart, third Earl of Bute (1713–1792), for whom *Stewartia* (*Stuartia*) was also named, British Prime Minister 1762–1763 but a better plantsman than politician, the unofficial director of Kew before Sir Joseph Banks. *B. monosperma*, 'Flame of the Forest', with big red flowers is one of the most conspicuous of Indian trees.

Bu'tia f. Vernacular name in Brazil for a species of palm (*B. capitata*) with numerous varieties.

Bu'tomus m. Flowering rush. From Gr. *bous*, ox; *temno*, to cut; in allusion to the sharp leaf margins which make the plant quite unsuitable for fodder. The baked root-stocks are said to be edible.

Butyrosper'mum n. From Gr. *boutyron*, butter; *sperma*, seed; in allusion to the fatty seeds.

buxbau'mii In honour of Johann Christian Buxbaum (1693–1730), German botanist who collected plants in the Near East.

buxifo'lius, -a, -um Box-leaved.

Bu'xus f. Box. The classical Latin name.

byzanti'nus, -a, -um From Istanbul (Constantinople), classically Byzantium.

C

Cabom'ba f. Fanwort, fish grass. Latinized version of the native Guiana name for these perennial aquatics.

cacaliafo'lius, -a, -um With leaves like *Cacalia*, a genus of virtually no garden interest.

caca'o Aztec name for the chocolate tree and still the name both for the tree and the chocolate in many countries. *Theobroma cacao* is the botanical name.

cachemi'ricus, -a, -um From Kashmir.

Cac'tus m. Name applied by the ancient Greeks to some spiny plant in no way connected with these American plants. Strictly *Cactus* was the name of one genus only, but it has been adopted loosely as a vernacular name for the whole family, comprising about 85 genera and 2,000 species.

cad'micus, -a, -um Metallic; like tin in appearance.

caerules'cens Becoming blue, bluish. Often incorrectly spelled 'coerulescens'.

caeru'leus, -a, -um Dark blue. Often incorrectly spelled 'coeruleus'.

Caesalpi'nia f. Ornamental trees and shrubs named in honour of Andrea Cesalpini (1524/25–1603), Italian botanist, philosopher, and physician to Pope Clement VIII, author of *De Plantis* (1583) and other works.

cae'sius, -a, -um Bluish-grey.

caespito'sus, -a, -um Growing in dense clumps; cespitose.

caf'fer, caf'fra, caffrum; caffro'rum Of South Africa; Kaffir; from Arabic *kafir*, unbeliever, pagan.

caini'to West Indian vernacular for the star apple (*Chrysophyllum cainito*).

Caio'phora f. Now known as *Blumenbachia*, q.v. Gr. *kaios*, a burn; *phoreo*, to bear; in allusion to the stinging hairs of these usually climbing or trailing plants.

cai'ricus, -a, -um From Cairo, Egypt.

ca'jan Adaptation of Malay vernacular name *katjang* for the pigeon pea.

cajanifo'lius, -a, -um With leaves like the pigeon pea.

Caja'nus m. Latinized version of the vernacular name *katjang* for the nutritious pigeon pea which is widely grown throughout the tropics.

In India it is called 'dall', and is cooked into a tasty potage to be eaten with curries.

cala′ba West Indian vernacular name of *Calophyllum calaba*.

calab′ricus, -a, -um From Calabria in Italy.

Cala′dium n. Latinized form of Malay plant-name *kaladi* used by Rumphius.

calamifo′lius, -a, -um With reedlike leaves.

Calamin′tha f. Gr. *kalaminthē*, Greek name for savory, from Gr. *kalos*, beautiful, and *minthē*, mint.

Ca′lamus m. Cane and rattan palms. Gr. *kalamos*, a reed.

Calandri′nia f. Rock purslane. Named in honour of Jean Louis Calandrini (1703–1758), professor of mathematics and philosophy at Geneva.

Calan′the f. Gr. *kalos*, beautiful; *anthos*, a flower. Chiefly terrestrial orchids.

Calathe′a f. Gr. *kalathos*, a basket; in allusion to the flower cluster of these perennial foliage plants, which looks like flowers in a basket.

calathi′nus, -a, -um Basketlike.

calcara′tus, -a, -um Spurred.

calca′reus, -a, -um Pertaining to lime.

Calceola′ria f. Slipper-wort. L. *calceolus*, a slipper; in allusion to the form of the flower.

calce′olus Diminutive of *calceus*, shoe.

calci′cola Growing in limy soil.

Caldcluvia f. In honour of Alexander Caldcleugh (fl. 1803–1858), traveller in South America; evergreen shrub.

Calen′dula f. Marigold. L. *calendae*, the first day of the month and also the day on which interest must be paid. The allusion is to the long flowering of some of the species. *C. officinalis* is the pot-marigold which was used in English country cooking to flavour thick soups. A single head thrown into the pan imparts a pleasing taste.

calendula′ceus, -a, -um Like *Calendula* in colour.

califor′nicus, -a, -um Of California.

Cali′meris f. Gr. *kallos*, beauty; *meris*, a part.

Cal′la f. Presumably from Gr. *kallos*, beauty. *C. palustris*, the water-arum, is the only species in the genus. It has nothing to do with the popularly named calla-lily which belongs to the genus *Zantedeschia*.

Callian′dra f. Gr. *kalli-*, beautiful; *andros*, a stamen. These flowering evergreen shrubs and trees have conspicuous stamens.

callian'thus, -a, -um Having beautiful flowers.

Callicar'pa f. Gr. *kalli-*, beautiful; *karpos*, fruit. The shrubs are sometimes called beauty-berry or French mulberry.

callicar'pus, -a, -um With beautiful fruit.

Calli'coma f. Gr. *kalli-*, beautiful; *kome*, hair; in allusion to the tufted heads of the flowers on these Australian evergreen shrubs or trees.

callimor'phus, -a, -um Beautifully shaped; of beautiful form.

Callir'hoe f. Poppy-mallow. Showy herbs named in honour of the daughter of a minor Greek deity, Achelous, a river god.

callista'chyus, -a, -um With beautiful spikes.

callistegioi'des Resembling *Calystegia*, a genus of *Convolvulaceae* which includes *C. sepium*, the vicious bindweed; from *calyx*, calyx; *stĕgon*, cover; in allusion to the large bracts covering the calyx.

Calliste'mon m. Bottlebrush tree. Gr. *kalli-*, beautiful; *stemon*, a stamen; in allusion to the beauty of the flowers of these ornamental shrubs.

Calli'stephus m. China or annual aster. From Gr. *kalli-*, beautiful; *stephanos*, a crown; from the appearance of the flowers.

callis'tus, -a, -um Very beautiful.

calli'triche f. Starwort. From Gr. *kalli-*, beautiful; *trix, trichos*, hair.

Calli'tris f. Cypress pine. From Gr. *kalli-*, beautiful; *treis*, three; in allusion to the beautiful three-fold arrangement of its parts, i.e. the leaves, cone-scales and lobes of the columella are in threes.

callizo'nus, -a, -um Beautifully banded or zoned.

callo'sus, -a, -um Thick-skinned; with callouses.

Callu'na f. Heather or ling. From Gr. *kalluno*, to cleanse or adorn. This is an allusion to the common use of heather for making brooms.

Calocar'pum n. See *Achras*.

Caloced'rus f. Gr. *kalos*, beautiful; *cedrus*, cedar.

Caloce'phalus f. Gr. *kalos*, beautiful; *cephalos*, a head; an allusion to the flower of these Australian herbs or shrubs.

caloce'phalus, -a, -um With beautiful head.

Calochor'tus m. Mariposa lily. Gr. *kalos*, beautiful; *chortos*, grass.

Caloden'drum n. Cape chestnut. Gr. *kalos*, beautiful; *dendron*, a tree.

Calome'ria f. From Gr. *kalos*, beautiful, good; *mĕris*, a part; see *Humea*.

Calonyc'tion n. Moonflower. From Gr. *kalos*, beautiful; *nyx, nyktos*, night; in allusion to the beauty of these night-blooming plants. This group is now included in *Ipomoea*.

Calo'phaca f. From Gr. *kalos*, beautiful; *phake*, a lentil. These are leguminous shrubs not unlike laburnum.

Calophyl'lum n. Calaba tree. From Gr. *kalos*, beautiful; *phyllon*, a leaf. The leaves of these evergreen, aromatic trees are beautifully veined.

calophyl'lus, -a, -um With beautiful leaves.

calophytum Beautiful plant.

Calopo'gon m. Gr. *kalos*, beautiful; *pōgōn*, a beard; in allusion to the fringed lip of this native North American bog-orchid.

calostro'tus, -a, -um Beautifully covered; with a beautiful cover.

Calotham'nus m. Gr. *kalos*, beautiful; *thamnos*, a shrub.

Calo'tropis f. Gr. *kalos*, beautiful; *trŏpis*, ship.

caloxan'thus, -a, -um Of beautiful yellow.

Calpur'nia f. African trees and shrubs named by the learned Ernst Meyer after T. Julius Calpurnius (1st cent. A.D.), a second-rate Latin poet who imitated Virgil. The genus is closely related to and may be said to imitate the genus *Virgilia*.

Cal'tha f.; In England, marsh-marigold or king-cup; in America, cowslip. From the Latin name for a plant with a yellow flower—possibly *Calendula*.

cal'vus, -a, -um Bald; hairless; naked.

calycan'thema Varietal epithet of the *Campanula* known as the cup-and-saucer Canterbury bell. Literally, blossoming in the calyx.

Calycan'thus m. Carolina allspice. Gr. *kalyx*, calyx; *anthos*, a flower. The calyx and the petals are the same colour in the flowers of these North American aromatic deciduous shrubs.

calyci'nus, -a, -um Calyx-like.

Calyco'tome Gr. *kalyx*, calyx; *tŏmē*, a part left after cutting; a tree stump. In these deciduous spiny shrubs the upper part of the calyx drops after the flower opens.

Caly'pso f. Bog-orchid named in honour of the nymph Calypso, daughter of Atlas, who entertained Odysseus for seven years but, even with the promise of immortality, could not overcome his longing to return home. In the end, Zeus had to send Hermes to bid her release him.

calyptra'tus, -a, -um Bearing a calyptra, any caplike covering of a flower or fruit.

Calyste'gia f. From Gr. *kalyx*, calyx; and *stĕgŏn*, cover; with reference to the large bracts concealing the calyx in these *Convolvulus*-like plants.

Caly'thrix f. Gr. *kalyx*, calyx; *thrix*, a hair. In these heathlike shrubs, the divisions of the calyx end in long bristling hairs.

camaldolulen'sis, -is, -e Of the Camaldoli garden near Naples, Italy.

ca'mara South American vernacular name for a species of *Lantana*.

Camaro'tis f. Gr. *kamarotos*, an arch; a reference to the arched shape of the lip of the flower of these epiphytic orchids.

Camas'sia f. Camas or bear-grass. From the North American Indian name, *quamash*. The bulbs formed an important part of the local diet.

cambessede'sii Epithet applied to a peony, named after Jacques Cambessèdes (1799–1863), a French botanist who published a flora of the Balearic Islands, where this peony occurs.

cam'bricus, -a, -um Of Wales (Cambria).

Camel'lia f. Evergreen flowering trees and shrubs named for Georg Josef Kamel (1661–1706), Jesuit pharmacist, born at Brno, Moravia; who botanized from 1688 onwards in Luzon in the Philippines and wrote an account of the plants there which was published in 1704 by his English correspondent John Ray under his Latinized name, Camellus. Commercially the most important species is the tea plant, *Camellia sinensis*.

camelliflo'rus, -a, -um With flowers like *Camellia*.

Camoen'sia f. Shrubby climbers named in honour of Luis de Camoens (1524–1580), celebrated Portuguese epic poet and author of *The Lusiads*.

campaniflo'rus, -a, -um With bell-shaped flowers.

Campa'nula f. Canterbury-bell. Bellflower. Harebell. Diminutive of late L. *campana*, a bell; in allusion to the form of the flower.

campanula'rius, -a, -um; campanula'tus, -a, -um Bell-shaped.

campanuloi'des Resembling *Campanula*.

campbel'lii Named in honour of Dr. Archibald Campbell, Superintendent of Darjeeling and Political Agent to Sikkim, who accompanied Joseph Hooker on his celebrated Sikkim journey in 1849.

campes'tris, -is, -e Of the fields or open plains.

camphora'tus, -a, -um Pertaining to or resembling camphor.

Camphoros'ma f. Gr. *kamphora*, camphor; *ŏsmĕ*, an odour. The shrub smells like camphor.

Campsi'dium n. From its likeness to *Campsis* or trumpet creeper.

Cam'psis f. Trumpet creeper. Gr. *kampe*, something bent. The stamens of these deciduous climbing plants are curved.

Camptoso'rus m. Gr. *kamptos*, curved; *sorus*, a group of fern spore cases. In these ferns the sori are curved.

campylocar'pus, -a, -um With curved or bent fruit.

campylo'gynus, -a, -um With curved or bent ovary.

camtschatcen'sis, -is, -e; camtscha'ticus, -a, -um From the Kamchatka Peninsula on the Siberian coast.

canaden'sis, -is, -e Canadian, but used by early writers also to cover the north-eastern United States.

canalicula'tus, -a, -um Channelled; grooved.

Canan'ga f. The Malayan vernacular name for these tropical trees, one of which yields the perfume known as macassar oil.

canarien'sis, -is, -e From the Canary Islands.

Canari'na f. The type species of these herbaceous perennials comes from the Canary Islands.

Canava'lia f. Jack-bean. Latinized version of the Malabar vernacular name *kanavali* for these climbing herbs or sub-shrubs.

can'byi In honour of William Marriott Canby (b. 1831), American botanist.

cancella'tus, -a, -um Cross-barred; latticed.

candelab'rum Like a branched candlestick.

can'dicans Shining or woolly-white.

candidis'simus, -a, -um Very white.

can'didus, -a, -um Shining or pure white.

Candol'lea f. Evergreen flowering shrubs from Australia named in honour of Augustin Pyramus de Candolle (1778–1841), professor of botany first at Montpellier, then Geneva, Switzerland, founder of a widely adopted system of classification, author of many important botanical works, notably the *Prodromus Systematis naturalis*.

Canel'la f. Wild cinnamon. From L. *canna*, a reed, in allusion to the rolled bark.

canes'cens With hairs off-white or ashy-grey colour.

cani'nus, -a, -um Pertaining to dogs; applied metaphorically to an inferior kind, e.g. a scentless as opposed to a scented species.

Canis'trum f. L. *canistra*, a flat basket; in allusion to the appearance of the inflorescence of these Brazilian plants.

Can'na f. From Gr. *kanna*, a reed. Tropical plants with spectacular flowers, from America and Asia.

cannabi′nus, -a, -um Like *Cannabis* or hemp.

Can′nabis f. Hemp. The Latin and Greek name for the plant which, besides producing fibres, is also the source of such narcotic drugs as hashish and marijuana. For this reason cultivation is often prohibited.

cannifo′lius, -a, -um With leaves like a *Canna*.

cantab′ricus, -a, -um From Cantabria, a district of northern Spain, or the Cantabrian mountains.

cantabrigien′sis, -is, -e Relating to Cambridge, England, to be distinguished from *cantabricus*, *cantianus*, and *cantuariensis* (qq. v.)

cantia′nus, -a, -um Relating to Kent, England.

Can′tua f. Latinized version of the Peruvian name for these evergreen flowering trees and shrubs.

cantuarien′sis, -is, -e Relating to Canterbury, Kent, England.

ca′nus, -a, -um Off-white; ash-coloured.

capen′sis, -is, -e From the Cape of Good Hope, South Africa.

capera′tus, -a, -um Wrinkled.

capilla′ceus, -a, -um; capilla′ris, -is, -e Resembling hair; very slender.

capilla′tus, -a, -um Furnished with fine hairs.

capillifor′mis, -is, -e Like hair.

capilli′pes Slender-footed.

capil′lus-ve′neris Venus' hair. Specific epithet of *Adiantum*.

capita′tus, -a, -um Growing in a dense head—referring to the flowers, the fruit, or the whole plant.

capitella′tus, -a, -um; capitel′lus, -a, -um; capitula′tus, -a, -um Having a small head.

cappado′cicus, -a, -um From eastern Asia Minor, the ancient Province of Cappadocia.

Cap′paris f. Caper. The ancient Greek name for these evergreen shrubs. The flowerbuds of *C. spinosa* are pickled and known as capers.

capreola′tus, -a, -um Provided with tendrils.

cap′reus, -a, -um Pertaining to goats.

capricor′nis, -is, -e
(*a*) Like a goat's horn.
(*b*) From or below the Tropic of Capricorn in the Southern Hemisphere.

caprifo′lium From L. *caper*, goat; *folium*, leaf.

Cap'sicum n. Peppers, both hot and sweet. From Gr. *kapto*, to bite. The true peppercorn, however, is *Piper nigrum*.

capsula'ris, -is, -e Having capsules.

caracasa'nus, -a, -um From Caracas, Venezuela.

Caraga'na f. Latinized version of the Mongolian name for one of the species of these ornamental shrubs or small trees.

Carda'mine f. Bitter cress, cuckoo flower, lady's smock. From the Greek name for a plant of the cress family.

cardaminifo'lius, -a, -um With leaves like *Cardamine*.

Cardian'dra f. Gr. *kardia*, a heart; *andros*, an anther; in allusion to the shape of the anthers of these ornamental shrubby plants.

cardina'lis, -is, -e Scarlet; cardinal red.

cardiobasis With heart-shaped base.

Cardiocri'num n. Gr. *kardia*, heart; *krinon*, lily; referring to the heart-shaped leaves.

cardiope'talus, -a, -um With heart-shaped petals.

Cardiosper'mum n. Gr. *kardia*, heart; *sperma*, a seed. The black seed of these climbing herbs has a heart-shaped spot.

cardua'ceus, -a, -um Resembling a thistle.

Carduncellus m. Diminutive of *Carduus*.

cardun'culus, -a, -um Resembling a small thistle.

carducho'rum Of the Kurds, eastern Asia Minor.

Ca'rduus m. Thistle. The classical Latin name.

Car'ex f. Sedge. The classical Latin name.

caribae'us, -a, -um Of the Caribbean area.

Ca'rica f. Papaya. The L. name *carica*, Gr. *karike*, a kind of fig, was transferred to the papaya on account of its fig-like leaves. The melon-like fruit, with its delicate and distinctive flavour, is a natural source of pepsin and is therefore a tenderizer. Tough meat wrapped in the rind overnight is much improved. In India pregnant women are forbidden the fruit because the black seeds are said to cause miscarriage. (See *carica*.)

car'ica Specific epithet for fig, *Ficus carica*. The word also refers to Caria, a district in Asia Minor, where figs seem to have been extensively cultivated.

caricifo'lius, -a, -um Sedge-leaved.

carici'nus, -a, -um Resembling sedge or *Carex*.

carina'tus, -a, -um Having a keel.

77

carinthia′cus, -a, -um From Carinthia in Austria.

Caris′sa f. Latinized version of the Indian vernacular name for these spiny shrubs with edible fruits.

carle′sii In honour of William Richard Carles (*c.* 1867–1900) of the British consular service in China, who collected plants in Korea between 1883 and 1885.

Carli′na f. Carline thistle. Medieval plant name derived from Carolus (Charles) and referring to the tradition that Charlemagne, having been shown the root of this thistle by an angel, used it successfully as a remedy for the plague which prevailed in his army.

Carludovi′ca f. Palmlike plants named in honour of Charles IV (1748–1819) and his queen, Louise (1751–1819), of Spain.

Carmichae′lia f. Shrubs, becoming leafless when fully grown. This genus, confined to New Zealand and Lord Howe Island, was discovered in 1769 by Joseph Banks and Daniel Solander. It remained unpublished until 1825 when Robert Brown named it in honour of Captain Dugald Carmichael (1722–1827), Scottish army officer and botanist who collected plants at the Cape of Good Hope and in Mauritius, Tristan da Cunha and India, but not New Zealand.

carmineus, -a, -um Carmine.

Carnegi′ea f. Giant cactus or saguaro, state flower of Arizona. Named in honour of Andrew Carnegie (1835–1919), steel magnate and philanthropist.

It is the only species in the genus. It was named *Cereus giganteus* by Engelmann in 1848 but placed in a genus of its own as *Carnegiea gigantea* by Britton and Rose in 1908.

On the occasion of one of Carnegie's visits to the Desert Laboratory, Tucson, Arizona, the dramatic announcement was made that the feature of his welcome was to be the magnificent plant recently named in his honour. Carnegie was much moved and looked forward with keen anticipation to seeing what he understood was a new discovery. On seeing the giant cactus blooming in profusion, it was natural for Carnegie to inquire how so enormous and conspicuous a plant could have remained unnamed. When the matter was explained, Carnegie lost interest in the festivities.

carnerosa′nus, -a, -um From the Carnerosa Pass in Mexico.

car′neus, -a, -um; carni′color Flesh-coloured; deep pink.

carnio′licus, -a, -um From Carniola in north-western Yugoslavia.

carno′sulus, -a, -um Somewhat fleshy.

carno′sus, -a, -um Fleshy.

carolinia'nus, -a, -um; carolinen'sis, -is, -e; caroli'nus, -a, -um From North Carolina or South Carolina.

Caro'ta f. Variant of Gr. *karŏtŏn*, carrot.

carpa'ticus, -a, -um (sometimes **carpathicus**) From the Carpathian Mountains between Czechoslovakia and Poland.

Carpente'ria f. An evergreen flowering shrub named in honour of Professor William M. Carpenter (1811–1848), Louisiana physician.

carpinifo'lius, -a, -um With leaves like hornbeam.

Carpi'nus f. Hornbeam. The classical Latin name. The English name is said to derive from the use of the wood for ox-yokes.

Carpobro'tus m. Gr. *karpos*, fruit; *brōtŏs*, edible. These succulents were formerly included in *Mesembryanthemum*.

Carpo'detus m. Gr. *karpos*, fruit; *detos*, bound; from the fact that the fruit of these evergreen shrubs or small trees is contracted in the middle.

Carrie'rea f. Deciduous trees with attractive foliage and flowers; named in honour of Elie Abel Carrière (1816–1896), French botanist and horticulturist, for many years editor of *Revue Horticole.*

Car'thamus m. Distaff thistle, false saffron. From an Arabic word meaning to paint—an allusion to the brilliant colour yielded by the flowers.

carthusiano'rum Of the monks of the Carthusian Monastery of Grande Chartreuse near Grenoble, France.

cartilagi'neus, -a, -um Resembling cartilage.

Ca'rum n. Latin form of Greek name *karŏn*. Caraway, *C. carvi*, from Caria, the name of the district in Asia Minor where it was much grown, is one of the few aromatics growing widely throughout the northern temperate zone where it is much used to flavour cakes and breads. It yields an aromatic oil.

Ca'rya f. Hickory. From Gr. *karya*, a walnut tree, the fruit of which was known as *karyon*, a word also applied to other nuts. According to legend, Carya, daughter of the King of Laconia, was changed by Bacchus into a walnut tree.

caryophyl'lus, -a, -um From Gr. *karya*, walnut; *phyllŏn*, a leaf; referring to the aromatic smell of walnut leaves, which led to the use of the name for the clove and thence to the clove pink (*Dianthus caryophyllus*).

caryopteridifo'lius, -a, -um With leaves like *Caryopteris*.

Caryop'teris f. Blue spiraea. From Gr. *karyon*, a nut; *pteron*, a wing. The fruit is winged.

Caryo'ta f. Gr. *karyon*, nut. *C. urens* is the toddy palm, widely grown throughout India and the East generally. It produces sap freely—as much as three gallons daily from a mature tree. This palm 'wine' quickly ferments and very soon turns into the highly intoxicating toddy, looking and tasting rather like inferior gasoline. The availability of the tree is the despair of Indian prohibitionists.

caryoti'deus, -a, -um *Caryota*-like.

caryotifo'lius, -a, -um With leaves like *Caryota* or fishtail palm.

cashmeria'nus, -a, -um From Kashmir.

Casimiro'a f. Mexican evergreen trees, named for Casimiro Gomez de Ortega (1740–1818), Spanish physician and botanist, director of the Madrid botanic garden from 1771 to 1801.

cas'picus, -a, -um; cas'pius, -a, -um From the shores of the Caspian Sea.

Cassan'dra f. Trojan princess with the gift of prophecy, whose warnings of calamity were disregarded but came true. The custom of naming genera of *Ericaceae* after figures in Greek mythology (e.g. *Andromeda*, *Cassiope*) led to its use for this genus also called *Chamaedaphne*.

Cas'sia f. Senna. The Greek name for a genus of leguminous plants which provide the senna leaves and pods that are important in pharmacy. Not to be confused with the cassia of commerce which is *Cinnamomum cassia* from which is derived an adulterant for true cinnamon. It is cheaper and looks the same, though coarser in flavour.

cas'sia Specific epithet of cinnamon. (See *Cassia*.)

cas'sine North American Indian name for dahoon, *Ilex cassine*.

Cassi'nia f. Evergreen shrubs, named after Count Henri de Cassini (1781–1832), French botanist specializing in the study of *Compositae*.

cassinoi'des Resembling *Ilex cassine*.

Cassi'ope f. In Greek mythology the mother of Andromeda; hence the use of the name for a genus of *Ericaceae* related to the genus *Andromeda*.

Casta'nea f. Sweet chestnut. Among the species are the American, Spanish, and Chinese chestnuts. This is the Latin name for these trees, after the town of Castania in Thessaly which was famous for them.

casta'neus, -a, -um Chestnut-coloured.

Castano'psis f. Gr., resembling chestnut; indicating its relationship to the genus *Castanea*.

Castanosper'mum n. Moreton Bay chestnut. From *Castanea*, chestnut; Gr. *sperma*, a seed; in allusion to the form and size of the seed.

Castille'ja f. Indian paint brush. Named for an 18th-century botanist, Domingo Castillejo, of Cadiz, Spain.

Castillo'a f. In honour of Juan Castillo y Lopez (d. 1794), Spanish botanist. One of these species, *C. elastica*, yields a milky juice which can be a commercial source of rubber when the market is high enough. This is the plant that yielded the rubber balls which astonished Columbus.

cas'tus, -a, -um Spotless; chaste.

Casuari'na f. So named because the long, drooping branches of this Pacific genus are supposed to resemble the feathers of a cassowary (*Casuarinus*). Called beefwood, sheoke, and Australian pine, these trees are features of many tropical beaches where alone they seem to grow well.

catacos'mus, -a, -um Gr. *katakŏsmus*, adorned.

Catal'pa f. The North American Indian name for these trees.

catalpifo'lius, -a, -um With leaves like catalpa.

Catanan'che f. Cupid's dart. Greek name for these herbs, traditionally the basis of love potions. The Greek name means a strong incentive.

Catap'pa f. Malayan name for the Indian almond.

Cata'ria f. Pertaining to cats.

catarrac'tae; catarrac'tarum Of a waterfall or waterfalls.

Catase'tum n. Gr. *kata*, downward; *seta*, a bristle. A reference to the two horns of the column or structure formed by the junction of stamens and pistils in some of the species of this genus of epiphytic orchids.

catawbien'sis, -is, -e From the region of the Catawba River in the Blue Ridge Mountains in the south-eastern United States.

ca'techu Vernacular name of various Asiatic plants including betel palm (*Areca catechu*), apparently from Malayan *caccu*.

Catesbae'a f. Lily thorn. In honour of Mark Catesby (1682–1749), of Sudbury, England, author of *A Natural History of Carolina*.

Ca'tha f. Latinized version of the Arabian name of this evergreen shrub with edible leaves, *qat*.

cathar'ticus, -a, -um Purgative; cathartic.

cathaya'nus, -a, -um; cathayen'sis, -is, -e From China.

Cathcar'tia f. A poppy named in honour of John Fergusson Cathcart (1802–1851), Calcutta judge and amateur botanist.

cat'jang Malayan vernacular name for *Vigna cylindrica*.

Cat'tleya f. In honour of William Cattley (d. 1832) of Barnet, England, a wealthy patron of botany and an ardent collector of rare plants, who employed young John Lindley to describe and draw his plants and had this genus named for him in return. For many people the name is synonymous with orchid, or an expensive, showy, purple-and-white bloom tied with silver ribbon.

caucas'icus, -a, -um From the Caucasus.

cauda'tus, -a, -um Having a tail: caudate.

caules'cens Having a stem.

cauliala'tus, -a, -um Having a winged stem.

cauliflo'rus, -a, -um Bearing flowers on the stem or trunk.

Caulophyl'lum f. Blue cohosh. From Gr. *kaulos*, a stem; *phyllon*, a leaf. The stem of this perennial seems to form a stalk for a single large leaf.

caulora'pus, -a, -um With a stem like a turnip, as in kohlrabi.

caus'ticus, -a, -um Burning or caustic to taste.

Cautle'ya f. In honour of Sir Proby Thomas Cautley (1802–1871), English military engineer, geologist and palaeontologist, responsible for large irrigation works in India.

cav'us, -a, -um Hollow.

Ceano'thus f. New Jersey tea. From the Greek name for a spiny plant, not this one. This genus of shrubs is more appreciated abroad than in its native North America.

Cecro'pia f. Milky-juiced trees named for Cēcrŏps, mythical Egyptian founder of Athens.

Ced'rela f. Diminutive of L. *cedrus*, a cedar; from a similarity in the appearance and fragrance of the wood. The genus includes West Indian cedar, much in demand for making cigar boxes.

Cedronel'la f. Diminutive of L. *cedrus*, cedar; probably from the fragrance of *C. triphylla* (balm of Gilead) which gives out a very sweet scent when crushed.

Ced'rus f. Cedar. The Latin name. The genus includes such true cedars as Atlas cedar, the cedar of Lebanon and deodar. The L. *cēdrus* and Gr. *kĕdros* may originally have been applied to species of *Juniperus*. The trees popularly called 'cedar' in the United States are generally *Juniperus* and *Thuja*. The name *citrus* is a variant of *cedrus*.

Cei'ba f. Latinized version of the South American name for the silk cotton tree which is the source of kapok, used for stuffing lifebelts and in the upholstery trade.

celastri'nus, -a, -um Resembling *Celastrus*.

Celas'trus f. Bittersweet. From the Greek name *kēlastrŏs* for an evergreen tree.

cele'bicus, -a, -um From the island of Celebes (Sulawesi), Indonesia.

Celmi'sia f. New Zealand daisy. Named after Celmisios, son of the Greek nymph Alciope, for whom a related genus was named.

Celo'sia f. Cock's comb. Gr. *kēlĕŏs*, burning; in allusion to the brilliant colour of some of the flowers.

Cel'sia f. Herbs and sub-shrubs named in honour of Olof Celsius (1670–1756), professor of theology at Uppsala and a supporter of Linnaeus in his impecunious student days.

Cel'tis f. Hackberry. The Greek name for another tree.

cem'bra Italian name of the Swiss stone pine, *Pinus cembra*.

ceni'sius, -a, -um From the area of Mont Cenis between France and Italy.

Centau'rea f. Knapweed, bachelor's button. From Gr *kentauros*, centaur. The genus includes sweet sultan and dusty miller.

Centaur'ium n. Centaury. From Gr. *centauros*, the half-man half-horse of Greek mythology. This herb was supposed to have excellent medicinal qualities, especially in healing wounds. Chiron the Centaur, famous for his knowledge of the uses of plants, is said to have used centaury to heal a wound in his foot, caused by one of the arrows of Hercules.

centifo'lius, -a, -um Many-leaved; with a hundred leaves or petals.

Centrade'nia f. Gr. *kĕntrŏn*, a spur; *adēn*, a gland. The anthers of these flowering evergreen shrubs have a spurlike gland.

centra'lis, -is, -e Central, in the middle.

centranthifo'lius, -a, -um With leaves like *Centranthus* or red valerian.

Centran'thus m. Red valerian. From Gr. *kĕntrŏn*, a spur; *anthŏs*, a flower; from the spurred flower. Often incorrectly spelled *Kentranthus*.

Centropo'gon m. Gr. *kĕntrŏn*, a spur; *pōgōn*, a beard; in allusion to the fringed stigma of these shrubs and sub-shrubs.

Centrose'ma n. Butterfly pea. From Gr. *kĕntrŏn*, a spur; *sēma*, a standard. The standard has a short spur behind it.

Ce'pa f. Latin for onion.

Cephae'lis f. Gr. *kĕphalē*, a head; from the close arrangement of the flowers of these shrubs and herbs, from one of which ipecacuanha is produced.

Cephalan'thus m. Button bush. Gr. *kĕphalē*, a head; *anthŏs*, a flower. The flowers of these shrubs are closely packed in globose clusters.

Cephala'ria f. Giant scabious. Gr. *kĕphalē*, a head. The flowers of these herbs are in round clusters.

Cephalocereus

Cephaloce'reus m. Old-man cactus. From Gr. *kĕphalē*, a head; *Cereus*, a genus of cactus. A woolly head is produced when the plant is of flowering size.

cephalo'nicus, -a, -um From the Grecian island of Cephalonia (Kefallinia) in the Ionian Sea.

Cephalosta'chium n. Gr. *kĕphalē*, a head; *stachys*, a spike.

Cephalota'xus f. Plum yew. From Gr. *kĕphalē*, a head; *Taxus*, yew; a reference to the appearance of the trees which resemble yews.

cephalo'tes Resembling a small head.

cepifo'lius, -a, -um With leaves resembling onion.

cera'ceus, -a, -um Waxy.

cera'micus, -a, -um Resembling pottery; or from Ceram or Serang, an island in the Moluccas.

cera'sifer, cerasi'fera, -um Bearing cherries or cherry-like fruit.

cerasifor'mis, -is, -e Shaped like a cherry.

cera'sinus, -a, -um Cherry-red.

cerastioi'des Resembling *Cerastium* or chickweed.

Ceras'tium n. Mouse-ear chickweed. From Gr. *kĕras*, a horn; in allusion to the horned shape of the seed capsule.

Ce'rasus f. Latin name for cherry, and used when the cherries are put in a genus of their own, but also now the specific epithet of sour cherry, *Prunus cerasus*.

ceratocau'lis, -is, -e With a stalk like a horn.

Cerato'nia f. Carob. From Gr. *kĕratōnia*, the name of the carob, from Gr. *kĕras*, horn, in allusion to the pods. The seed pods, sometimes sold for food, are called St. John's bread and are supposed to be the locusts which John the Baptist ate with wild honey in the wilderness. The seeds were used as weights and from them derives the carat, the jewellers' weight for gold and precious stones.

Ceratope'talum n. Gr. *kĕras*, a horn; *petalon*, a petal. In one species of these Australian trees the petals look like stag's horns.

Ceratophyl'lum n. Hornwort. From Gr. *kăras*, a horn; *phyllon*, a leaf; from the antler-like appearance of the leaves of these submerged plants.

Ceratop'teris f. Gr. *kĕras*, a horn; *pteris*, a fern; from the horned appearance of this water fern.

Ceratostig'ma n. Gr. *kĕras*, a horn and stigma; from the hornlike excrescence on the stigma of the flower.

Ceratoza'mia f. Gr. *kĕras*, a horn; *Zamia*, the name of another genus,

84

which it resembles except for the fact that here the cones of these Mexican foliage plants have horned scales.

Cercidiphyl'lum n. Gr. *kerkis*, redbud or Judas tree; *phyllon*, a leaf; from the resemblance of the leaves to those of one species of the genus *Cercis*.

Cerci'dium n. Resembling *Cercis* or redbud.

Cer'cis f. Redbud, Judas tree. The Greek name *kerkis* for a European species, probably a poplar, but also applied to *Cercis siliquastrum*, which through confusion between Judas and Judaea is now traditionally the tree on which Judas hanged himself. The flowers are produced along the branches and trunk before the leaves appear.

Cercocar'pus m. Gr. *kerkos*, a tail; *karpos*, a fruit. The fruit of these trees and shrubs has a tail-like plume.

cerea'lis, -is, -e Pertaining to agriculture. Ceres was the goddess of farming.

cerefo'lius, -a, -um Waxen-leaved.

Ce'reus m. L. *cereus*, a wax taper; in allusion to the shape of some of the species of this cactus.

ce'reus, -a, -um Waxy.

ce'rifer, ceri'fera, -um Wax-bearing.

Cerin'the f. Honeywort. From Gr. *kērinos*, waxen. It was once supposed that bees obtained wax from these flowers, of which they are very fond.

cerinthoi'des Resembling *Cerinthe* or honeywort.

ceri'nus, -a, -um Waxy.

cer'nuus, -a, -um Drooping; nodding.

Cerope'gia f. Gr. *kēros*, wax; *pēgē*, a fountain; in allusion to the waxen appearance of the flower clusters on these vines.

Cero'pteris f. A synonym of *Pityrogramma*.

Ceroxy'lon n. Wax palm. From Gr. *kēros*, wax; *xylon*, wood. The trunks of these feather palms are coated with wax.

cerris Classical Latin name for the Turkey or mossy-cup oak, and now the specific epithet (*Quercus cerris*).

Ces'trum n. Bastard jasmine. Greek name for some other plant than these ornamental shrubs.

Chaenome'les f. Japanese quince. From Gr. *chaino*, to gape; *mēlon*, an apple; an allusion to the erroneous belief that the fruit is split.

Chaenorhi'num n. From Gr. *chaino*, to gape, and *Antirrhinum*; an allusion to the open mouth of the corolla which distinguishes the genus from *Antirrhinum* and *Linaria*.

Chaeno'stoma n. Gr. *chaino*, to gape; *stŏma*, a mouth; an allusion to the form of the corolla of these African herbs and sub-shrubs.

Chaerophyl'lum n. Bulbous chervil. From Gr. *chairo*, to please; *phyllon*, a leaf. The foliage is fragrant.

chaetomal'lus, -a, -um From Gr. *chaite*, loose flowing hair; *mallŏs*, flock of wool.

chalcedo'nicus, -a, -um From Chalcedon, the classical name for what is now Kadekoy, a district in Asia Minor opposite Istanbul.

chalmersia'nus, -a, -um In honour of James B. Chalmers (1841–1901), traveller and author of works on New Guinea where he was killed by cannibals.

chamae- n. Gr. prefix indicating dwarfness or low growth, from *chamai*, on the ground.

Chamaeba'tia f. From Gr. *chaimaibatos*, a dwarf bramble.

Chamaece'reus m. Gr. *chamai*, on the ground; *cereus*, a genus of cactus; an allusion to its prostrate habit.

Chamaecy'paris f. False cypress. From Gr. *chamai*, dwarf; *kuparissos*, cypress. Often sold as *Retinospora*, the genus includes such specimen and timber trees as Port Orford and Alaska cedar, Hinoki cypress, and southern white cedar.

Chamaedaph'ne f. Gr. *chamai*, dwarf; *daphne*, laurel.

Chamaedo'rea f. Gr. *chamai*, on the ground; *dorea*, a gift. The shining bright-coloured fruits of these palms are very easily reached.

chamaedrifo'lius, -a, -um With leaves like *Chamaedrys* or ground oak.

chamae'iris Specific epithet for an iris meaning dwarf iris.

Chamaeli'rium n. Devil's bit. This North American perennial is also sometimes called blazing star. From Gr. *chamai*, dwarf; *lirion*, a lily.

Chamaeran'themum n. Gr. *chamai*, dwarf; *Eranthemum*.

Chamae'rops f. Dwarf fan-palm. From Gr. *chamai*, dwarf; *rhŏps*, a bush.

Chambeyro'nia f. Palms named for Captain Chambeyron, who commanded the ship in which the French botanist and doctor of medicine, Eugene Vieillard (1819–1896), explored the coasts of New Caledonia where this genus was found.

Chamelau'cium n. Meaning and derivation of the name for these Australian evergreen shrubs is obscure.

chapma'nii In honour of Alvin Wentworth Chapman (1809–1899), American botanist, author of *Flora of the Southern United States*.

charan'tia A pre-Linnaean name and now the specific epithet for *Momordica charantia*.

Chardi'nia f. In honour of Jean Chardin, later Sir John Chardin (1643–1713), French jeweller, traveller in Asia Minor, Iran, Caucasus and India, Huguenot refugee in England where in 1681 Charles II knighted him.

charian'thus, -a, -um With elegant flowers.

Chari'eis f. Greek word for elegant—in allusion to the beauty of the flowers of this South African herb.

Cha'rira f. Sanskrit name for a kind of pepper.

Chasman'the f. From Gr. *chasme*, yawning, gaping; *chasmamai*, to yawn; *anthos*, flower. A genus of S. African *iridaceae*.

chasmanthus, -a, -um With wide-open or gaping flowers.

chatha'micus, -a, -um From the Chatham Islands in the South Pacific.

Cheilan'thes f. Lip ferns. From Gr. *cheilos*, a lip; *anthos*, a flower; in allusion to the form of the membranous covering (indusium) of the spore-bearing parts.

cheilan'thus, -a, -um With a flower furnished with a lip.

Cheiran'thus m. Wallflower. The meaning is obscure although some suppose that it may derive from Arabic *keiri* and others from Gr. *cheir*, a hand; *anthos*, a flower; and thus refer to the custom of carrying these old-fashioned, sweetly scented favourites in the hand as a bouquet.

chelidonioi'des Resembling *Chelidonium* or celandine.

Chelido'nium n. Celandine. From Gr. *chĕlidōn*, a swallow. The tradition is that the herb flowers when the swallows arrive.

Chelo'ne f. Turtlehead. Greek word *chĕlōnē* for a tortoise; in allusion to the appearance of the flower of these North American perennial herbs.

Chenopo'dium n. Goosefoot. From Gr. *chēn*, a goose; *pŏdion*, a little foot; from the shape of the leaf of these plants, mostly weeds but including some which are edible, such as good-King-Henry.

Chiapa'sia f. Named after the state of Chiapas, Mexico, where this cactus grows.

Chiastophyl'lum n. From Gr. *chiastos*, arranged diagonally, cross-wise; *phyllon*, leaf; referring to the pairs of opposite leaves.

chilen'sis, -is, -e From Chile.

Chilio'trichum n. From Gr. *chilii*, thousand; *trix*, hair. With reference to the numerous bristles of the pappus.

chiloen'sis, -is, -e From the island of Chiloe off the Chilean coast.

Chilopsis

Chilo'psis f. Gr. *cheilos*, a lip; *opsis*, resembling. The calyx of this low-growing tree or shrub has a distinct lip.

chimae'ra Specific epithet of uncertain application. The word means a monster; and, figuratively, a wild fancy.

Chima'phila f. Pipsissewa. From Gr. *cheima*, winter weather; *phileo*, to love. These small perennials are evergreen.

Chimonan'thus m. Wintersweet. Gr. *cheimōn*, winter; *anthos*, flower. These shrubs flower in winter.

Chimonobambu'sa f. From Gr. *cheimōn*, winter; *Bambusa*, bamboo.

chinen'sis, -is, -e Chinese; also spelled *sinensis*.

Chiococ'ca f. Snowberry. Gr. *chiōn*, snow; *kŏkkŏs*, a berry. The berries of one species of these evergreen shrubs, mostly climbers, are snow-white.

Chio'genes f. Creeping snowberry. Gr. *chiōn*, snow; *genos*, offspring; in allusion to the snow-white berries.

Chionan'thus m. Fringe-tree. Gr. *chiōn*, snow; *anthos*, a flower. These shrubs have snowy white flowers.

chionan'thus, -a, -um Having snow-white flowers.

Chionodo'xa f. Gr. *chiōn*, snow; *doxa*, glory. Among the very earliest of spring flowers, they often bloom when snow is still on the ground. The type-species, discovered by Boissier, was flowering among melting snow high in the Tmolus mountains of Asia Minor and named *C. luciliae* in honour of his wife.

Chiono'phila f. Gr. *chiōn*, snow; *phileo*, to love. It grows at high elevations in the Rocky Mountains.

chiotil'la Mexican name for a cactus, *Escontria chiotilla*.

Chiri'ta f. Nepalese vernacular name. Asiatic evergreens.

Chiro'nia f. South African plants in the gentian family named for Chiron, a centaur, son of Phyllira and Saturn. One of the beings half-man and half-horse, he was skilled in the medicinal use of plants.

chirophyl'lus, -a, -um With leaves resembling a hand.

Chlidan'thus m. From Gr. *chlidē*, luxury, costly ornament; *anthos*, flower.

chloo'des Grass-green.

chloracan'thus, -a, -um With green spines.

Chlorae'a f. Gr. *chloros*, green. Flowers of some species in this terrestrial fleshy-rooted genus of orchids are green.

chloran'thus, -a, -um With green flowers.

chlorifo'lius, -a, -um With leaves like *Chlora* (i.e. *Blackstonia*), a genus of little horticultural interest.

Chlo'ris f. Finger-grass. Ornamental flowering grasses named in honour of Chloris, the Greek goddess of flowers.

chlorochi'lon With a green lip.

Chloroco'don m. Gr. *chlŏrŏs*, green; *kŏdŏn*, a bell; from the shape and colour of the flower of this climber from Natal.

Chloro'galum n. Gr. *chlŏrŏs*, green; *gala*, milk; the sap of these bulbous plants from California is green.

Chloro'phora f. Gr. *chlŏrŏs*, green; *phoreo*, to bear. One species of these trees produces a green dye.

Chloro'phytum n. From Gr. *chlŏrŏs*, green; *phyton*, a plant. Some of these herbaceous plants are grown for their foliage.

chlo'rops With a green eye.

Choisy'a f. Mexican orange. Evergreen Mexican shrubs, named in honour of Jacques Denis Choisy (1799–1859), Swiss botanist and professor of philosophy at Geneva.

Chordospartium n. From Gr. *chŏrdē*, string; *Spartium*. Presumably in allusion to the slender branches.

chorda'tus, -a, -um Cordlike.

Chori'sia f. Floss silk tree. Named in honour of Ludwig (Louis) Choris (1795–1828), a botanical artist who accompanied the noted Russian navigator Otto von Kotzebue on his second scientific expedition to the Pacific, 1823–1826.

Chorize'ma n. Contraction of *chorizonema* from Gr. *chŏris*, separate; *nēma*, thread; referring to the separate filaments of the stamens. This genus of evergreen flowering shrubs was discovered by the French botanist Labillardière in Australia in 1792.

Chrozo'phora f. From Gr. *chrōzō*, to tinge, dye, stain; *phorus*, carrying.

Chrysalidocar'pus m. Gr. *chrysos*, gold; *karpos*, fruit; in allusion to the golden fruit of one species. *C. lutescens* is the *Areca* of florists, properly cane palm.

Chrysan'themum n. Gr. *chrysos*, gold; *anthos*, flower. Comprising about 150 species, the genus includes both the hothouse chrysanthemum and the common white daisy of the fields.

chrysanthoi'des f. Resembling *Chrysanthemum*.

chrysan'thus, -a, -um With golden flowers.

chry'seus, -a, -um Golden.

Chrysoba'lanus f. Gr. *chrysos*, gold; *balanos*, an acorn; in allusion to the yellow fruit of some species of these shrubs and trees.

chrysocar'pus, -a, -um With golden fruit.

Chryso'coma f. Gr. *chrysos*, gold; *kŏmē*, hair.

chryso'comus, -a, -um With golden hairs.

chrysodo'ron A golden gift.

Chryso'gonum n. Golden star. Gr. *chrysos*, gold; *gonu*, a knee.

chryso'lepis, -is, -e With golden scales.

chrysoleu'cus, -a, -um Gold-and-white.

chryso'lobus, -a, -um Golden-lobed.

chrysophyl'lus, -a, -um With golden leaves.

chrys'ops Gr. *chrysos*, golden; *ōps*, eye, face.

Chrys'opsis f. Golden aster. Gr. *chrysos*, gold; *opsis*, appearance. The flowers are golden.

chryso'stomus, -a, -um Golden-mouthed.

Chrysotham'nus m. Gr. *chrysos*, gold; *thamnos*, shrub.

Chus'quea f. A local South American name for these bamboos.

Chy'sis f. From Gr. *chysis*, melting. The pollen masses in these orchids appear to be fused into a whole.

Cibo'tium n. Scythian lamb (see *barometz*). Gr. *kibōtŏs*, a small box; from the appearance of the sori of these tree ferns.

Ci'cer f. Chick pea or garbanzo. The classical Latin name.

Cicer'bita f. Italian name, apparently of medieval origin, for sow-thistle, but now for a genus of blue-flowered composites related to *Lactuca*.

cichoria'ceus, -a, -um Resembling chicory or *Cichorium*.

Cicho'rium n. Chicory, endive. The Latinized version of the Arabic name for one species.

Cicu'ta f. Water hemlock. The Latin name for poison hemlock (*Conium maculatum*) from which the brew was made for Socrates. Both plants are equally poisonous and have no garden interest.

cicuta'rius, -a, -um Resembling water hemlock or *Cicuta*.

cicutifo'lius, -a, -um With leaves like water hemlock or *Cicuta*.

cilia'ris, -is, -e; cilia'tus, -a, -um Fringed with hairs on many leaves and petals; ciliate, from L. *cilium*, eyelid, then eyelash.

cili'cicus, -a, -um From Cilicia, a classical region of Southern Turkey.

ciliica'lyx With a fringed calyx.

cilio'sus, -a, -um Fringed or ciliate.

Cimici'fuga f. Bugbane. Black cohosh. From L. *cimex*, a bug; *fugo*, to drive away; from the use of *C. foetida* as an insect repellent. The genus includes snakeroot, regarded as useful in cases of snakebite because of the form and shape of the root.

Cincho'na f. Named in honour of Countess de Chinchon, wife of the Viceroy of Peru, who, legend has it, was cured in 1638 of malaria by use of what was then called Peruvian bark. It is the source of quinine, the commercial production of which is now centred in Java. During World War II it was an important part of Japanese strategy to cut off the world from supplies of quinine. Fortunately Western medicine quickly provided an effective synthetic without which the campaigns in malarial areas could not have been fought.

cin'ctus, -a, -um Girdled; girt around.

Cinera'ria f. From L. *cinereus*, ash-coloured; from the grey down on the leaves of these herbs and sub-shrubs.

cinerariifo'lius, -a, -um With leaves like *Cineraria*.

cineras'cens Becoming ashy-grey.

cine'reus, -a, -um Ash-coloured.

cinnabari'nus, -a, -um Cinnabar-red; vermilion.

cinnamo'meus, -a, -um Brown like cinnamon.

cinnamomifo'lius, -a, -um With leaves like cinnamon.

Cinnamo'mum n. From the classical Greek name for cinnamon. The genus includes the camphor tree and also cassia which is widely grown throughout the Far East as an adulterant of true cinnamon, *C. zeylanicum*.

Cipu'ra f. Derivation of the name of these South American bulbous plants is unknown.

circina'lis, -is, -e; circina'tus, -a, -um Coiled.

cirra'tus, -a, -um; cirrho'sus, -a, -um Equipped with tendrils.

Cirrha'ea f. L. *cirrus* or *cirrhus*, tendril; in allusion to the prolonged tendril-like rostellum of this orchid.

Cirrhope'talum n. Presumably from *cirrus* or *cirrhus*, curl, tendril; *petalum*, petal. In allusion to the long lateral segments of the flowers of these orchids.

Cir'sium n. Plumed thistle. From the Greek for a kind of thistle. The genus includes a great many pernicious weeds and very few plants of value to the gardener.

cismonta'nus, -a, -um On this side of the mountain—meaning the south side of the Alps, the side on which Rome stands.

cisplati'nus, -a, -um From the neighbourhood of the River Plate in South America.

Cis'sus f. From Gr. *kissos*, ivy; an allusion to the climbing habits of these shrubs.

cistiflo'rus, -a, -um *Cistus*-flowered.

cistifo'lius, -a, -um With leaves like *Cistus*.

Cis'tus m. Rock-rose. From the Greek name for the plant.

Cithare'xylum n. Gr. *kithara*, a lyre; *xylon*, wood; a Latinization of the English name 'fiddlewood'.

citra'tus, -a, -um Resembling *Citrus*.

citrifo'lius -a, -um With leaves resembling *Citrus*.

citriniflo'rus, -a, -um With lemon-yellow flowers.

citri'nus, -a, -um Lemon-coloured or resembling citron.

citriodo'rus, -a, -um Lemon-scented.

citroi'des Resembling *Citrus*.

Citro'psis f. *Citrus* plus Gr. *opsis*, resembling. These small trees are somewhat like *Citrus*.

Citrul'lus m. From *Citrus*, from the appearance of the fruit. Watermelon is *C. vulgaris* (*C. lanatus*).

Cit'rus f. Latin name for the citron (*Citrus medica*), the fruit of which was substituted in ancient Jewish ritual for the cone of the cedar (*Cedrus libani*), but applied by Linnaeus and earlier authors to the whole genus which includes oranges, lemons, limes, and grapefruit.

Cladan'thus m. From Gr. *klados*, a branch; *anthos*, a flower. The flowers of this annual herb occur at the ends of the branches.

Cladas'tris f. Kentucky yellow-wood. Gr. *klados*, a branch; *thraustos*, fragile; from the brittle nature of the twigs.

Cla'dium n. Gr. *kladion*, a small branch.

Cladotham'nus m. From Gr. *klados*, a branch; *thamnos*, shrub. In allusion to its much-branched habit.

clandesti'nus, -a, -um Hidden.

Clar'kia f. North American annuals named in honour of Captain William Clark (1770–1838), who, with Captain Meriwether Lewis, made the first transcontinental expedition, crossing the Rocky Mountains in 1806.

Clause'na f. A genus of trees named in honour of Peder Claussøn (1545–1614), a Danish priest who wrote on topography and natural history.

clau'sus, -a, -um Closed; shut.

cla'va-her'culis Club of Hercules. A specific epithet of *Zanthoxylum*.

clava'tus, -a, -um Club-shaped.

clavella'tus, -a, -um Shaped like a small club.

clave'nae In honour of Niccolo Chiavera (d. 1617), Italian apothecary.

Clavi'ja f. Evergreen trees and shrubs named in honour of Don Jose de Viera y Clavijo (1731–1813), who translated the works of Buffon into Spanish.

cla'vus A club.

Clayto'nia f. Little spring-blooming perennials named in honour of John Clayton (1686–1773), who came to Virginia from England in 1705. He corresponded with the botanical great of the day—Linnaeus, Gronovius, Kalm, and John Bartram—as well as with Benjamin Franklin and Thomas Jefferson. Collinson (see *Collinsonia*), the English Quaker botanist, described him as the greatest botanist in America. The common spring-beauty, sometimes called good-morning spring (*C. virginica*), is a species of the genus.

Cleistocac'tus m. From Gr. *kleistos*, closed; plus *cactus*. The flowers hardly open.

clemati'deus, -a, -um Like *Clematis*.

Cle'matis f. Greek name for various climbing plants.

Clematocle'thra f. From the generic names *Clematis* and *Clethra*.

clementi'nae In honour of Clementine, wife of the plant-collector George Forrest.

Cleo'me f. Derivation unknown. Spider flower is *C. hasslerana* or *C. spinosa*.

Cleroden'drum n. Gr. *klēros*, chance; *dendron*, a tree. Supposed to be an allusion to the variable medicinal qualities of these shrubs, trees, and climbers.

Cle'thra f. Sweet pepper bush. White alder. *Klēthra*, the Greek name for alder; the leaves are similar to those of *Clethra*.

clethroi'des Resembling *Clethra*.

Cley'era f. In honour of Andreas Cleyer (d. 1697 or 1698), German doctor and soldier in service of Dutch East India Company in Java and Japan.

Clian'thus m. Glory Pea. Gr. *klĕŏs*, glory; *anthos*, a flower; in allusion to the brilliantly coloured flowers of these shrubs.

Cliffortia

Cliffor'tia f. In honour of George Clifford (1685–1760), Dutch merchant-banker of English descent, the plants of whose magnificent garden at De Hartekamp were described by Linnaeus in his *Hortus Cliffortianus* (1738).

Clifto'nia f. An evergreen shrub named in honour of Dr. Francis Clifton, English physician, who died in 1736.

Clinto'nia f. Bear-tongue. Low-growing herbaceous plants named for Governor De Witt Clinton (1769–1828), of New York, who promoted the construction of the Erie Canal.

Clito'ria f. Butterfly pea. From *L. clitoris*, an anatomical term; an allusion to a characteristic of the flower.

Cli'via f. Kaffir lily. Bulbous evergreen plants named after Lady Charlotte Florentina Clive, Duchess of Northumberland (d. 1866), granddaughter of Robert Clive, and the first to flower the type-species.

clivo'rum Of the hills.

clusia'nus, -a, -um; clu'sii In honour of Carolus Clusius (Charles de l'Ecluse, 1526–1609), celebrated Flemish botanist.

clypea'tus, -a, -um Resembling a Roman shield.

clypeola'tus, -a, -um Somewhat shield-shaped.

Clyto'stoma n. Trumpet flower. Gr. *klytos*, beautiful; *stŏma*, a mouth.

Cneo'rum n. Spurge olive. From the Greek name *knĕŏron* for some shrub resembling olive.

Cni'cus m. Blessed thistle. From Gr. *knēkos*, a thistle.

coarcta'tus, -a, -um Pressed or crowded together.

Coba'ea f. Tendril-climbers named in honour of a Jesuit Father, Bernardo Cobo (1572–1659), Spanish missionary and naturalist in Mexico and Peru.

cocci'fera, -um Berry-bearing; being a host to coccoid insects, the gravid females of which resemble berries.

cocci'neus, -a, -um Scarlet.

Cocci'nia f. L. *coccineus*, scarlet; from the colour of the fruits.

Cocco'loba f. Seaside grape. From Gr. *kokkŏlŏbis*, ancient name of a kind of grape also called *balisca* and transferred to these plants in allusion to the fruit.

Coccothri'nax f. From Gr. *kŏkkŏs*, a berry, and *Thrinax*, a related genus of fan palms. They have berrylike fruits.

Coc'culus m. Moonseed, snail-seed. From the diminutive of Gr. *kŏkkŏs*, a berry; in allusion to the fruits of these evergreen climbers or shrubs.

Cochemi'ea f. From the name of an Indian tribe which once lived in lower California, home of these plants of the *Cactus* family.

cochenilli'fera Cochineal-bearing. Specific epithet of the cochineal fig, *Nopalea cochenillifera*. Cochineal is a dyestuff made from the dried bodies of the females of a scale insect found in Mexico and Central America. It is used as a reddish or purple colouring for foodstuffs and is the source of carmine.

Cochlea'ria f. Scurvy grass. L. *cochlear*, a spoon; from the shape of the basal leaves of these herbs.

cochlearifo'lius, -a, -um With leaves like *Cochlearia*.

cochlea'ris, -is, -e Spoon-shaped.

cochlea'tus, -a, -um Spiral like a snail shell.

Cochleoste'ma n. From Gr. *kŏchlŏs*, spiral shell; *stēma*, penis, stamen. In allusion to the spirally twisted anthers.

Cochlio'da f. From Gr. *kŏchlŏs*, snail-shell; in allusion to the snail-shell-like thickening on the lip of the type-species of this genus of orchids.

Cochlosper'mum n. From Gr. *kŏchlŏs*, snail-shell or other spiral shell; *sperma*, seed; in allusion to the shape of the seeds of these tropical trees.

cockburnia'nus, -a, -um In honour of members of the Cockburn family domiciled in China.

Co'cos f. The coconut palm has become so widespread through human activity over many centuries that its original home, probably in the Malaysian region, is uncertain. It has spread itself by natural means over the sea coasts of the tropics wherever its seedlings could find a footing on a sandy shore unmolested by wild pigs, rats and crabs. One of the important sources of vegetable oil, it is an indispensable article of diet in many places. Sailing vessels in many parts of the world are rigged with coconut-fibre ropes, the nut shells are used as drinking and other cups, the leaves are used for thatching, while the trunks make a handsome timber commercially called porcupine wood.

Codia'eum m. Croton. Latinized version of the Ternate vernacular name, *kodiho*.

Codono'psis f. From Gr. *kōdōn*, a bell; *ŏpsis*, resembling; from the shape of the corolla of these herbs.

coelesti'nus, -a, -um; coeles'tis, -is, -e Sky-blue.

Coe'lia f. Gr. *kŏilŏs*, hollow; from an erroneous concept that the pollen masses of these epiphytic orchids were hollow.

coe'licus, -a, -um Celestial; heavenly.

coeli-ro′sa Specific epithet meaning rose of the sky.

Coeloglos′sum n. From Gr. *kŏilŏs*, hollow; *glōssa*, tongue. Referring to the lip of these orchids.

Coelo′gyne f. Gr. *kŏilŏs*, hollow; *gynē*, female; referring to the depressed stigma of these epiphytic orchids.

coerules′cens; coeruleus, -a, -um The first, bluish, becoming blue; the second, blue.

Co′ffea f. From *kahwah*, the Arabic name of the beverage. A genus of some forty species of shrubs or small trees native to tropical Asia and Africa. The coffee we drink is prepared from the seeds which are enclosed in a bright scarlet berry.

coggygria A false rendering of the Gr. *kŏkkugia*, name for the smoke-tree (see *Cotinus*).

cogna′tus, -a, -um Closely related to.

Co′ix f. Job's tears. From the Greek name for a reedlike plant.

Co′la f. From the African vernacular name. Cola or goora nuts are widely used in a variety of soft drinks. The nuts are said by Africans to increase endurance.

Col′chicum n. Autumn crocus. Said by ancient authors to be especially abundant in Colchis, the Black Sea region of Georgia, Caucasus.

col′chicus, -a, -um From Colchis.

Coleone′ma n. From Gr. *kŏlĕŏs*, a sheath; *nema*, a thread; referring to the filaments of the stamens which are folded in the petals of these South African evergreen shrubs.

Coleotry′pe f. From Gr. *kŏlĕŏs*, a sheath; *trypē*, hole. In allusion to the young growths breaking through a hole in the leaf-sheath instead of emerging from the top of the sheath.

Co′leus m. Gr. *kŏlĕŏs*, a sheath; an allusion to the manner in which the stamens are enclosed. Called simply the foliage plant in large areas of the United States, the leaves tend to be very colourful. The plants were once widely used for ornamental ribbon bedding, sometimes on a very large scale; a large and pretentious garden used many thousands of plants annually in its formal bedding arrangements. The fashion for this kind of thing has waned and few gardeners remain as exponents of the art. This genus is considered insufficiently distinct from *Plectranthus*.

Colle′tia f. Spiny shrubs named in honour of Philibert Collet (1643–1718), French botanist.

collettia′nus, -a, -um; collettii. In honour of General Sir Henry Collett (1836–1901) of the Indian Army, author of *Flora Simlensis* (1902).

Collin'sia f. Blue-eyed Mary. Annuals named in honour of Zaccheus Collins (1764–1831), Vice-President of the Philadelphia Academy of Natural Sciences.

Collinso'nia f. Horse balm, horse weed. Coarse herbs named in honour of Peter Collinson, 18th-century English Quaker merchant, who corresponded widely with men such as Linnaeus, John Bartram and John Clayton (see *Claytonia*) in America, and helped to introduce many North American plants into British gardens by way of his garden at Mill Hill.

colli'nus, -a, -um Pertaining to hills.

Collo'mia f. From Gr. *kŏlla*, glue. The seeds of these herbs are muci-laginous when wet.

Coloca'sia f. West Indian kale, taro root, elephant's ear. From the Gr. *kŏlŏkasia*, used for the root of *Nelumbo nucifera*, according to Dioscorides.

colora'tus, -a, -um Coloured.

Colquhou'nia f. Asian shrubs named in honour of Sir Robert Colquhoun (d. 1838), British Resident in Nepal in 1819, patron of the Calcutta Botanic Garden in the early part of the 19th century. The name is pronounced *ko-hoo-nia*.

colubri'nus, -a, -um Snakelike; shaped like a snake.

columba'rius, -a, -um Dovelike or pertaining to doves.

columbia'nus, -a, -um From British Columbia or the Columbia River separating Washington and Oregon, U.S.A.

columel'laris, -is, -e Pertaining to a small pillar or pedestal.

colum'naris, -is, -e Columnar; in the shape of a column.

Colum'nea f. Tropical American herbs and shrubs named in honour of Fabius Columna (Fabio Colonna, 1567–1640); author of the first botanical book with copper-plate illustrations, published in Naples 1592.

colur'na Classical name, now the specific epithet of the hazelnut.

Colu'tea f. Bladder senna. From the Greek name *kŏlutea* for these shrubs.

Colvil'lea f. A red-flowered tree named in honour of Sir Charles Colville (1770–1843), Scottish soldier, governor of Mauritius 1828–1834.

Coma'rum n. From Gr. *kŏmaros*, a name of the strawberry-tree, transferred to *Potentilla comarum* on account of the strawberry-like appearance of the fruiting heads.

coma'tus, -a, -um Furnished with a tuft, sometimes of hair.

com'beri In honour of Harold F. Comber (1898–1969), English gardener and plant collector.

Combre'tum n. Latin name for a climbing plant, but not this genus.

Comesper'ma n. From Gr. *kŏmē*, hair; *sperma*, a seed; in allusion to the tufts of hair on the seeds of these herbs and shrubs.

Commeli'na f. Day-flower. Named for the Dutch botanists Commelin (Commelijn), Johan (1629–1692) and his nephew Caspar (1667–1731), by Plumier and adopted by Linnaeus, who while in Holland evidently learned of Caspar's son Caspar, the author of a Latin poem in 1715 but otherwise unknown. To quote Linnaeus, 'Commelina has flowers with three petals, two of which are showy, while the third is not conspicuous, from the two botanists called Commelin, for the third died before accomplishing anything in botany.'

Commerso'nia f. In honour of Philibert Commerson (1727–1773), French doctor and naturalist on Bougainville's voyage of 1766–1769. Commerson remained in Mauritius to make further biological investigations and never returned to France. Among the new plants collected by him was *Bougainvillea*.

commix'tus, -a, -um Mixed together; mingled.

commu'nis, -is, -e Common; general; growing in company.

commuta'tus, -a, -um Changed or changing.

como'sus, -a, -um Furnished with a tuft, sometimes of sterile flowers, as in *Muscari comosum*, or conspicuous bracts, as in *Eucomis comosa*.

compac'tus, -a, -um Compact; dense.

Comparet'tia f. In honour of Andrea Comparetti (1745 or 1746–1801 or 1802), Italian botanist, professor of medicine at Padua.

complana'tus, -a, -um Flattened; levelled.

comple'xus, -a, -um Encircled; embraced; complex.

complica'tus, -a, -um Complicated; complex.

compo'situs, -a, -um Compound.

compres'sus, -a, -um Compressed; flattened.

Compto'nia f. Sweet fern. Named for Henry Compton (1632–1713), Bishop of London and patron of botany. Compton began his career as an officer of the Life Guards. As Bishop of London he was suspended by King James II for his strongly expressed religious beliefs. At the time of the Revolution of 1688, he girded on sword and pistols and personally escorted Princess Anne, soon to be Queen, out of London in his own coach.

comp'tus, -a, -um Adorned; ornamented.

Conan'dron n. From Gr. *kōnŏs*, a cone; *anēr, andrŏs*, man, hence stamen. So named because of the way in which the appendages of the anthers of this herb are arranged in a cone shape around the style.

conca′tenans Linking together.

conca′vus, -a, -um Hollowed out; basin-shaped.

conchifo′lius, -a, -um With leaves like sea shells.

concin′nus, -a, -um Neat; elegant; well-made.

con′color Of the same colour throughout.

condensa′tus, -a, -um; conden′sus, -a, -um Crowded together.

confertiflo′rus, -a, -um With the flowers crowded together.

confer′tus, -a, -um Crowded.

confor′mis, -is, -e Conforming to the type. Similar in shape or other ways to related species.

confu′sus, -a, -um Confused; uncertain; apt to be taken for another species.

conges′tus, -a, -um Arranged very closely together; congested.

conglomera′tus, -a, -um Crowded together, conglomerate.

congola′nus, -a, -um From the Congo.

Conico′sia f. From Gr. *kōnicŏs*, cone-shaped, conical. A genus of S. African plants related to *mesembryanthemum*.

co′nifer, coni′fera, -um Cone-bearing.

Coniogram′me f. Gr. *kŏniŏs*, dusty; *grammē*, a line; in reference to the way the spore cases are arranged along the veins of these ferns.

Coni′um n. Poison hemlock. The Greek name for both the plant and the poison derived from it. The latter was administered to those condemned in Athens to die for certain offences against the state. Plato's description of Socrates taking the poison remains one of the most moving passages in literature.

conjuga′tus, -a, -um; conjuga′lis -is, -e Joined in pairs; wedded.

conjunc′tus, -a, -um Joined.

conna′tus, -a, -um United; twin; having opposite leaves joined together at their bases.

conoi′deus, -a, -um Conelike.

Conophy′tum n. Cone-plant. From Gr. *kŏnŏs*, a cone; *phytŏn*, a plant; an allusion to the shape of these succulents.

cono′pseus, -a, -um Gnat-like, from Gr. *kŏnōps*, gnat, mosquito.

conradi′nae In honour of Conradine, wife of the German botanist B. A. E. Koehne (1848–1918).

consangui'neus, -a, -um Related.

Conso'lida f. Medieval name of a wound-healing herb, apparently from L. *consolido*, make firm.

conso'lidus, -a, -um Solid; stable.

consper'sus, -a, -um Scattered.

conspi'cuus, -a, -um Conspicuous.

constric'tus, -a, -um Constricted.

conti'guus, -a, -um
(*a*) Near together—so close as to touch one another.
(*b*) Closely related.

continenta'lis, -is, -e Continental.

contor'tus, -a, -um Twisted; bent irregularly; contorted.

contrac'tus, -a, -um Drawn together; contracted.

controver'sus, -a, -um Doubtful; controversial.

Convalla'ria f. Lily-of-the-valley. From L. *convallis*, a valley, the old name being *Lilium convallium*.

convallaroi'des Resembling lily-of-the-valley.

convalliodo'rus, -a, -um Scented like lily-of-the-valley.

convolvula'ceus, -a, -um Similar to *Convolvulus*.

Convol'vulus m. Bindweed. From L. *convolvo*, to twine around.

conyzoi'des Resembling *Conyza*, a genus of no garden interest.

coo'peri In honour of Thomas Cooper (1815–1913), English gardener, who collected in South Africa from 1859 to 1862, introducing many notable plants. Father-in-law of the Kew botanist N. E. Brown (1849–1934) who also worked in his youth for W. W. Saunders and thereby became interested in South African plants.

Coope'ria f. Evening-star. Night-flowering bulbous plants named in honour of Joseph Cooper, gardener to Earl FitzWilliam at Wentworth, Yorkshire, *c.* 1830.

Copai'fera f. From *copaiba*, the Brazilian name for a balsam, plus L. *fero*, to bear. The trees yield copal, a commercial gum used in making varnish.

copalli'nus, -a, -um Gummy; resinous.

Coperni'cia f. Carnauba palm, named for Copernicus (Nicolaus Koppernigk, 1473–1543), Polish astronomer, physician, economist, and theologian, most famous for having propounded and proved the heliocentric theory which placed the sun as the centre of our solar system around which the planets, including the earth, revolve. Thus, once and for all, he upset the Ptolemaic system which had been accepted for 1,500 years.

Copiapo'a f. From the province of Copiapo, Chile, where these cacti grow.

Copros'ma f. Gr. *kŏprŏs*, dung; *ŏsmē*, a smell; in allusion to the fetid smell of many of these small trees and shrubs when bruised.

Cop'tis f. Gold thread. Mouth root. From Gr. *kopto*, to cut; a reference to the deeply cut leaves of these perennial herbs.

coralliflo'rus, -a, -um With flowers of coral-red.

coralli'nus, -a, -um Coral-red.

coralloden'dron Specific epithet of the coral-tree (*Erythrina*).

Cor'chorus m. Greek name *kŏrchŏros* of obscure derivation. Jute is *C. capsularis*.

corda'tus, -a, -um Heart-shaped.

Cor'dia f. Trees and shrubs, named in honour of Euricius Cordus (1486–1535) and his son Valerius (1515–1544), German botanists and pharmacists.

cordifo'lius, -a, -um With heart-shaped leaves.

cordifor'mis, -is, -e Heart-shaped.

Cordy'line f. Gr. *kordyle*, a club. Often offered as *Dracaena* to which this genus of flowering trees and shrubs is closely related.

corea'nus, -a, -um From Korea.

Core'ma n. Gr. *korēma*, a broom; from the habit of these heathlike shrubs.

Coreo'psis f. Tickseed. From Gr. *koris*, a bug; *opsis*, like. The seed of these herbs looks like a bug or tick.

coria'ceus, -a, -um Thick and tough; leathery.

Corian'drum n. Coriander. The Greek name, *kŏriandrŏn*, for the plant which has a long history of use as a flavouring. Today coriander seed is used in such diverse things as gin, confectionery, bread, curry powder, and so on. When unripe the seed has an unpleasant odour which disappears when dry.

Coria'ria f. L. *corium*, leather. Some species of these shrubs and herbs are used in tanning.

coria'rius, -a, -um Like leather.

coridifo'lius, -a, -um With leaves like *Coris*.

corifo'lius, -a, -um With leathery leaves.

coriophyl'lus, -a, -um With leaves like *Coris*, a plant of no garden interest.

coris Specific epithet for a St. John's Wort (*Hypericum*), from a similarity to *Coris*.

cor'neus, -a, -um Horny.

cornicula'tus, -a, -um With small horns.

cor'nifer, corni'fera, -um; cor'niger, corni'gera, -um Bearing or being furnished with a horn.

cornubien'sis, -is, -e Of Cornwall, south-west England.

cornuco'piae Shaped like a cornucopia or horn of plenty.

Cor'nus f. Dogwood or cornel, bunchberry. The Latin name for the cornelian cherry (*Cornus mas*). This genus is easily broken up into smaller groups treated as genera by many botanists, e.g. *Benthamidia*, *Cornus*, proper (*Macrocarpium*) and *Swida* (*Thelycrania*).

cornu'tus, -a, -um Horned or horn-shaped.

Coro'kia f. From the Maori name for these evergreen shrubs.

corolla'tus, -a, -um Like a corolla.

coromandelia'nus, -a, -um From the Coromandel Coast, the south-eastern part of the Indian peninsula south of the Kistna River.

cor'onans; corona'tus, -a, -um Crowned.

corona'rius, -a, -um Used for garlands, or pertaining to garlands.

Coronil'la f. Crown vetch. Diminutive of L. *corona*, a crown; in reference to the umbels of these shrubs and herbs.

coronopifo'lius, -a, -um With leaves like *Cŏrōnŏpus*, a rather weedy plant.

Coro'zo f. Native South American name for this palm.

Corpuscula'ria f. From Latin *corpusculum*, a little body.

Cor'rea f. Australian shrubs named in honour of José Francesco Correa da Serra (1751–1823), Portuguese botanist.

corruga'tus, -a, -um Corrugated; wrinkled.

cor'sicus, -a, -um From Corsica.

Cortade'ria f. Pampas grass. From the Argentinian name.

cortico'sus, -a, -um With bark like a cork tree; heavily barked.

Cortu'sa f. Perennial herbs named in honour of Jacobi Antonio Cortusi (1513–1593), director of the botanic garden at Padua.

cortusoi'des Resembling *Cortusa*.

corus'cans Glittering; coruscating.

Coryan'thes f. Helmet flower or bucket orchid. From Gr. *kŏrys*, a helmet; *anthos*, a flower.

corya'nus, -a, -um In honour of Reginald R. Cory (1871–1934), coal-owner, benefactor by bequest of the Cambridge Botanic Garden and the Lindley Library, Royal Horticultural Society.

Cory'dalis f. Greek word meaning a lark; the flowers of these annual and perennial herbs have spurs like those of larks.

corylifo'lius, -a, -um With leaves like *Corylus* or hazel.

Corylo'psis f. Gr. *kŏrylŏs*, hazel; *ŏpsis*, like; from the resemblance of these deciduous flowering shrubs to *Corylus* or hazel.

Co'rylus f. Hazel. The Greek name. Now the name for a large genus of shrubs and trees which includes cob-nut and filbert.

corym'bifer, corymbi'fera, -um Corymb-bearing. A corymb is a cluster of flowers in which the outer stalks or pedicels are longer than the inner ones. The blossom of the sweet cherry is an example.

corymbiflo'rus, -a, -um With flowers produced in a corymb.

corymbo'sus, -a, -um Provided with corymbs; corymbose.

coryne'phorus, -a, -um Carrying a club-shaped organ of some sort.

coryno'calyx With a calyx like a club.

Corynocar'pus m. New Zealand laurel. From Gr. *kŏrynē*, a club; *karpos*, fruit; from the shape of the fruit of this tree.

Co'rypha f. Gr. *koryphe*, a summit or hilltop; with reference to the terminal crown of leaves of this palm.

coryphae'us, -a, -um Gr. *koryphaios*, leading.

Coryphan'tha f. From Gr. *koryphe*, a summit; *anthos*, a flower. The flowers are produced on the top of these globular or cylindrical cacti.

cosmophyl'lus, -a, -um With leaves like *Cosmos*.

Cos'mos m. Gr. *kosmos*, beautiful. These annual and perennial herbs have a variety of colours.

costa'tus, -a, -um Ribbed.

Cos'tus m. Spiral-flag. The Latin name derived from an Oriental name for an imported aromatic root.

cotinifo'lius, -a, -um With leaves like *Cotinus* or smoke-tree.

cotinoi'des Resembling *Cotinus* or smoke-tree.

Co'tinus m. Smoke-tree. From Gr. *kŏtinŏs* for the wild-olive but of uncertain application here.

Cotonea'ster m. L. *cotoneum*, quince; *-aster*, suffix indicating incomplete resemblance, hence applied to a wild or inferior kind. From a similarity of the leaves to quince in some species of these handsome-fruited shrubs and small trees.

Co'tula f. Greek word meaning a small cup. The bases of the leaves of these small herbs form cups.

Cotyle'don f. Gr. *kŏtylēdōn*, from *kŏtylĕ*, a cavity or small cup; in reference to the cuplike leaves of some species of these succulent plants.

coul'teri In honour of Thomas Coulter (1793–1843), Irish botanist.

Couta'rea f. Latinized version of the vernacular name in Guyana for these evergreen trees and shrubs.

Cowa'nia f. Flowering evergreen shrubs named in honour of James Cowan (d. 1823), London merchant who travelled in Mexico and Peru and introduced many plants.

cowania'nus, -a, -um In honour of John Macqueen Cowan (1892–1960), forester and botanist, authority on the genus *Rhododendron*.

Cram'be f. Seakale. The Greek name for cabbage.

Craspe'dia f. Gr. *kraspĕdon*, a fringe; in allusion to the feathery pappus of these perennial herbs.

crassicau'lis, -is, -e Thick-stemmed.

crassifo'lius, -a, -um Thick-leaved.

cras'sipes Thick-footed or thick-stemmed.

crassius'culus, -a, -um Somewhat thick.

Cras'sula f. Diminutive of L. *crassus*, thick; from the somewhat thick leaves of these shrubs and herbs.

cras'sus, -a, -um Thick; fleshy.

crataegifo'lius, -a, -um With leaves like *Crataegus* or hawthorn.

crataegoi'des Resembling hawthorn.

Crataegomes'pilus f. Graft hybrids of *Crataegus* and *Mespilus*.

Cratae'gus f. Hawthorn. The Greek name for the tree. From *kratos*, strength; an allusion to the strength and hardness of the wood.

Crataemes'pilus f. Sexual hybrids of *Crataegus* and *Mespilus*.

cre'ber, cre'bra, -um Thickly clustered; close; frequent.

crebiflo'rus, -a, -um With thickly clustered flowers.

Cremantho'dium n. From Gr. *krĕmao*, hang, and *anthodium*, flower-head or capitulum (from *anthōdēs*, flower-like), referring to the drooping flower-heads of these Sino-Himalayan herbs.

crenatiflo'rus, -a, -um With flowers cut in rounded scallops.

crena'tus, -a, -um Cut in rounded scallops; crenate.

crenula'tus, -a, -um Somewhat scalloped.

crepida'tus, -a, -um Shaped like a sandal or slipper.

Cre'pis f. Hawk's-beard. A large genus of herbs with dandelion-like flowers. Gr. *krēpis*, a boot; but why is not clear.

cre'pitans Rustling; crackling.

Crescen'tia f. The Calabash tree, bearing the big fruits from which bowls and dippers are made in hot countries. Named in honour of Pietro Crescenzi (1230–1321), Italian author of a work on country life.

creta'ceus, -a, -um Pertaining to chalk; chalky.

cretensis, -is, -e; cre'ticus, -a, -um From the island of Crete.

cri'niger, crini'gera, -um Bearing hairs; from L. *crinis*, hair.

crini'tus, -a, -um Furnished with long, generally weak, hairs.

Crinoden'dron n. From Gr. *krinŏn*, lily; *dendron*, tree; in allusion to the flowers of these Chilean evergreen shrubs.

Crinodon'na f. Generic name given to a hybrid group between *Crinum* and *Amaryllis* (*Belladonna*).

Cri'num n. Gr. *krinon*, a lily. Bulbous plants of the *Amaryllis* family.

crispa'tus, -a, -um; cris'pus-, -a, -um Finely waved; closely curled.

cris'ta-gal'li A specific epithet meaning cock's comb.

crista'tus, -a, -um Having tassel-like tips, crested.

crithmifo'lius, -a, -um With leaves like *Crithmum*.

Crith'mum n. Samphire. Gr. *krithe*, barley; in allusion to the form of the seed of this perennial herb.

croca'tus, -a, -um; cro'ceus, -a, -um Saffron-coloured; yellow.

Crocos'mia f. Gr. *krŏkŏs*, saffron; *osme*, a smell. The dried flowers when placed in warm water have a strong smell of saffron.

crocosmiiflo'rus, -a, -um *Crocosmia*-flowered.

Cro'cus m. Gr. *krŏkŏs*, saffron, derived from Semitic *karkom*. The genus includes both the common and the saffron crocus. The saffron crocus is widely grown in Mediterranean countries and elsewhere for the stigmas and part of the styles which are dried to produce a yellow dye, the royal colour of ancient Greece. Saffron is also used in various types of cookery. It is an almost essential ingredient of bouillabaisse and also of Spanish rice to which it adds a pleasant flavour and colour. It is fragrant and was once used on a considerable scale by Roman emperors to scatter in the streets when they went abroad. In some countries it is used in embalming.

Crossan'dra f. Gr. *krossos*, a fringe; *aner*, male. The anthers are fringed.

Crotala'ria f. Rattle-box. Gr. *krŏtalon*, a rattle, clapper; with reference to the seeds rattling in the inflated pods.

Cro'ton m. Gr. *krŏtōn*, a tick, from the appearance of the seeds of these plants of solely economic interest. *C. tiglium* is the source of croton oil, a powerful cathartic.

crotonifo'lius, -a, -um With leaves like *Croton*.

Cro'wea f. In honour of James Crowe (1750–1807), an English surgeon, born at Norwich, who studied mosses, fungi, and willows, of which he had a large living collection.

Crucianel'la f. Crosswort. From the diminutive of L. *crux*, a cross; the leaves of these herbs being crosswise.

crucia'tus, -a, -um In the form of a cross.

cru'cifer, cru'cifera, -um Bearing a cross.

cruen'tus, -a, -um Blood-coloured, gory.

Crupi'na f. Latinized from a Belgian or Dutch vernacular name.

crus-gal'li A specific epithet meaning cock's spur.

crusta'tus, -a, -um Encrusted.

crux-an'drae Specific epithet meaning St. Andrew's cross.

crux-maltae Specific epithet meaning Maltese cross.

Cryptan'tha f. From Gr. *krypto*, to hide; *anthos*, a flower. The flowers of these herbs are hidden.

Cryptan'thus m. From Gr. *krypto*, to hide; *anthos*, a flower. The flowers of these Brazilian plants are hidden among the bracts.

Cryptoco'ryne f. From Gr. *krypto*, to hide; *kŏrynē*, club.

Cryptogram'ma n. Rockbrake. From Gr. *krypto*, to hide; *gramma*, writing. In these ferns the spore cases are concealed.

Cryptome'ria f. Japanese cedar. Gr. *krypto*, to hide; *meris*, a part. All parts of the flower of this ornamental evergreen tree are concealed.

Cryptoste'gia f. Rubber vine. From Gr. *krypto*, to hide; *stĕgo*, to cover. The scales in the throat of the flower cover the anthers.

Cryptostem'ma n. Gr. *krypto*, to hide; *stĕmma*, a crown. The hairs of the seed cases cover the scales of the pappus in these herbs.

crystal'linus, -a, -um Crystalline; transparent.

Ctenan'the f. From Gr. *kteis, ktĕnos*, a comb; *anthos*, a flower; from the arrangement of the bracts of these Brazilian herbs.

ctenoi'des Resembling a comb.

cube'ba Specific epithet for *Piper cubeba* from the pre-Linnaean name for cubeb.

Cucuba'lus A Latin plant name mentioned by Pliny.

cuculla'ris, -is, -e; cuculla'tus, -a, -um Hoodlike; having sides or apex curved inwards to resemble a hood.

cucumerifo'lius, -a, -um Having leaves like cucumber.

cucumeri'nus, -a, -um Resembling cucumber.

Cu'cumis m. Cucumber, melon. The ancient Latin name.

Cucur'bita f. Latin name for a gourd. The genus also includes such plants as squash and pumpkin.

cultra'tus, -a, -um; cultrifor'mis, -is, -e Shaped like a knife blade.

Cumi'num n. Cumin. Greek name for this aromatic herb of ancient use.

cunea'tus, -a, -um Wedge-shaped, generally with the narrow end down.

cuneifo'lius, -a, -um With wedge-shaped leaves.

cuneifor'mis, -is, -e In the form of a wedge.

Cu'nila f. Dittany. The Latin name for one of the mints.

Cunningha'mia f. Decorative coniferous evergreens named for James Cunninghame (d. *c.* 1709), East India Company surgeon at Amoy, China, who, between the years 1698–1702, sent home large collections of plants and plant drawings from China, Malacca, etc.

Cupa'nia f. Ornamental evergreen trees, resembling *Blighia*, named in honour of Francesco Cupani (1657–1711), Italian monk and author of works on Sicilian plants.

Cu'phea f. Gr. *kyphos*, curved; from the shape of the seed capsule of these herbs and sub-shrubs.

cuprea'tus, -a, -um; cupreus, -a, -um Coppery; copper-coloured.

cupressifo'lius, -a, -um With leaves like cypress.

cupressifor'mis, -is, -e; cupressi'nus, -a, -um; cupressoi'des Resembling cypress.

Cupres'sus f. Cypress. The Latin name for the Italian cypress tree (*C. sempervirens*).

curassa'vicus, -a, -um From Curaçao, Caribbean Sea.

Curcu'ligo f. L. *curculio*, a weevil. The ovary of these palmlike foliage plants is beaked like a weevil.

Curcu'ma f. Turmeric. The Latinized version of the Arabic name. The stout rootstocks are the source of turmeric which is both a dyestuff and a condiment. It is an important ingredient of curry to which it imparts the yellow colour. The name *crocus* has the same derivation.

Curto'nus m. From Gr. *kyrtos*, swelling, hunch-backed, bent.

cur'tus, -a, -um Shortened.

curva'tus, -a, -um Curved.

curvifo'lius, -a, -um With curved leaves.

Cuscu'ta f. Medieval Latin name for dodder.

cuscutifor'mis, -is, -e Resembling *Cuscuta* or dodder.

cuspida'tus, -a, -um With a sharp stiff point; cuspidate.

cuspidifo'lius, -a, -um With leaves having a sharp stiff point.

Cyanan'thus m. Gr. *kyanos*, blue; *anthos*, a flower; from the colour of the flower of these herbaceous perennials.

cyanan'thus, -a, -um With blue flowers.

Cyanel'la f. Gr. *kyanos*, blue.

cya'neus, -a, -um Blue.

cyanocar'pus, -a, -um With blue fruit.

cyanophyl'lus, -a, -um With blue leaves.

Cyano'tis f. Gr. *kyanos*, blue; *ous*, an ear; an allusion to the colour and form of the petals of these *Tradescantia*-like herbs.

cy'anus Old name of a dark blue substance now the specific epithet for the cornflower.

Cya'thea f. Tree fern. From Gr. *kyatheion*, a little cup; in allusion to the spore cases.

cyatheoi'des Resembling *Cyathea*.

Cyatho'des f. From Gr. *kyathōdēs*, cup-like; in allusion to the cup-shaped, toothed disc of these evergreen shrubs.

Cybiste'tes m. Gr. *kybistētēs*, one who tumbles; referring to the dry mature fruiting heads being tumbled and blown about by the wind.

Cy'cas f. Greek name for a kind of palm. Although palmlike these are not palms. Sago-palm is *C. revoluta*.

Cyc'lamen n. Sowbread. The Greek name. They are regarded as a favourite food for swine in the South of France, Sicily, and Italy. It is claimed that a diet of cyclamen bulbs adds particular flavour to the pork products of the Périgord area.

cyclami'neus, -a, -um Resembling *Cyclamen*.

Cyclanthe'ra f. Gr. *kyklos*, a circle; *anthera*, anther; from the arrangement of the anthers of these herbs.

Cyclan'thus m. Gr. *kyklos*, a circle; *anthos*, a flower; from the arrangement of the flowers of these herbs.

cyclocar'pus, -a, -um With fruit arranged in a circle.

Cyclo'phorus m. Japanese fern. From Gr. *kyklos,* a circle; *phoreo,* to bear; from the circular shape of the spore cases.

Cyclo'pia f. From Gr. *kyklŏs,* circle; *pous,* foot. Referring to the circular base of the calyx.

cyc'lops Gigantic.

Cydis'ta f. From Gr. *kydistos,* most glorious; in reference to the spectacular flowers of these ornamental vines.

Cydo'nia f. Quince. The Latin name for this shrub or small tree, derived from the town of Cydon in Crete.

cylindra'ceus, -a, -um; cylindricus, -a, -um Long and round, cylindrical.

cylindrosta'chyus, -a, -um With a cylindrical spike.

Cymba'laria f. Ivy-leaved toadflax. From Gr. *kymbalon,* Latin *cymbalum,* cymbal, referring to the leaf shape of some species of these herbs.

Cymbi'dium Gr. *kymbē,* a boat. There is a hollow recess in the lip of these orchids.

cymbifor'mis, -is, -e Boat-shaped.

Cymbopo'gon m. Gr. *kymbē,* a boat; *pōgōn,* a beard. A rather technical allusion to the form of the spikelets of these oil-producing grasses.

cymo'sus, -a, -um Furnished with cymes or flower clusters in which the flower in the centre opens first, the remainder of the flowers open in succession outwards to the periphery (as in *Phlox*).

Cynanast'rum n. Gr. *kyanos,* blue; *astrum,* incompletely resembling. The name was apparently suggested by resemblance to the genus *Cyanella.*

Cynan'chum n. Gr. *kyōn,* a dog; *ancho,* to strangle. Some of the species of these twining herbs and sub-shrubs are poisonous.

Cy'nara f. Artichoke and cardoon. The Latin name for these herbs with spiny involucres.

cynaroi'des Resembling *Cynara.*

Cyno'ches n. From Gr. *kyknŏs,* swan; *auchē,* neck; referring to the arched slender column of the male flowers of these epiphytic South American orchids.

Cy'nodon n. Bermuda grass. From Gr. *kyōn,* a dog; *ŏdŏus,* a tooth.

Cynoglos'sum n. Hound's tongue. From Gr. *kyōn,* a dog; *glōssa,* a tongue; an allusion to the texture of the leaves of these weedy herbs.

Cyparis'sias f. Latin name of a kind of spurge, from Gr. *kyparissos,* cypress.

Cypel'la f. Gr. *kypellon*, a goblet; from the form of the flowers of these small bulbous plants.

Cype'rus m. Galingale. Greek word meaning sedge. The genus includes *Papyrus*, from which the Egyptians made a writing material by cutting the pithlike tissues into strips, laying them crosswise and uniting them under pressure.

Cyphoman'dra f. Gr. *kyphos*, a tumour; *aner*, male; with reference to a thickening on the connective of the anther.

Cypripe'dium n. Lady's slipper, moccasin flower. From Gr. *Kypris*, Latin *Cypria*, Venus (Aphrodite) who was worshipped especially on Cyprus; *pĕdilon*, a slipper, unfortunately latinized as *pedium* by Linnaeus, from the shape of the flower.

cyp'rius, -a, -um Cyprian, relating to the island of Cyprus.

Cyril'la f. Leatherwood. A shrub named in honour of Dominico Cirillo (1734–1790), physician and professor of botany at Naples, executed by the Bourbons for his liberal opinions.

Cyrtanthe'ra f. From Gr. *kyrtos*, arched. In allusion to the curved anthers.

Cyrtan'thus m. Gr. *kyrtos*, arched; *anthos*, a flower; referring to the curved tube of the flower of these South African bulbs.

Cyrto'mium n. Gr. *kyrtos*, arched; from the habit of growth of these ferns.

Cyrtopo'dium n. Gr. *kyrtos*, arched; *pous*, a foot; from the form of the lip of the flower of these terrestrial orchids.

Cyrto'stachys m. Gr. *kyrtos*, arched; *stachys*, a spike. The flower spikes of these palms are curved.

Cysto'pteris f. Bladder fern. From Gr. *kystis*, a bladder; *ptĕris*, a fern.

cythe'reus, -a, -um From the Greek island of Kythera.

cytisoi'des Resembling *Cytisus* or broom.

Cy'tisus m. Broom. Gr. *kytisos*, used by the Greeks for several kinds of woody *Leguminosae*.

D

Daboe'cia f. St. Dabeoc's heath. Named after a rather obscure Irish saint. Misspelled *Daboecii* by Linnaeus.

dacrydioi'des Resembling *Dacrydium*.

Dacry'dium n. Gr. *dakrydion*, a small tear, diminutive of *dacry*, tear. The trees, called pines in New Zealand, exude drops of resin.

dacty′lifer, dactyli′fera, -um Fingerlike; furnished with fingers.

Dac′tylis f. Cock's foot grass. From Gr. *dactylŏs*, a finger; referring to the finger-like appearance of the inflorescence.

dactyloi′des Resembling fingers.

Dactylorhi′za f. From Gr. *daktylŏs*, finger; *rhiza*, root; in allusion to the finger-like tubers of these orchids which contrast with the rounded tubers of *Orchis* proper.

Daemo′norops f. From Gr. *daimon*, Latin *daemon*, evil spirit, demon; *rhŏps*, a shrub; in allusion to the stem of these climbing feather palms. Like cats, they climb by means of sharp recurved hooks on the main stem and leaf stalks.

Da′hlia f. Pronunciation is largely a matter of taste; probably the most convenient course would be to use 'day-lia' when referring to the garden dahlias in the English-speaking world but 'dah-lia' when referring to species of the genus *Dahlia*, from which the garden forms have been derived by hybridization. These herbaceous tuberous-rooted perennials from Mexico were named in honour of Dr. Anders Dahl (1751–1789), Swedish botanist and pupil of Linnaeus, who sought unsuccessfully to retain the Linnaean collections in Sweden.

dahu′ricus, -a, -um; dau′ricus, -a, -um; davu′ricus, -a, -um From Dahuria (Dauria, Davuria), region of south-east Siberia and north-east Mongolia.

Da′is f. Greek word meaning a torch; an allusion to the form of inflorescence, the flowers of these shrubs being arranged in clusters.

Dalber′gia f. A genus of trees, shrubs, and climbers named for the brothers Nils Dalberg (1736–1820), Swedish botanist and royal physician, and Carl Gustav Dalberg (*fl.* 1754–1775), Swedish officer, who owned an estate in Surinam, whence he sent specimens to Linnaeus.

Da′lea f. Herbs and small shrubs named in honour of Dr. Samuel Dale (1659–1739), English botanist and author.

dalecar′licus, -a, -um From the province of Dalecarlia, or Dalarna, in Sweden.

Dalecham′pia f. Tropical shrubs named for Jacques Dalechamps (1513–1588), French physician and botanist.

dalhou′siae In honour of Lady Dalhousie, wife of James A. B. Ramsay (1812–1860), 1st Marquess of Dalhousie, Governor-General of India.

Dalibar′da f. An herbaceous perennial named in honour of Thomas François Dalibard (1703–1779), French botanist, author of a *Flora of Paris* (1749).

dalma′ticus, -a, -um From Dalmatia.

damasce'nus, -a, -um From Damascus, Syria.

Damaso'nium n. Greek name for water-plantain (*Alisma*).

Danaë f. Alexandrian laurel. Named for Dănăē, in Greek mythology the daughter of Acrisius, King of Argos. Having been warned that she would bear a son who would eventually kill him, he shut her up in a brazen tower only to have Zeus descend in a shower of gold with the result that she gave birth to Perseus. Acrisius placed both in a wooden box and threw them into the sea. By one of those twists of fate which make up Greek tragic plots, Perseus grew up to kill his grandfather accidentally when practising with the discus.

danaeifo'lius, -a, -um With leaves like *Danaë*.

Da'phne f. Greek name for the bay tree or laurel (*Laurus nobilis*) later transferred to the present genus. According to mythology it was named after a nymph changed by the gods into a bay tree to save her from pursuit by Apollo, but the name itself may be derived from an Indo-European root meaning 'odour'.

Daphniphyl'lum n. Gr. *daphnē*, laurel; *phyllon*, a leaf. The leaves of these evergreen shrubs and small trees resemble laurel.

daphnoi'des Resembling *Daphne*.

Darlingto'nia f. An insectivorous herbaceous plant named in honour of Dr. William Darlington (1782–1863), Philadelphia physician and botanist who published the posthumous botanical work of his friend and fellow physician, William Baldwin (1779–1819), who died at Franklin on an exploring trip up the Missouri River.

Darwi'nia f. Australian evergreen shrubs named in honour of Dr. Erasmus Darwin (1731–1802), physician, man of science and writer of poems on scientific subjects, grandfather of Charles Darwin (1809–1882).

dasy- A prefix in many Greek compound words meaning thick. Thus:

dasyacan'thus	thick-spined
dasyan'thus	shaggy-flowered
dasycar'pus	with hairy fruit
dasy'cladus	shaggy-branched
dasyphyl'lus	shaggy-leaved
dasyste'mon	with hairy stamens

Dasyli'rion n. Gr. *dasys*, thick; *lirion*, lily; in allusion to the character of these succulents.

Datis'ca f. The derivation of the name of these herbaceous perennials is obscure.

Datu'ra f. Angel's trumpet, thorn-apple. From an Indian vernacular name. Some of the species are violently narcotic and poisonous,

including Jimson weed (*D. stramonium*) after wilting. *D. metel* is common in India and provides a convenient source of poison.

Daubento'nia f. Herbs or shrubs named for a French naturalist, Louis Jean Marie Daubenton (1716–1799).

daucifo'lius, -a, -um With leaves like carrot or *Daucus*.

daucoi'des Resembling carrot or *Daucus*.

Dau'cus m. Carrot. The Latin name, Gr. *daukos*.

Daval'lia f. Hare's-foot fern. Named for Edmond Davall (1763–1798), Swiss botanist of English origin, a friend of James Edward Smith, to whom he bequeathed his herbarium.

Davi'dia f. A handsome deciduous tree named in honour of Abbé Armand David (1826–1900), French missionary in China, 1862–1873, who collected many plants and animals. He discovered what is now known as Père David's Deer (*Elaphurus davidianus*) in the grounds of the Imperial Summer Palace at Peking; it survives only in captivity.

davi'dii In honour of Armand David.

Davie'sia f. In honour of Hugh Davies (*c.* 1739–1821), Welsh clergyman, author of *Welsh Botanology* (1813).

dawsonia'nus, -a, -um Species named for Jackson T. Dawson (1841–1916), first superintendent of the Arnold Arboretum in Boston, Massachusetts.

dealbatus, -a, -um Whitened; covered with opaque white powder.

Dea'mia f. A cactus named for Charles Clemon Deam (1865–1953), American druggist and botanist in Indiana.

deb'ilis, -is, -e Weak; frail.

Debregea'sia f. A genus of shrubs or trees, named after Prosper Justin de Brégeas (b. 1807), French naval officer, who commanded the corvette *La Bonite* on a voyage of exploration to the Far East (1836–1837).

Decais'nea f. Ornamental shrubs named in honour of Joseph Decaisne (1807–1882), director of the Jardin des Plantes, Paris, eminent botanist and horticulturist.

decan'drus, -a, -um With ten stamens.

decape'talus, -a, -um With ten petals.

decaphyl'lus, -a, -um With ten leaves.

deci'duus, -a, -um Deciduous.

deci'piens Deceptive; cheating.

declina'tus, -a, -um Bent downwards.

De'codon n. Swamp loosestrife. Gr. *deka*, ten; *ŏdŏus*, a tooth. The calyx of this shrubby perennial has ten teeth.

decolo'rans Discolouring; staining.

decompo'situs, -a, -um More than once divided.

decora'tus, -a, -um; deco'rus, -a, -um Decorative; becoming; comely.

decor'ticans Shedding bark.

decuma'nus, -a, -um Very large; immense; literally the tenth part, but acquiring its meaning of importance from association with the main entrance of Roman camps, called the Decumana Porta, because the tenth part of a Roman division was quartered near it.

Decuma'ria f. L. *decimus*, ten; from the number of parts of the flower of these ornamental climbing shrubs.

decum'bens Trailing with tips upright; decumbent.

decur'rens Running down the stem; decurrent; e.g. a leaf base that merges with the stalk below it.

Deerin'gia f. Climbing herbs or sub-shrubs named for George Charles Deering (*c*. 1695–1749), physician, who was born in Saxony, studied at Leiden and Paris and practised medicine in London and later Nottingham. He was the author of a catalogue of plants found at Nottingham.

defl'exus, -a, -um Bent abruptly downward.

defor'mis, -is, -e Misshapen, deformed.

dehis'cens Dehiscent, namely the splitting or opening of a seed-pod for the release of seeds or the opening of an anther to discharge pollen. The function can sometimes be relatively violent and both visible and audible as in the popping of broom pods on a hot day. In the case of the castor-oil plant the seeds are thrown several feet and it is important to harvest in time.

Deinan'the f. From Gr. *dĕinŏs*, wondrous; *anthē*, flower.

dejec'tus, -a, -um Debased.

delec'tus, -a, -um Chosen.

delicatis'simus, -a, -um Most delicate.

delica'tus, -a, -um Delicate.

delicio'sus, -a, -um Delicious.

De'lonix f. Royal poinciana, flamboyant, gul mohur, fancy-anna, possibly the most gorgeous tree in cultivation. From Gr. *dēlŏs*, evident, conspicuous; *ŏnux*, a claw; referring to the long-clawed petals.

Delosper'ma n. From Gr. *dēlŏs*, evident; *sperma*, seed. In allusion to the seeds exposed in the unroofed chambers of the capsule. A genus formerly included in *Mesembryanthemum*.

Delo'stoma n. Gr. *dēlos*, evident; *stŏma*, a mouth. The flower of these Peruvian shrubs has a wide mouth.

del'phicus, -a, -um Of Delphi, Greece.

delphinan'thus, -a, -um With flowers like *Delphinium*.

delphinen'sis, -is, -e Of Dauphiné, France.

delphinifo'lius, -a, -um With leaves like *Delphinium*.

Delphin'ium n. Larkspur. The Greek name, *dělphiniŏn*, derived from the word *dělphis* for a dolphin in allusion to the form of the flower of annual species (now placed in *Consolida*) found in Greece.

deltoi'des; deltoi'deus, -a, -um Triangular, like the Greek letter *Δ* (delta).

Dema'zeria (incorrectly, **Desmazeria**) f. A genus of grasses named after Jean Baptiste Joseph Henri Desmazières (1796–1862), French botanist.

demer'sus, -a, -um Living under water; submerged.

demis'sus, -a, -um Hanging down; weak.

dendri'cola A dweller on trees.

Dendro'bium n. Gr. *dĕndrŏn*, a tree; *bios*, life. These tree-perching orchids are, next to the cattleyas, the most popular greenhouse orchids.

Dendroca'lamus m. Giant bamboo. From Gr. *dĕndrŏn*, a tree; *calamus*, a reed. Some of the species grow to a height of 100 feet. Among them is the male bamboo from which surf-casting and split-cane fishing rods are made.

Dendrochi'lum n. Gr. *dĕndrŏn*, a tree; *cheilos*, a lip. The flowers of these tree-perching orchids have conspicuous lips.

dendroi'deus, -a, -um Treelike.

Dendrome'con f. Tree-poppy. From Gr. *dĕndrŏn*, a tree; *mēcōn*, a poppy.

dendro'philus, -a, -um Tree-loving.

Denmo'za f. Name an anagram of Mendoza, the province of Argentina to which this cactus is native.

Dennstae'dtia f. Hayscented fern. In honour of August Wilhelm Dennstedt, German botanist of the early 19th century.

densa'tus, -a, -um; densus, -a, -um Compact; dense.

dens-ca'nis Specific epithet meaning dog's tooth.

densiflor'us, -a, -um Densely flowered.

densifo'lius, -a, -um Densely leaved.

Denta'ria f. Toothwort. From L. *dens*, a tooth; in allusion to the toothlike scales on the roots of this early-flowering herb which led to the supposition that it might be good for toothache.

denta'tus, -a, -um; den'tifer, denti'fera, -um; dento'sus, -a, -um Toothed; furnished with teeth.

denticula'tus, -a, -um Slightly toothed.

denuda'tus, -a, -um Bare; naked.

deoda'ra Specific epithet from the North Indian name for the deodar (*Cedrus deodara*).

depaupera'tus, -a, -um Imperfectly developed; dwarfed.

depen'dens Hanging down.

depres'sus, -a, -um Flattened; pressed down.

Der'ris f. Gr., a leather covering; in allusion to the tough seed pods of these woody climbers. The insecticide, derris, is made from the powdered tuberous root. In the East Indies the powder is used to poison fish by stupefying them.

Descham'psia f. In honour of Louis August Deschamps (1765–1842), French surgeon-naturalist who investigated the natural history of Java.

Descurai'nia f. In honour of François Descourain (1658–1740), French pharmacist.

deser'ti Of the desert.

deserto'rum Of the deserts.

Desfontai'nea f. A Chilean evergreen shrub named for René Louiche Desfontaines (1750–1833), French botanist, professor at the Jardin des Plantes, Paris; see *Fontanesia*.

Desman'thus m. Gr. *děsmē*, a bundle; *anthos*, a flower. The flowers of these herbs or shrubs are collected together in spikes resembling bundles.

Desmo'dium n. Tick-trefoil. From Gr. *děsmos*, a band or chain. The stamens are joined together in these herbs and sub-shrubs.

desmoncoi'des Resembling the genus *Desmoncus*.

Desmon'cus m. Gr. *děsmos*, a band or chain; *onkos*, a hook. The ends of the leaves of these spiny feather palms are furnished with hook-like tips.

desquama'tus, -a, -um With scales removed; scoured; bereft of scales.

deter'sus, -a, -um Wiped clean.

deton'sus, -a, -um Bare; shorn.

deu'stus, -a, -um Burned.

Deut'zia f. Very ornamental shrubs. After Johan van der Deutz (1743–1788), Dutch friend and patron of Carl Thunberg, Swedish botanist and pupil of Linnaeus.

diabo'licus, -a, -um Devilish; applied, for example, to a plant with a two-horned fruit.

diacan'thus, -a, -um Furnished with two spines.

diac'ritus, -a, -um Separated; excellent.

Dia'crium n. From Gr. *di*, double; *akris*, peak, point; referring to the two horn-like points on the lip of these orchids.

diade'ma Gr. *dĭadēma*, royal head-dress, crown, diadem.

dian'drus, -a, -um Furnished with two or twin stamens.

Dianel'la f. Flax lily. Diminutive of Diana, goddess of the chase.

dianthiflo'rus, -a, -um *Dianthus*-flowered.

Dian'thus m. Pink, carnation. From Gr. *Di*, of Zeus or Jove; *anthos*, flower. These small herbs have long been held in the highest esteem.

Diapen'sia f. Classical Greek plant name adopted by Linnaeus for these dwarf, evergreen, tufted perennials.

dia'phanus, -a, -um Transparent.

diap'repes Distinguished.

Dia'scia f. Gr. *di*, two; *askos*, a sac. The flowers of these herbs have two spurs.

Dicen'tra f. From Gr. *dis*, twice; *kentron*, a spur. The flowers of these perennial plants, which include bleeding heart, have two spurs.

Dichorisan'dra f. Gr. *dis*, twice; *chōri*, separate; *aner*, male; in allusion to two lateral stamens which spread outwards and away from the converging four other stamens.

dicho'tomus, -a, -um Forked in pairs; repeatedly dividing into two branches.

Di'chroa f. Gr. *dis*, twice; *chrŏa*, colour.

dichroan'thus, -a, -um With flowers of two distinct colours.

dichro'mus, -a, -um; di'chrous, -a, -um Of two distinct colours.

Dickso'nia f. Tree ferns named after James Dickson (1738–1822), British botanist and nurseryman.

Dicli'ptera f. Gr. *diklis*, double-folding; *pteron*, wing. In allusion to the two wing-like divisions of the valves of the capsule.

dicoc'cus, -a, -um Furnished with two berries or nuts.

Dicranostig'ma n. Glaucous herbs of the poppy family. From Gr. *dikranos*, two-branched; plus *stigma*.

Dictam'nus m. Greek name for a Cretan origanum, probably named after Mount Dikte, later transferred to this perennial herb which is variously called dittany, gas plant, burning bush, and fraxinella. The leaves and stem, which have a strong odour, give out a volatile oil which, on a hot, still day, can be ignited with a match.

dictyophyl'lus, -a, -um With leaves having a conspicuous network of veins and veinlets. Similarly, *dictyocarpus*, with netted fruit.

Dictyosper'ma n. From Gr. *dictyon*, a net; *sperma*, seed; in a technical allusion to the seed of these *Areca*-like palms.

dictyo'tus, -a, -um Made in net-fashion; netted; latticed.

Didis'cus m. Gr. *di*, two; *diskos*, disc.

Didymocar'pus m. From Gr. *didymos*, double; *karpos*, fruit. The capsule breaking into two parts.

di'dymus, -a, -um Twin or in pairs; two-fold.

Dieffenba'chia f. Dumb cane. Foliage plants named in honour of J. F. Dieffenbach, 1790–1863, in charge of the gardens of the royal palace of Schönbrunn at Vienna around 1830.

dieffenba'chii Specific epithet of a curious monotypic umbelliferous genus, *Coxella*, confined to the Chatham Islands where the first botanical collections were made by Ernst Dieffenbach, naturalist to the New Zealand Company, 1839–1841.

dielsia'nus, -a, -um; dielsii In honour of Friedrich Ludwig Emil Diels (1874–1945), director of Berlin–Dahlem botanic garden and museum.

Diera'ma n. Wandflower. Gr. *dĭĕrāma*, meaning a funnel; from the shape of the perianth of these South African iridaceous plants.

Diervil'la f. Bush honeysuckle. Named for M. Dierville, a French surgeon who travelled in Canada, 1699–1700, and who introduced *D. lonicera*.

diffor'mis, -is, -e Of unusual form in relation to the normal run of the genus.

diffu'sus, -a, -um Spreading; diffuse.

Digita'lis f. Foxglove. From L. *digitus*, a finger. The flowers of these herbaceous plants are like the fingers of a glove. *D. purpurea* is the source of the digitalis used in cardiac medicine.

digita'tus, -a, -um Shaped like an open hand; digitate.

dilata'tus, -a, -um; dila'tus, -a, -um Spread out; expanded.

Dille'nia f. A genus of tall tropical, evergreen trees named by Linnaeus for Johann Jacob (John James) Dillenius (1684–1747), German botanist and physician. He came to England in 1721 from Giessen to help William Sherard, produced a magnificent book *Hortus Elthamensis* (1732), illustrating and describing the plants of James Sherard's garden at Eltham near London, and became Sherardian professor of botany at Oxford in 1734 where in 1741 he published his celebrated *Historia Muscorum*. To quote Linnaeus, '*Dillenia* of all plants has the showiest flower and fruit, even as Dillenius made a brilliant show among botanists.'

Dillwy'nia f. Evergreen flowering shrubs named in honour of Lewis Weston Dillwyn (1778–1855), British botanist.

dimidia'tus, -a, -um Divided into two dissimilar or unequal parts.

Dimorphothe'ca f. Cape marigold. From Gr. *dis*, twice; *morphe*, a shape; *theka*, fruit; in allusion to the two kinds of achenes (*cypselae*) found in one and the same fruiting head of these South African herbs and sub-shrubs; those of the disc and ray differ in shape.

dimor'phus, -a, -um Having two forms of leaf, flower, or fruit on the same plant.

Dinteran'thus m. In honour of Professor Kurt Dinter (1868–1945), German botanist who made extensive collections in South-west Africa from 1897 to 1906. *Juttadintera* commemorates his wife.

di'odon Having two teeth.

dioi'cus, -a, -um Dioecious, i.e. having the male reproductive organs borne on one plant and the female on another.

Diona'ea f. Venus flytrap. One of the Greek names for Venus. Found only in North and South Carolina, the plant has leaves each divided into two hinged, valvelike segments, the inner faces of which have three sensitive 'trigger' hairs. When an insect lands inside and touches any two of the hairs, the valves close like jaws, the interlocking stiff hairs around the outside edge of the leaves holding the victim. The trap reopens when the insect is digested.

Diony'sia f. From *Dĭŏnȳsŏs*, the earlier Greek name of Bacchus.

Dio'on n. From Gr. *dis*, twice; *ōŏn*, an egg. With reference to the paired seeds; originally spelled *Dion*.

Diosco'rea f. Yam. Named in honour of Pedanios Dioscorides, 1st-century Greek physician and herbalist, who compiled *Materia Medica* which dealt with medicinal herbs and which has gone into countless editions through the centuries.

Dios'ma f. Gr. *dios*, divine; *ŏsme*, fragrance. The leaves of these heath-like shrubs when crushed are very fragrant.

diosmifo'lius, -a, -um With leaves like *Diosma*.

Dios'pyros f. Persimmon, ebony. Gr. *diŏspyrŏs* from *dios*, divine; *pyros*, wheat; a name transferred to this genus with edible fruits.

Dio'tis f. From Gr. *di*, two; *ŏus*, *ōtŏs*, an ear. The lobes of the corolla of this perennial are shaped like ears.

Dipel'ta f. Gr. *di*, two; *pelte*, a shield; referring to the form of the bracts on these ornamental deciduous shrubs.

dipe'talus, -a, -um Having two petals.

Diphylle'ia f. Gr. *di*, two; *phyllon*, a leaf. The leaves on this perennial herb usually have two deep lobes.

diphyl'lus, -a, -um Having two leaves or leaflets.

Diplade'nia f. Gr. *diplŏŏs*, double; *aden*, a gland. There are two glands on the ovary of these climbers.

Diplarrhe'na f. From Gr. *diplŏŏs*, double; *arrēn*, male; referring to the two fertile stamens of these Australian perennials.

Dipla'zium n. Gr. *diplasios*, double. The indusium, the covering over the spores, is double in these ferns.

Diploglot'tis f. Gr. *diplŏŏs*, double; *glottis*, a tongue. The inner scale of the petals is double on this Australian tree.

diplostephioi'des Resembling *Diplostephium*, a genus of little garden interest.

diplo'trichus, -a, -um Gr. *diplŏŏs*, double; *thrix*, a hair. With two kinds of hair, that is long and short hairs intermixed.

dipsa'ceus, -a, -um Resembling the teasel.

Dip'sacus m. Teasel. Gr. *dipsakos*, from *dipsa*, thirst; the leaf base holds water. The plants are cultivated for the fruiting heads which are covered with stiff, hooked points. These heads are used in the woollen industry to raise the nap on cloth. So far, no man-made device has been found as effective.

Dipteracanth'us m. Gr. *dipteras*, two-winged; *Acanthus*. In allusion to the two-leaved peduncle of these plants of the *Acanthus* family.

dipterocar'pus, -a, -um Having two-winged fruit.

Diptero'nia f. Gr. *di*, two; *pteron*, a wing; in allusion to the winged fruit of this ornamental Chinese tree.

dip'terus, -a, -um Two-winged.

dipyre'nus, -a, -um Having two seeds or kernels.

Dir'ca f. Leatherwood. From Gr. *Dirke*, a fountain. These early-blooming North American shrubs grow in moist places. The tough, flexible shoots can be used as tying material or for baskets, while rope can be made from the bark.

Di'sa f. In allusion to the mythical Queen Disa of Sweden who was commanded to come before the King of the Sveas neither naked nor clothed and who accordingly came wrapped in a fishing net; the upper lip of the flower of the type-species has a somewhat netted appearance.

Disca'ria f. Gr. *diskos*, a disc. The flower of these spiny shrubs has a large fleshy disc.

discifor'mis, -is, -e In the shape of a disc.

Discocac'tus m. Gr. *diskos*, a disc, and *Cactus*. From the shape of these plants which are globe-shaped or flattened endwise.

discoi'deus, -a, -um Discoid; without rays.

disco'lor Of two different and, usually, distinct colours.

Disocac'tus m. From Gr. *dis*, twice; *ĭsŏs*, equal; *cactus*; the sepals and petals of these cacti are equal in number.

dis'par Unequal; dissimilar from the normal state of the genus.

disper'sus, -a, -um Scattered.

Dis'porum n. Fairybells. From Gr. *dis*, two; *spora*, seed; in allusion to the two ovules in each chamber of the ovary.

dissec'tus, -a, -um Deeply cut; divided into deep lobes or segments.

dissi'milis, -is, -e Unlike the normal for the genus.

dissitiflo'rus, -a, -um With flowers in loose heads, not compact.

dis'situs, -a, -um Lying apart, well-spaced.

dista'chyus, -a, -um With two spikes.

dis'tans Widely separated.

distichophyl'lus, -a, -um With leaves arranged in two ranks.

dis'tichus, -a, -um In two ranks.

Dis'tictis f. Gr. *dis*, twice; *stiktos*, spotted. The flattened seeds look like two rows of spots in the capsule of these climbing shrubs.

distor'tus, -a, -um Misshapen; of grotesque form.

Disty'lium n. Gr. *dis*, two; *stylos*, a style. These ornamental trees and shrubs have two styles.

disty'lus, -a, -um Having two styles.

Diu'ris f. From Gr. *di-* two; *oura*, tail; referring to the two tail-like lateral sepals of these terrestrial Australian orchids.

diur'nus, -a, -um Day-flowering.

divarica'tus, -a, -um Spreading; growing in a straggling manner.

diver'gens Spreading out widely from the centre.

diversi'color Diversely coloured. Also:

diversiflo'rus, -a, -um	diversely flowered
diversifo'lius, -a, -um	diversely leaved
diversifor'mis, -is, -e	of diverse forms

divi'sus, -a, -um Divided.

di'vus, -a, -um Belonging to the gods.

Dizygothe'ca f. False aralia. From Gr. *dis*, twice; *zygos*, a yoke; *thēkē*, a case. The anthers of these ornamental trees and shrubs have twice the normal number of lobes.

Docy'nia f. Anagram of *Cydonia*. These trees are closely related to *Cydonia*, quince.

dodecan'drus, -a, -um Having twelve stamens.

Dodeca'theon n. Shooting star, American cowslip. From Gr. *dodeka*, twelve; *thĕŏs*, god; the name given by Pliny to a spring flower, believed to be the primrose, which was under the care of the twelve principal gods.

Dodonae'a f. A genus of trees and shrubs, chiefly Australian, named in honour of Rembert Dodoens (1517–1585), celebrated Flemish royal physician and herbalist, who became professor of medicine at Leiden in 1582.

dodonaeifo'lius, -a, -um *Dodonaea*-leaved.

dolabra'tus, -a, -um; dolabrifor'mis, -is, -e Hatchet-shaped.

Dolichan'dra f. Gr. *dolichos*, long; *aner*, male; in allusion to the long anthers of these climbing shrubs.

Do'lichos m. Greek for long; used by the ancient Greeks for long-podded beans, whence transferred to these climbing plants closely related to beans.

Dolichothe'le f. Gr. *dolichos*, long; *thēlē*, a nipple; an allusion to the elongated tubercles or small above-ground tubers of these cacti.

dolo'sus, -a, -um Deceitful; appearing like some other plant.

Dombey'a f. Ornamental evergreen trees or shrubs named in honour of Joseph Dombey (1742–1794), French botanist, who travelled in Chile and Peru with Ruiz and Pavon.

domes'ticus, -a, -um Frequently used as a house plant; domesticated.

Do'nax m. Gr. *dŏnax*, a kind of reed.

Doo'dia f. Decorative ferns named in honour of Samuel Doody (1656–1706), London apothecary, curator of the Chelsea Physic Garden from 1691.

Dore'ma n. Gr. *dōrēma*, gift.

Dori'tis f. From Gr. *dŏry*, a spear; the lip of these orchids being spear-shaped.

doronicoi'des Resembling *Doronicum*.

Doro'nicum n. Derivation obscure. Called leopard's bane, this genus of perennial herbs was considered useful in destroying and warding off wild beasts. *D. pardalianches*, meaning to strangle leopards, is reputedly poisonous.

Dorothean'thus m. In honour of Dorothea Schwantes, mother of the German specialist on succulent plants, Dr. Martin Heinrich Schwantes (1881–1960), archaeologist and professor at Kiel.

Dorste'nia f. Herbs and small shrubs named for Theodor Dorsten (1492–1552), German botanist, professor of medicine at Marburg.

Doryan'thes f. Spear lily. From Gr. *dŏry*, a spear; *anthos*, a flower. The flower stem of this Australian desert plant is very long, sometimes as much as 20 feet.

Dory'cnium n. The ancient Greek name *dŏryknĭon* for a *Convolvulus*, later transferred to these shrubby plants from the Mediterranean region.

Doryo'pteris f. From G. *dŏry*, a spear; *pteris*, a fern; from the form of the fronds of these small ferns.

doshongen'sis, -is, -e From the Doshong Pass, south-eastern Tibet.

Dougla'sia f. Tufted perennial herbs, named for David Douglas (1798–1834), Scottish collector for the Horticultural Society of London, who introduced many Pacific Coast plants of great merit into European gardens. He came to Oregon in 1825 for a two-year botanical exploration, returning in 1830. He was killed in Hawaii; by a strange accident he fell into a wild-cattle trap where he was gored to death by a bull already trapped.

dougla'sii In honour of David Douglas.

Dovya'lis f. Woody plants of southern Africa. Derivation obscure. Sometimes incorrectly spelled *Doryalis*. *D. caffra*, the Kei apple, is much used in South Africa as a hedge plant and has orange-like fruits.

Downin'gia f. Annual herbs, mostly Californian, named in honour of Andrew Jackson Downing (1815–1852), American landscape gardener and pomologist. He was the founder of a great tradition of American landscape design and his book *A Treatise on the Theory and Practice of Landscape Design* has run through many editions, with re-editing at least up to 1921. His ideas were free and naturalistic and had great influence on subsequent practice.

Downing was a very notable pomologist. His book *The Fruits and*

Fruit Trees of America was the standard work on the subject in America for at least sixty years. With his brother Charles he founded the Downing Nursery at Newburgh, New York.

Doxan'tha f. Cat's claw, North American climbing shrub. From Gr. *dŏxa*, glory; *anthos*, a flower.

Dra'ba f. Whitlow-grass or nailwort. The classical Greek name *drabē*, for a cruciferous plant, probably *Cardaria draba*. This spring-flowering tufted herb was supposed to have value in poulticing whitlows.

drabifo'lius, -a, -um Having leaves like *Draba* or whitlow-grass.

Dracae'na f. The accepted derivation of the name for these ornamental treelike and shrublike plants is from Gr. *drakaina*, a female dragon. The colour dragon's blood can be obtained from *D. draco*, the dragon tree. At least one authority (Pulteney), however, suggests that the genus was named by Clusius in honour of Sir Francis Drake, whom he met in 1581 and from whom he received some of his plant discoveries from the New World. Clusius, living in Flanders, which was then occupied by the Spanish, and through the time of the Armada, would naturally have regarded Drake as someone worth commemorating.

A number of the plants popularly described and sold as *Dracaena* are in fact *Cordyline*.

dracaenoi'des Resembling *Dracaena*.

dra'co Specific name meaning a dragon.

Dracoce'phalum n. Dragonhead. From Gr. *draco*, a dragon; *cephale*, a head; from the shape of the flower of these herbaceous plants.

dracoce'phalus, -a, -um Dragon-headed.

dracunculoi'des Resembling *Artemisia dracunculus* or tarragon.

Dracun'culus m. Dragon arum. Latin name for another plant, not this evil-smelling *Arum*-like plant.

dracun'culus Latin word meaning a small dragon; specific epithet for *Artemisia dracunculus* or tarragon.

drepanophyl'lus, -a, -um With sickle-shaped leaves.

Dri'mia f. Gr. *drimys*, acrid, pungent.

Drimio'psis f. From *Drimia*, name of a related genus; *opsis*, resemblance.

Dri'mys f. Greek word meaning acrid; from the taste of the bark of these evergreen trees and shrubs related to the magnolia family.

Drosan'themum n. From Gr. *drŏsŏs*, dew; *anthos*, flower. With reference to the papillae on the flowers.

Dro'sera f. Sundew. From Gr. *drŏserŏs*, dewy; in allusion to the gland-tipped hairs on the leaves which give a dewy appearance to these carnivorous plants.

dru'cei In honour of George Claridge Druce (1850–1932), Oxford pharmacist and celebrated amateur botanist who was a keen student of British plants and the author of several floras.

drupa'ceus, -a, -um; dru'pifer, drupi'fera -um Bearing fleshy fruits or drupes such as peach, cherry, or plum.

Dryan'dra f. Australian evergreen shrubs, named for Jonas Dryander (1748–1810), Swedish botanist, librarian (1782–1810) to Sir Joseph Banks, part-author of *Hortus Kewensis* 1789, published under W. Aiton's name, and author of the *Catalogus Bibliothecae Historico-Naturalis J. Banks* in 5 volumes.

Dry'as f. Mountain avens. From Gr. *dryas*, a wood nymph or dryad, to whom the oak was sacred. The leaves of at least one species of these tufted evergreen plants, *D. octopetala*, resemble oak leaves.

drynarioi'des Resembling *Drynaria*, a fern with fronds which look like oak leaves.

dryophyl'lus, -a, -um Oak-leaved.

Dryo'pteris f. American shield fern. Wood fern. From Gr. *dryŏptĕris*, from *drys*, oak; *pteris*, a fern.

Dry'pis f. Gr. *drypis*, a kind of thorny shrub mentioned by Theophrastus.

du'bius, -a, -um Doubtful in the sense of not conforming to pattern.

Duboi'sia f. In honour of Charles Du Bois (1656–1740), merchant, treasurer to the Honourable East India Company, who amassed a big herbarium now at Oxford.

Duches'nea f. Indian or mock strawberry. Trailing herbs named in honour of Antoine Nicolas Duchesne (1747–1827), a French horticulturist, who published in 1766 a remarkable *Histoire naturelle des Fraisiers* dealing with the history and kinds of strawberry, to which this plant is related.

Dudley'a f. In honour of William Russel Dudley (1849–1911), first professor of botany at Stanford University, California.

dulcama'ra Latin for bittersweet. The specific epithet for climbing nightshade (*Solanum dulcamara*).

dul'cis, -is, -e Sweet.

dumeto'rum Of hedges or of bushy places.

dumo'sus, -a, -um Bushy, shrubby.

dunen'sis Of sand dunes.

duplex

du'plex; duplica'tus, -a, -um Double, duplicate.

dura'bilis, -is, -e Durable, lasting.

dura'cinus, -a, -um With hard fruit or berries.

Duran'ta f. Tropical American woody plants named for Castore Durante (*c.* 1529–1590), Papal physician and botanist in Rome.

Du'rio m. From the Malayan name, durian, which is regarded by some people as among the most delicious of fruits. The fruit, which is larger than a coconut, has a spiky outside rind which encloses an evil-smelling pith in which the pulp-covered seeds are embedded. This pulp is the edible part. It is creamy white, delicate, and sweet. The durian is a feature of Malaysian bazaars in June. It is popularly regarded as an aphrodisiac with many earthy sayings attached to it.

durius'culus, -a, -um Somewhat hard.

dur'ra Native Egyptian name for a fodder sorghum, *Sorghum vulgare* var. *durra*, fed to camels and other beasts.

du'rus, -a, -um Hard. Thus, *durior*, harder.

Duva'lia Dwarf succulent herbs named for Henri Auguste Duval (1777–1814), French medical man who published in 1809 a pamphlet on succulent plants in the Alençon garden containing descriptions of the genera *Haworthia* and *Gasteria*; this is reprinted in *The Cactus Journal*, 7, no. 4 (June 1939).

Dy'ckia f. Succulent plants named for Prince Joseph Salm-Reifferscheid-Dyck (1773–1861), author of outstanding books on succulents.

dy'eri In honour of Sir William Turner Thiselton-Dyer (1843–1928), Director of the Royal Botanic Gardens, Kew, from 1885 to 1905, distinguished not only as a botanist and an administrator but also as a classical scholar interested in the interpretation of ancient Greek plant-names.

dys- Greek prefix meaning 'bad, ill, hard, with difficulty', i.e. 'destroying the *good* sense of a word or increasing the *bad*' and used in the formation of many generic names.

Dyschoris'te f. Gr. *dys*, with difficulty; *choristos*, separated; referring to the valves of the capsule which adhere tenaciously.

dysente'ricus, -a, -um Relating to dysentery; used for plants serving as remedies for dysentery.

Dyso'xylum n. Gr. *dys*, bad; *xylon*, wood. The wood of some species of these trees has an unpleasant smell.

E

ebena'ceus, -a, -um Resembling ebony.

E'benus f. Gr. *ĕbĕnŏs*, apparently of Egyptian origin for another plant, not this genus of shrubs and sub-shrubs.

e'benus, -a, -um Ebony-black.

Eberlan'zia f. In honour of Father Eberlanz of Lüderitz Bay, Southwest Africa, collector of succulent plants *c.* 1926.

eboracen'sis, -e Of York, England.

ebractea'tus, -a, -um Without bracts.

e'bulus Latin and Greek name of the dwarf elder or danewort, *Sambucus ebulus*.

ebur'neus, -a, -um Ivory-white.

Ecbal'lium n. Gr. *ekballein*, to cast out. The seeds of this trailing plant, called squirting cucumber, are violently ejected in a squirt of fluid when ripe.

Eccremocar'pus m. Glory flower. From Gr. *ĕkkrĕmēs*, hanging; *karpos*, fruit. The fruit is a slender, hanging pod.

Echeve'ria f. Succulent plants named for Athanasio Echeverriay Godoy, botanical artist who accompanied the expedition under Sessé and Mociño to New Spain (Mexico) (1787–1797), and thus must have done many of the 1,200 drawings of Mexican plants copied within 10 days in 1816 by the citizens of Geneva for A. P. de Candolle.

Echidno'psis f. Gr. *echidna*, a viper; *opsis*, like; in allusion to the slender, somewhat prostrate stems of these succulents.

Echina'cea f. Cornflower (in the United States), purple cone flower. From Gr. *ĕchĭnos*, a hedgehog; in allusion to the prickly scales of the receptacle of these perennial herbs. The thick, black roots have a pungent flavour and are edible.

echina'tus, -a, -um Covered with prickles like a hedgehog.

Echinocac'tus m. Gr. *ĕchĭnos*, a hedgehog. These cacti are like hedgehogs in appearance.

echinocar'pus, -a, -um With prickly fruit like a hedgehog.

Echinoce'reus m. Gr. *ĕchĭnos*, a hedgehog. The spiny fruit differentiates this genus from *Cereus*.

Echino'chloa f. Gr. *ĕchĭnos*, a hedgehog; *chlŏē*, grass.

Echinocystis

Echinocys'tis f. Prickly cucumber. Gr. *ĕchīnos*, a hedgehog; *kystis*, a bladder; from the prickly fruit.

Echino'dorus m. Gr. *ĕchīnos*, a hedgehog; *dŏrŏs*, bag; referring to the spiny achenes.

Echinofossulocac'tus m. From L. *ĕchīnos*, a hedgehog; *fossula*, a little ditch; *cactus*. With reference to the spines and furrows of these cacti.

Echinomas'tus m. Gr. *ĕchīnos*, a hedgehog; *mastos*, a breast. The tubercles of this cactus are spiny.

Echino'panax m. Gr. *ĕchīnos*, a hedgehog; *Panax*, the generic name for American ginseng. This ornamental shrub is densely covered with spines.

Echi'nops m. Globe thistle. From Gr. *ĕchīnos*, a hedgehog; *ōps*, appearance; in allusion to this herb's handsome spiny, globe-shaped flower heads of metallic blue.

Echino'psis Sea-urchin cactus. From Gr. *ĕchīnŏs*, a sea-urchin (also hedgehog); *ŏpsis*, resemblance; from the resemblance of these round, spiny cacti to a sea-urchin.

echinose'palus, -a, -um With prickly sepals.

echinosper'mus, -a, -um With spiny seeds.

echioi'des Resembling viper's bugloss or *Echium*.

Echi'tes f. Gr. *ĕchis*, a viper. This shrub is poisonous and of twining habit.

Echi'um n. Viper's bugloss. The Greek name *ĕchīŏn*. This genus of rough herbs and shrubs was supposed to discourage serpents, while the root drunk with wine was regarded both as good for snake-bite and also as a protection against future bites.

eclec'teus, -a, -um Picked out; selected.

ecornu'tus, -a, -um Without horns.

Edgeworthia f. Ornamental shrubs named in honour of Michael Pakenham Edgeworth (1812–1881) of the East India Company's Service, a keen amateur botanist who collected many new plants in India.

Edraian'thus m. Gr. *hedraios*, sitting; *anthos*, flower. The flowers of these perennials are sessile in a cluster at the top of the main stalk.

edu'lis, -is, -e Edible.

effu'sus, -a, -um Loosely spreading, straggling, spread out.

Ehre'tia f. A genus of trees and shrubs named in honour of Georg Dionysius Ehret (1708–1770), celebrated German botanical artist, brother-in-law of Philip Miller, the English author of the great *Gardeners Dictionary*.

Eichhor'nia f. Water-hyacinth. A genus of aquatic plants named for J. A. Fr. Eichhorn (1779–1856), Prussian minister of education.

eich'leri Of Wilhelm Eichler, a German who found *Tulipa eichleri* in the Caucasus, or August Wilhelm Eichler (1839–1887), German professor of botany, successively at Graz, Kiel and Berlin.

elaeagnifo'lius, -a, -um With leaves resembling *Elaeagnus*.

Elaeag'nus f. A large genus of shrubs and trees. Often said to be from Gr. *elaia*, the olive tree; *agnos*, the Greek name for *Vitex agnus-castus* or chaste-tree; but, as it was applied by Theophrastus to a willow (*Salix caprea*) and was probably *hĕleagnos*, it is most likely from *hĕlōdēs*, marshy; *hagnŏs*, pure, hence white; in allusion to the white fluffy masses of fruits.

Ela'eis f. African oil palm. From Gr. *ĕlaia*, the olive tree. Oil is obtained from the fruit of these feather palms. It is of great commercial importance in West Africa and other tropical areas. Palm-nut oil, less desirable but still valuable, is obtained from the kernels of the nuts within the fruits. Palm oil, locally called chop oil, is prized for cooking on the West African coast.

Elaeocar'pus m. Gr. *ĕlaia*, olive; *karpos*, fruit; from the appearance of the round fruit of these handsome evergreens which encloses a nut with a rough shell.

Elaeoden'dron n. Gr. *ĕlaia*, olive; *dĕndrŏn*, a tree. The fruit of these small trees or shrubs is like an olive with an oily seed.

Elaphoglos'sum n. From Gr. *elaphos*, a stag; *glossa*, a tongue; an allusion to the shape of the fronds in some species. The elephant-ear fern is *E. crinitum*, with broadly oblong fronds.

elas'ticus, -a, -um Elastic; producing an elastic substance.

elater'ium Specific epithet of *Ecballium elaterium*, squirting cucumber. Both words have the implication of driving with force—an allusion to the way in which the seeds are thrown out.

Elati'ne f. Ancient Greek plant-name *ĕlatinē*.

ela'tior Taller.

Elatoste'ma n. From Gr. *ĕlatŏs*, driving, striking, doctile; used by the authors J. R. and G. Foster in the sense of 'elastic'; *stēma*, penis, stamen; in allusion to the stamens springing up.

ela'tus, -a, -um Tall.

e'legans; elegan'tulus, -a, -um Elegant.

elegantis'simus, -a, -um Very elegant.

129

elephan'tidens Large-toothed.

elephan'tipes Like an elephant's foot.

elephan'tum Of the elephants; monstrous, big.

Eletta'ria f. Cardamon. From *elettari*, the vernacular name in Malabar, south-western India. The cardamon of commerce is the seed of this perennial herb. In India and the East generally it is often chewed as a post-prandial carminative to aid digestion.

Eleu'sine f. From the city of Eleusis in Greece where the temple of Ceres stood. This genus of coarse grasses includes various cereals such as African millet.

elisabe'thae Of Elisabeth.

ellacombia'nus, -a, -um After Canon Henry Nicholson Ellacombe (1822–1916) of Bitton, Gloucestershire, England, celebrated gardener and author of *Plant-lore and Garden-craft of Shakespeare*.

Elliot'tia f. A handsome shrub named after Stephen Elliott (1771–1830), botanist at Charleston, South Carolina, U.S.A.

ellipsoida'lis, -is, -e Elliptic in outline but solid.

ellip'ticus, -a, -um Elliptic; longer than wide, broadest at the middle, with curved sides.

Elo'dea f. Waterweed or ditchmoss. From Gr. *hĕlōdēs*, marshy, bred in marshes.

elonga'tus, -a, -um Lengthened; elongated.

Elshol'tzia f. Aromatic herbs and sub-shrubs named in honour of Johann Sigismund Elsholtz (1623–1688), Prussian naturalist and physician.

E'lymus m. Wild rye. From the Greek name *ĕlymos*, millet.

emargina'tus, -a, -um With a shallow notch at the end as though a piece had been removed.

Emboth'rium n. Gr. *en*, in; *bŏthrĭŏn*, a little pit; an allusion to the position of the anthers on these evergreen shrubs or trees.

eme'rus Latinized form used by Cesalpino in 1583 of the Italian vernacular name *emero* for *Coronilla emerus*.

eme'ticus, -a, -um Causing vomiting; emetic.

Emi'lia f. Tassel flower. Probably named for some individual but just who is commemorated in these weedy herbs is not known.

e'minens Eminent; prominent.

Emmenan'the f. California golden bells. Gr. *ĕmmĕnēs*, lasting; *anthos*, a flower. The corolla lasts for a long time on these bushy annuals.

Emmeno'pterys f. Gr. *ĕmmĕnēs*, enduring; *ptĕryx*, a wing. The calyx lobes of this ornamental tree are large and winged.

emo'di Of the Himalaya, the Emodi Montes in classical geography.

empetrifo'lius, -a, -um With leaves like crowberry or *Empetrum*.

Em'petrum n. Crowberry. Gr. *ĕmpĕtrŏn*, from *en*, on; *petros*, a rock.

Ence'lia f. Derivation obscure. South American herbaceous composites.

Encephalar'tos m. Gr. *ĕn*, in; *cĕphalē*, a head; *artos*, bread. The inner parts of the top of the trunks of these primitive palmlike trees are farinaceous and edible.

Endy'mion m. Gr. *Endymiōn*, in Greek mythology a beautiful youth beloved by Juno, now applied to a genus which includes the English bluebell or wild hyacinth.

engelman'nii Of Engelmann. Specific epithet of Engelmann's spruce and oak (*Picea* and *Quercus engelmannii*) as well as of other less well-known plants. Named for Georg Engelmann (1809–1884), German-born St. Louis physician who had great knowledge of plants. He did some exploring but his greater contribution lay in his encouragement of others. He was first to call attention to the immunity of American grape stocks to phylloxera. This virtually saved the European wine industry.

Enkian'thus m. Gr. *enkyos*, pregnant; *anthos*, a flower. In the type-species (*E. quinqueflorus*) of this genus of Eastern Asiatic shrubs the corolla has marked swellings at the base as if, to quote Loureiro, 'the flowers are gravid with other flowers'. The flower gives the appearance of being a flower within a flower.

enneacan'thus, -a, -um With nine spines.

enneaphyl'lus, -a, -um With nine leaves or leaflets.

ensa'tus, -a, -um Sword-shaped.

ensifo'lius, -a, -um With sword-shaped leaves.

ensifor'mis, -is, -e Quite straight and with a sharp point like a sword or the leaf of an iris.

Enta'da f. A southern Indian (Malabar) name used by van Rheede. The large brown seeds, often called 'sea beans', of the West Indian species *Entada gigas* float across the Atlantic Ocean and are cast up on the shores of north-west Europe.

Ente'lea f. Gr. *ĕntĕlēs*, complete, perfect. The stamens of this shrub or small tree are all fertile.

entomo'philus, -a, -um Insect-loving. Applied as a specific epithet to certain plants that by means of colour, smell, or nectar attract the insects which carry their pollen to other plants.

Eomecon

Eome'con f. Snow poppy. Gr. *ēōs*, dawn, the east; *mēkōn*, a poppy; a Far Eastern genus.

E'pacris f. Gr. *epi*, upon; *akris*, a summit; the upper regions of hills are often the habitat of some species of these heathlike shrubs.

Ephe'dra f. Greek name *ĕphĕdra* for the common mare's tail (*Hippuris*) which it resembles. Certain species of these low evergreen shrubs are the source of ephedrine which, as *ma-huang*, the Chinese have used in medicine for centuries.

Epiden'drum n. Gr. *ĕpĭ*, upon; *dendron*, a tree. These are epiphytic or tree-perching orchids.

Epiga'ea f. Ground-laurel. Mayflower or trailing arbutus. From Gr. *ĕpĭ*, upon; *gaia*, the earth; a reference to the creeping habit of these evergreen, woody-stemmed plants.

epiga'eus, -a, -um Growing close to the ground or, sometimes, growing on land and not in water.

Epilo'bium n. Willow-herb. From Gr. *ĕpĭ*, upon; *lobos*, a pod. The petals of these herbs surmount the podlike ovary.

Epime'dium n. Barren wort. The Greek name *ĕpĭmēdĭon* for a plant quite different from these dainty herbs.

Epipac'tis f. Greek plant name adopted for this genus of terrestrial orchids.

Epiphroni'tis f. Name applied to orchid hybrids resulting from crossing *Epidendrum* and *Sophronitis*.

Epiphyl'lum n. Gr. *ĕpĭ*, upon; *phyllon*, a leaf. The flowers of these cacti are borne on flattened green leaf-like stems once thought to be leaves.

Epis'cia f. From Gr. *ĕpiskios*, shaded; from the natural habitat of these herbaceous perennials.

Epithelan'tha f. Gr. *ĕpĭ*, upon; *thēlē*, a nipple; *anthos*, a flower. The flowers of these globular cacti are borne on tubercles.

eques'tris, -is, -e Pertaining to horses or horsemen.

equi'nus, -a, -um Of horses.

equisetifo'lius, -a, -um With leaves like horsetail or *Equisetum*.

Equise'tum n. Horsetail. L. *equus*, a horse; *seta*, a bristle. Of more historical than garden interest, the genus links modern plants with some of the most ancient orders of vegetation. In the carboniferous period, the earth was covered with immense forests of gigantic woody *Equisetum*-like plants, the fossilized remains of which comprise most of our coal measures. It was called scouring rush in the 17th-century because the presence of silica gave a fine finish in scouring pewter. In modern times it is occasionally used to put a really fine finish on cabinet work.

Eragros'tis f. Love-grass. From Gr. *erōs*, love; *agrostis*, grass. The reason for the name is obscure.

Eran'themum n. Gr. *ēranthĕmon*, an old name transferred to this genus of tropical shrubs and sub-shrubs.

Eran'this f. Winter aconite. From Gr. *ēr*, spring; *anthos*, a flower. The plant is one of the earliest blooming of the spring flowers.

Ercil'la f. A Peruvian twining shrub named for Alonso de Ercille (1533–1595) of Madrid.

erec'tus, -a, -um Erect; upright.

Erema'ea f. Gr. *ĕrēmaiŏs*, lonely; the flowers are borne singly at the ends of shoots.

Ereman'thus m. Gr. *ĕrēmos*, solitary; *anthos*, flower; each head has only one flower.

Ere'mia f. Gr. *ĕrēmŏs*, solitary. There is but one seed in each chamber of the fruit of these heathlike shrubs.

Eremocit'rus f. Australian desert kumquat. From Gr. *ĕrēmia*, desert; *Citrus*. There is but one species of this shrub or small tree.

Eremo'phila f. From Gr. *ĕrēmia*, desert; *philos*, friend; i.e. desert-lover; Robert Brown collected the two original species in a sandy desert.

Eremo'stachys f. From Gr. *ĕrēmia*, desert; *Stachys*, woundwort; another genus of *Labiatae*.

Eremu'rus m. Desert-candle or foxtail lily. From Gr. *ĕrēmia*, desert; *oura*, a tail; from the appearance of the flower spike of these giant desert- and steppe-inhabiting liliaceous plants.

Ere'psia f. Possibly from Gr. *ĕrĕpo*, to cover with a roof; or Gr. *ĕrĕpsis*, a roof. In allusion to the stamens hidden by the staminodes.

eri- Gr. prefix indicating woolliness (the medieval Latin epithet *erigena* means, however, 'Irish-born'). Thus:

eriacan'thus woolly-spined
erianthe'rus with woolly anthers
erian'thus with woolly flowers
eriogy'nus with a woolly ovary

E'ria f. Gr. *ĕriŏn*, wool. The flowers and pedicels of these orchids are often woolly.

Erian'thus m. Plume-grass. From Gr. *ĕriŏn*, wool; in allusion to the silvery woolly effect of the inflorescence.

Eri'ca f. Heath. Latin *ĕrīcē*; Gr. *ĕrĕikē*. The Mediterranean *E. arborea* *ericaeus*

is the source of brier (or briar) for tobacco pipes which are turned from the roots which are sometimes very large and solid. (French *bruyère*, heath.)

ericae'us -a, -um Relating to *Erica* or heath.

ericifo'lius, -a, -um With leaves like *Erica* or heath.

ericoi'des Resembling *Erica* or heath.

Erige'nia f. Harbinger-of-spring. From Gr. *ēr*, spring; *genia*, born; *Erigĕnĕia* being another name for Aurora, a reference to this small North American plant's early flowering.

eri'genus, -a, -um Irish-born. From Old Irish *Eriu*, *Erin*, Ireland.

Eri'geron m. Fleabane. Possibly from Gr. *ēri*, early; *gĕrōn*, an old man.

erina'ceus, -a, -um Resembling a hedgehog.

E'rinus m. Greek name *ĕrinos*, used by Discorides for a plant like basil, transferred to this genus.

Eriobot'rya f. Gr. *ĕriŏn*, wool; *botrys*, a cluster of grapes; from the woolly, clustered panicles of this evergreen shrub or small tree.

eriobotryoi'des Resembling *Eriobotrya*.

eriocar'pus, -a, -um Woolly-fruited. Also:
erioce'phalus	woolly-headed
erio'phorus	wool-bearing
erio'spathus	with a woolly spathe
erio'stachys	with a woolly spike
erioste'mon	with a woolly stamen

Erioce'phalus m. Gr. *ĕriŏn*, wool; *cĕphale*, a head. The heads of these scented shrubs become woolly.

Erio'gonum n. Gr. *ĕriŏn*, wool; *gŏny*, a knee. The stems of these herbs or sub-shrubs are downy at the nodes.

Erio'phorum n. Cottongrass. From Gr. *ĕriŏn*, wool; *phŏros*, bearing. The seed heads of these sedgelike plants are cottony.

Eriophyl'lum n. Gr. *ĕriŏn*, wool; *phyllon*, a leaf. The leaves of these perennial herbs are woolly.

Eriop'sis f. Gr. resembling *Eria*, another genus of epiphytic or tree-perching orchids.

Erioste'mon m. Gr. *ĕriŏn*, wool; *stēmōn*, a stamen. The stamens of these evergreen shrubs are woolly.

Eritri'chium n. Gr. *ĕriŏn*, wool; *thrix*, hair. These tufted perennials' herbs are woolly.

Erlan'gea f. Shrubs or herbs, mostly African, named in honour of the

University of Erlangen in Bavaria, where the author, C. H. Schulz Bipontinus, studied botany in 1825–1826.

ernes'tii In honour of Ernest Henry Wilson (1876–1930); see *Sinowilsonia*.

Ero'dium n. Heron's bill. From Gr. *erōdios*, a heron. The carpels of these pretty plants resemble the head and beak of a heron.

Ero'phila f. From Gr. *ĕr*, spring; *philos*, loving.

ero'sus, -a, -um Jagged, as if irregularly gnawed or bitten off.

erra'ticus, -a, -um Unusual; sporadic; departing from the normal of the genus.

erro'menus, -a, -um Gr. *ĕrrōmĕnŏs*, in good health, vigorous.

erubes'cens Blushing; becoming red.

Eru'ca f. Latin name for *Eruca sativa* or rocket-salad, grown for its oil-rich seed and sometimes for use as a salad plant.

erucas'trum; erucoi'des Resembling *Eruca* or rocket-salad.

Eryn'gium n. Sea-holly, eryngo. The Greek name, *ēryggiŏn*, for *Eryngium campestre*. The roots of sea-holly (*E. maritimum*) are sweet and aromatic and once were well regarded for their supposed aphrodisiac effect when candied. Cookery books up to the 18th century carried recipes for their preparation.

Ery'simum Blistercress. Gr. *ĕrysimŏn*. Some species of these herbs are said to produce blisters.

Erythe'a f. These fan-palms are named for one of the three Hesperides who lived far away in the West on the border of the ocean where the sun set, guarding the golden apples which Earth had given to Hera. They were sisters but no one quite knew who the parents were; some thought they were Erebus and Night, others Atlas and Hesperis.

Erythri'na f. Coral tree. From Gr. *ĕrythrŏs*, red; an allusion to the colour of the flower.

erythro- As a prefix in compound words it signifies red. Thus:

erythroca'lyx	with a red calyx
erythrocar'pus	having red fruit
erythroce'phalus	red-headed
erythro'podus	having a red foot or stem
erythro'pterus	with red wings
erythroso'rus	having red spore cases

Erythro'nium n. Dog's-tooth violet, trout-lily, adder's tongue. Gr. *ĕrythrŏniŏn*, from *ĕrythrŏs*, red, originally the name for another plant.

Erythro'xylon n. From Gr. *ĕrythrŏs*, red; *xylŏn*, wood. Some species have red wood.

E. coca, the only source of cocaine, is a bush cultivated in the tropics for that product. In Bolivia and Peru, where it is native, the inhabitants chew the leaves both for their euphoric effect and to sustain themselves in feats of endurance.

Escallo'nia f. Shrubs and small trees named for Señor Escallon, a Spanish traveller in South America.

Eschschol'zia f. California poppy. Named in honour of Johann Friedrich Eschscholtz (1793–1831) of Tartu (Dorpat), U.S.S.R., who accompanied Otto von Kotzebue on his first expedition round the world (1815–1818). He wrote a number of botanical works.

Escoba'ria f. A genus of cacti named for two Mexicans, Romulo and Numa Escobar.

Escon'tria f. A treelike cactus named after Don Blas Escontria, a Mexican of distinction.

esculen'tus, -a, -um Related to eating; edible.

Espo'stoa f. In honour of Nicolas E. Esposto, botanist at Lima, Peru.

estria'tus, -a, -um Without stripes.

etrus'cus, -a, -um From Tuscany, the classical Etruria, in Italy.

etubero'sus, -a, -um Without tubers.

Euan'the f. From Gr. *euanthēs*, blooming, well-flowered. With reference to the showy flowers.

eucalyptoi'des Resembling *Eucalyptus*.

Eucalyp'tus f. Gum tree. Blue gum. From Gr. *eu*, well; *kalypto*, to cover, as with a lid; an allusion to the united calyx-lobes and petals forming a lid or cap which is shed when the flower opens. Among the largest trees in the world (highest recorded, 326 feet) they have very small and even minute seeds. The timber is durable and is much used in Australia for shipbuilding, engineering, and flooring.

Euchari'dium n. Gr. *eucharis*, pleasing. These pretty annuals are sometimes referred to as *Clarkia*.

Eu'charis f. Amazon lily. Greek word meaning pleasing, charming; from the fragrant beauty of the flowers of these bulbous plants.

Euchlae'na f. Gr. *eu*, good; *chlaina*, cloak, covering; referring to the stigma of these grasses.

euchlo'rus, -a, -um Fresh green, from Gr. *eu*, good; *chlōrŏs*, green.

Eucni'de f. Gr. *eu*, good; *knidē*, stinging nettle.

Eu'comis f. From Gr. *eu*, good; *kome*, hair; implying a beautiful head, from the tufted crown of the flower spike of these bulbous plants.

Eucom'mia f. Gr. *eu*, good; *kommi*, gum. Consisting of a single species

(*E. ulmoides*) it is interesting as the only hardy rubber-producing tree. Extraction, however, is difficult and not commercially feasible, yet when the leaves are torn gently across, threads of rubber remain strong enough to support the torn portion.

Eucry'phia f. Gr. *eu*, well; *kryphios*, covered. The sepals of these white-flowered trees and shrubs, cohering at the tips, form a cap.

eudo'xus, -a, -um Of good repute.

Euge'nia f. Clove tree, rose apple. Ornamental trees and shrubs named in honour of Prince Eugene of Savoy (1663–1736). Most authorities suggest that he was a patron of botany, which is possible but not very probable in the light of his preoccupation with war. With him it was a passion which he pursued with rare energy, imagination, and resolution. Wounded thirteen times in fifty years in the field, he did not spend two years together without fighting, almost invariably with success. Among his best-known campaigns were those he fought as an ally of the Duke of Marlborough.

eugenioi'des Resembling *Eugenia*.

Eulo'phia f. From Gr. *eu*, good; *lŏphŏs*, crest; referring to the crest on the lip of these orchids.

Euo'nymus f. Spindle-tree. The ancient Greek name *euōnymŏn dĕndron* whence the Latin *euonymus*, meaning 'of good name', referred ironically to its being poisonous to animals.

eupatorioi'des Like *Eupatorium*.

Eupato'rium n. Joe-Pye-weed, thoroughwort, boneset. The Greek name for these herbaceous and shrubby plants, commemorating Mithridates Eupator, King of Pontus, who is said to have discovered that one species was an antidote against poison. The English names indicate a traditional use in medicine.

Euphor'bia f. Spurge. Classically supposed to have been named for one Euphorbus, physician to the King of Mauretania.

euphorbioi'des Resembling spurge or *Euphorbia*.

Eupho'ria f. Gr. *eu*, well; *phoreo*, to carry. The white, sweet, sub-acid fruits, native to India, travel well. The vernacular name for these trees has been anglicized to Longan and Linkang.

Euphra'sia f. Eye bright. Gr. *euphrasia*, delight, mirth.

Eupritchar'dia f. Gr. *eu*, good; *Pritchardia*, a related genus of ornamental Pacific palms, named for William Thomas Pritchard, author of *Polynesian Reminiscences* (1866).

Eupte'lea f. Gr. *eu*, good; *ptĕlĕa*, an elm; in reference to the edible fruit of these shrubs or small trees.

europae′us, -a, -um European.

Euro′tia f. Gr. *euros*, mould; in reference to the grey-white downy or mouldy appearance of these herbs or sub-shrubs.

Eu′rya f. Gr. *euru*, broad.

Eury′ale f. Named after one of the three Gorgons, the monstrous daughters of the sea god, who had venomous snakes for hair and of whom only Medusa was mortal. The allusion here is to the very thorny, prickly character of these large and handsome water-lilies. The seeds are roasted and eaten in India.

Eu′scaphis f. Gr. *eu*, good; *scaphis*, a vessel; from the character of the seed pod of this ornamental woody plant.

Eu′stoma n. Prairie gentian. From Gr. *eu*, good; *stŏma*, a mouth, or (less literally) a pretty face; in allusion to the showy flowers.

Eu′strephus m. Gr. *eu*, well; *strepho*, to twine; from the character of these Australian climbers.

Euta′xia f. Greek word meaning modesty; from the nature of these Australian flowering shrubs.

Euter′pe f. Named for one of the nine Muses, goddesses of the liberal arts. Euterpe had charge of music. The genus includes the cabbage palm (*E. oleracea*), the terminal bud of which is eaten fresh as a salad or pickled.

evec′tus, -a, -um Extended, springing forwards.

ever′tus, -a, -um Overturned.

Evo′dia f. Gr. *euōdia*, a sweet scent. The leaves of these shrubs are fragrant.

Evol′vulus m. L. *evolvo*, to untwist or unravel. These plants of the convolvulus family do not twine.

ewer′sii In honour of Joseph Philipp Gustav Ewers (1781–1830), German statesman in Russian employ who promoted Ledebour's botanical exploration in Siberia.

E′xacum n. Variant of the Gallic name *exacon* for *centaurium* transferred to this genus of the gentian family.

exalta′tus, -a, -um Very tall; lofty.

exara′tus, -a, -um Engraved; furrowed.

exaspera′tus, -a, -um Roughened.

excava′tus, -a, -um Hollowed out.

excel′lens Excelling; excellent; distinguished.

excel′sior Taller.

excel′sus, -a, -um Tall.

exci'sus, -a, -um Cut away; cut out.

excortica'tus, -a, -um Without bark or cortex; stripped of bark.

exi'guus, -a, -um Very little; meagre; poor.

exi'mius, -a, -um Out of the ordinary; distinguished.

exitio'sus, -a, -um Pernicious; destructive.

Exochor'da f. Pearl-bush. From Gr. *exo*, outside; *chorde*, a cord; referring to a technical characteristic of the ovary; there are fibres outside the placenta.

exole'tus, -a, -um Mature; dying away.

exonien'sis, -is, -e From Exeter, Devon, England.

exo'ticus, -a, -um From a foreign country.

expan'sus, -a, -um Expanded.

explo'dens Bursting suddenly; exploding.

exscul'ptus, -a, -um Dug out.

exser'tus, -a, -um Protruding.

exsur'gens Rising up.

exten'sus, -a, -um Extended.

exu'dans Exuding.

F

faasse'nii Of J. H. Faassen, Dutch nurseryman, of Tegelen, Netherlands.

Fa'ba f. Latin for broad bean (*Vicia faba*); more appreciated in Europe than in the United States where the climate is generally too hot and dry.

faba'ceus, -a, -um Resembling the broad bean.

Fabia'na f. False heath. Named in honour of Archbishop Francisco Fabian y Fuero (1719–1801) of Valencia, Spain, a promoter of botanical study.

fagifo'lius, -a, -um Beech-leaved.

Fagopy'rum n. Buckwheat. From *Fagus*, beech; *pȳrŏs*, wheat.

Fa'gus f. Beech. The Latin name.

falca'tus, -a, -um; falcifor'mis, -is, -e Sickle-shaped; falcate.

falcifo'lius, -a, -um With sickle-shaped leaves.

falcinel'lus, -a, -um Resembling a small sickle.

falco'neri In honour of Hugh Falconer (1808–1865), Scottish doctor, geologist and botanist in India from 1830 to 1855.

fal'lax Deceptive; false.

Fallu'gia f. In honour of Abbot V. Fallugi of Vallombrosa, *c.* 1840.

farge'sii In honour of Paul Guillaume Farges (1844–1912), French missionary and naturalist in Central China.

farina'ceus, -a, -um Yielding starch; mealy, like flour.

farino'sus, -a, -um Mealy; powdery.

farleyen'sis, -is, -e From Farley Hill Gardens in Barbados, British West Indies.

farnesia'nus, -a, -um From the gardens of the Farnese Palace, Rome.

farre'rae In honour of Mrs Farrer, wife of an Honourable East India Company captain.

farre'ri Specific epithet of *Gentiana farreri* and other plants. In honour of Reginald John Farrer (1880–1920), English plant hunter and botanist, who specialized in Alpines; he collected in the European Alps, Upper Burma and the mountain regions of Kansu where he discovered this particular gentian among other good things. He wrote several books on his travels and on rock-garden plants, possibly the best known being *The English Rock Garden*.

fascia'tus, -a, -um Bound together.

fascicula'ris, -is, -e; fascicula'tus, -a, -um Clustered or grouped together in bundles.

fascina'tor Bewitcher, enchanter.

fastigia'tus, -a, -um Having branches close together and erect, often forming a column; fastigiate.

fastuo'sus, -a, -um Proud.

Fatshe'dera f. Name of intergeneric hybrid of *Fatsia* and *Hedera*.

Fat'sia Latinized adaptation of an obsolete or misrendered Japanese name, *Fatsi*, for *Fatsia japonica*.

fa'tuus, -a, -um Simple or foolish but, in this relationship, insipid or, merely, not good.

Fauca'ria f. L. *faux*, a gullet or throat. The paired leaves with tooth-like edges of these succulents resemble the wide open mouth of a carnivore, hence the local South African name of tiger's chaps.

febri'fugus, -a, -um Fever-dispelling.

Fe'dia f. Meaning of the name for this valerian-like annual is obscure.

Feijo'a f. A small tree or shrub with pineapple- or guava-flavoured fruit. Named for Don de Silva Feijoa, 19th-century Brazilian botanist.

Feli'cia f. Blue daisy. After Felix, a German official at Regensburg, who died in 1846.

fe'mina A female.

Fendle'ra f. In honour of August Fendler (1813–1883), German plant-collector in North and Central America.

fenestra'lis, -is, -e Pierced with openings resembling windows.

Fenestra'ria f. From L. *fenestra*, a window; in allusion to transparent areas on the leaves.

fen'nicus, -a, -um Finnish, of Finland.

fe'rax Fruitful.

Ferocac'tus m. From L. *ferox*, ferocious, savage; in allusion to the horrible spines with which this cactus is armed.

Fero'nia f. Wood-apple. A spiny tree named in honour of the Roman nymph who presided over woods and groves.

Feroniel'la f. Diminutive of *Feronia* to which it is closely related.

fe'rox Ferocious; very thorny.

Ferra'ria f. In honour of Giovanni Battista Ferrari (1584–1655) of Siena, Italian botanist, author of *Flora* (1633) and *Hesperides* (1646).

fer'reus, -a, -um Pertaining to iron or, sometimes, iron-hard.

ferrugi'neus, -a, -um Rust-coloured; rusty.

fer'tilis, -is, -e Fruitful; producing numerous seeds.

Fe'rula f. Giant fennel. The classical Latin name. The word also means the rod used to chastise schoolboys, and also slaves, for minor offences. The giant fennel has tall sticklike stems. The true fennel used for flavouring and as a vegetable belongs to the genus *Foeniculum, q.v.*

ferulifo'lius, -a, -um With leaves resembling giant fennel or *Ferula*.

fe'rus, -a, -um Wild.

festa'lis, -is, -e; festi'nus, -a, -um Festive; gay; bright.

Festu'ca f. Fescue. Latin word meaning a grass stalk.

fibrillo'sus, -a, -um; fibro'sus, -a, -um Fibrous; composed of fibres.

Fica'ria f. Medieval plant-name probably from *ficus*, fig, with reference to the tubers somewhat resembling little figs of this plant now included in *Ranunculaceae*.

ficifo'lius, -a, -um With leaves like a fig.

ficoi'des; ficoi'deus, -a, -um Resembling a fig or *Ficus*.

fic'tus, -a, -um False; feigned; hence *fictolacteum*, for a species of *Rhododendron* confused with *R. lacteum*.

Fi'cus f. Latin name for *Ficus carica*, the edible fig. The genus includes some quite extraordinary species such as the banyan (*F. benghalensis*) of India which, starting from a single main trunk, sends down aerial roots which themselves become trunks, thus extending the original tree over areas which may be very large. *F. elastica* is possibly the most familiar in temperate climates. Naturally quite a large tree and once an important source of rubber, it is now grown as the common household rubber plant, and most commercial rubber is obtained from the genus *Hevea* and from synthetics. *F. carica*, a native of western Asia, has been widely grown for its fruit from remote antiquity. It was probably introduced early into Britain but was lost, as was the vine, possibly due to climatic changes. It was reintroduced by Cardinal Pole in 1525 and planted in the garden of the Archbishop at Lambeth. Another celebrated tree was brought from Aleppo in 1648 by Dr. Pocock, Regius Professor of Hebrew at Oxford, and planted in the garden at Christ Church College where it survived, producing fine fruit, until 1833.

Fila′go f. Medieval Latin plant name from *filum*, thread.

filamento′sus, -a, -um; fila′rius, -a, -um Furnished with filaments or threads.

fili- As a prefix, signifying threadlike. Thus:
filicau′lis   With a threadlike stem
fi′lifers   Thread-bearing
filifor′mis   Threadlike
fi′lipes   With threadlike stalks

filicifo′lius, -a, -um With leaves resembling fern.

filici′nus, -a, -um; filicoi′des Resembling fern. From L. *filix*, a fern.

Filipen′dula f. Meadow-sweet. Dropwort. L. *filum*, a thread; *pendulus*, hanging. The root tubers in some species hang together with threads.

filipen′dulus, -a, -um Resembling meadow-sweet or *Filipendula*.

fimbriat′ulus, -a, -um With a small fringe.

fimbria′tus, -a, -um Fringed.

Firmia′na f. Tropical trees named for Karl Josef von Firmian (1716–1782), Governor of Lombardy when that province was part of the Austrian Empire, and a patron of the Padua botanic garden.

fir′mus, -a, -um Strong.

fis′silis, -is, -e; fissura′tus, -a, -um; fis′sus, -a, -um Cleft; split.

fistulo′sus, -a, -um Hollow, like a pipe.

Fitto′nia f. Ornamental-leaved perennials named in honour of Elizabeth and Sarah Mary Fitton (d. 1866), authors of *Conversations on Botany* (1817).

Fitzroy'a f. Evergreen trees or shrubs named for Captain (later Vice-Admiral) Robert Fitzroy (1805–1865), Royal Navy, who commanded the five-year surveying expedition of H.M.S. *Beagle*. Charles Darwin was aboard as official naturalist and, on the voyage, laid the foundation for his life's work on the origin of species. Fitzroy became a celebrated meteorologist and instituted a system of storm warnings that led to daily weather reports.

flabella'tus, -a, -um Like an open fan.

flabel'lifer, flabelli'fera, -um Fan-bearing.

flabellifor'mis, -is, -e Fan-shaped.

fla'ccidus, -a, -um Feeble; weak; soft.

Flacour'tia f. Shrubs and small trees named for Etienne de Flacourt (1607–1660), director of the French East India Company.

flagella'ris, -is, -e; flagellifor'mis, -is, -e Whiplike; having long, thin, supple shoots like whips.

flagel'lum A whip or flail.

flam'meus, -a, -um Flame-coloured or flamelike.

flam'mula A small flame. Application obscure but possibly alluding to a burning taste. The pre-Linnaean name for *Clematis flammula*.

flavanthe'rus, -a, -um With yellow anthers.

fla'vens Being yellow.

flave'olus, -a, -um Yellowish.

flaves'cens Becoming yellow, yellowish.

fla'vidus, -a, -um Yellowish.

flavis'simus, -a, -um Deepest yellow.

flavoru'fus, -a, -um Yellow-red.

fla'vus, -a, -um Pure yellow. Also:
flavi'comus   yellow-haired
flavispi'nus   yellow-spined

Flemin'gia f. Shrubs and sub-shrubs named in honour of John Fleming (1747–1829), President of the Medical Board of Bengal, Hon. East India Company. This genus is also called *Moghania*.

flet'cheri In honour of Harold Roy Fletcher, director of the Royal Horticultural Society's Gardens, Wisley, then Regius Keeper of Royal Botanic Garden, Edinburgh. Retired 1970.

flexicau'lis, -is, -e Having a pliant stem.

fle'xilis, -is, -e Pliant; limber.

flexuo'sus, -a, -um Tortuous; zigzag.

floc'ciger, flocci'gera, -um Bearing flocks of wool.

flocco'sus, -a, -um Woolly.

flocculo'sus, -a, -um Somewhat woolly.

flo're-al'bo With white flowers.

florenti'nus, -a, -um From Florence, Italy.

flo're-ple'no With double flowers. This phrase is now usually replaced by the epithets *duplex*, *pleniflorus* or *plenus*.

floribun'dus, -a, -um; flo'ridus, -a, -um; flo'rifer, -a, -um Free-flowering; producing abundant flowers.

florida'nus, -a, -um From Florida.

florin'dae Specific epithet of a *Primula* discovered by Kingdon Ward in Tibet and named after his first wife.

flos Latin word meaning a flower, as in:
   flos-cu'culi   cuckoo flower (*Lychnis*)
   flos-jo'vis   Jove's flower (*Lychnis*)

flu'itans Floating.

fluminen'sis, -is, -e From Rio de Janeiro (Flumen Januarii), Brazil.

fluvia'lis, -is, -e; fluvia'tilis, -is, -e Growing in a river or running water.

foe'mina Feminine.

foenicula'ceus, -a, -um; foenicula'tus, -a, -um Resembling fennel or *Foeniculum*.

Foeni'culum n. Fennel. The Latin name for this traditional salad and potherb which, in Italian, is *finocchio*.

foe'tidus, -a, -um Bad-smelling. Also *foetidissimus*, very bad-smelling.

folia'ceus, -a, -um Leaflike.

folia'tus, -a, -um Provided with leaves.

foliolo'tus, -a, -um; foliolo'sus, -a, -um Furnished with leaflets.

folio'sus, -a, -um Full of leaves; leafy.

follicula'ris, -is, -e Bearing follicles. These are dry, one-chambered fruits splitting along only one suture, e.g. peony, milkweed.

Fontane'sia f. Named for René Louiche Desfontaines (1750–1833), author of *Flora Atlantica*; see *Desfontainea*.

fonta'nus, -a, -um; fontina'lis, -is, -e Pertaining to springs or fountains; growing in fast-running water.

Forestie'ra f. Shrubs named for Charles Le Forestier, 18th-century French physician and naturalist.

formosa'nus, -a, -um From the island of Formosa (Taiwan).

formo'sus, -a, -um Handsome; beautiful. Also *formosissimus*, very handsome.

forres'tii In honour of George Forrest (1873–1932), Scottish plant-collector who between 1904 and 1932 made immense collections of seeds and specimens in western China.

Forsy'thia f. Golden-bell. Ornamental shrubs named in honour of William Forsyth (1737–1804), Scottish superintendent of the Royal Gardens of Kensington Palace and author, among other works, of *A Treatise on the Culture and Management of Fruit Trees* which in its day was probably the most widely read work on the subject. From 1789 until his death Forsyth was a controversial figure in horticultural circles as a result of his claims for the concoction known as 'Forsyth's plaister'. This mixture, made up of lime, dung, wood-ashes, soap-suds, sand, urine, and so on, was not very different from others of the day designed to heal wounds in the bark of trees. Forsyth claimed that his 'plaister' was capable 'of curing defects in growing trees' and would restore oak trees 'where nothing remained but the bark'. As a result of these claims the British government, involved in the Napoleonic wars and in urgent need of sound timber for the Navy, paid him the generous gratuity of £1,500. Thomas A. Knight, later President of the Horticultural Society of London for many years, took active exception to statements made either by or on behalf of Forsyth, such as that holes in trees could be filled with the 'plaister' and the trees themselves 'brought to such a degree of soundness that no one can know the new wood from the old'. Nevertheless Forsyth found plenty of supporters. Forsyth died before the argument was satisfactorily resolved, but in due course the 'plaister' was thoroughly discredited.

fortu'nei Specific epithet after Robert Fortune (see *Fortunella*).

Fortunel'la f. Kumquat. Evergreen shrubs named for Robert Fortune (1812–1880), Scottish horticulturist and collector in China who ranks among the great plantsmen. With the exception of *Rhododendron fortunei*, which he found wild, and one or two other less notable plants, most of his finds were in Chinese gardens and nurseries. In his day the difficulties of travel were such that he had small chance of going further afield. With a brief interval as curator of the Chelsea Physic Garden, he was in China practically continuously from 1843 to 1861. He learned Chinese and earned the Chinese name of Sing Wah.

In 1848, he was engaged by the Hon. East India Company to bring tea plants to India. Despite the justifiable secretiveness of the Chinese, he was able to investigate the growing and manufacture of tea and eventually exported the plants, Chinese workers, and implements which established the industry, first in India and then in Ceylon. It has since, of course, grown to enormous commercial importance. In 1858, the U.S. government employed Fortune to bring tea-plants to America for trial, with a view to the eventual establishment of an industry. Although there had been previous attempts—as early as 1772 in Georgia, between 1848 and 1852 by Junius Smith in the mountains of South Carolina, in Alabama and Florida with the use of slave labour—all had failed. Within a couple of years. Fortune's plants, imported in Wardian cases, numbered some 35,000 growing under glass in the Propagation Garden in Washington, D.C. After several attempts, however, which lasted into the 1870's, the project of a United States tea industry was finally abandoned.

Fortune wrote several readable books on his adventures and, having started as a poor garden apprentice, he died reasonably well off as a result of his plant sales and of his flourishing business in curios.

Fothergil'la f. Dwarf alder. Ornamental shrubs named for Dr. John Fothergill (1712–1780), Quaker physician of Stratford, Essex, England, who specialized in growing American plants.

Fouquie'ra f. Candlewood. Small trees or shrubs named for Pierre Edouard Fouquier, 19th-century French physician.

fourcroi'des Resembling the genus *Furcraea*.

fournie'ri In honour of Eugène Pierre Nicolas Fournier (1834–1884), French botanist.

foveola'tus, -a, -um Slightly pitted.

Fraga'ria f. Strawberry. From L. *fraga*, strawberry; deriving from *fragrans*, fragrant, in allusion to the perfume of the fruit.

fragariflo'rus, -a, -um With flowers like those of *Fragaria*, or else red as crushed strawberry.

fragarioi'des Resembling strawberry or *Fragaria*.

fra'gilis, -is, -e Brittle; easily broken; or, sometimes, wilting quickly.

fra'grans Fragrant.

fragrantis'simus, -a, -um Very fragrant.

Frai'lea f. In honour of Manuel Fraile (b. 1850), Spanish gardener for many years in charge of the cactus collection of the U.S. Department of Agriculture, Washington.

Franco'a f. Chilean perennial herbs named in honour of Francisco Franco, 16th-century physician of Valencia, Spain.

Franke'nia f. Heathlike perennials and sub-shrubs named for Johan Frankenius (1590–1661), professor of anatomy and botany at Uppsala, Sweden.

Frankli'nia f. Named in honour of Benjamin Franklin (1706–1790), American printer, scientist, philosopher and statesman, this remarkable tree was discovered in Georgia in 1764 and exterminated in the wild by 1803 through excessive collecting by nurserymen in its only known locality.

Fra'sera f. Glabrous herbs named for John Fraser (1750–1811), Scottish collector of North American plants who also kept a nursery in Chelsea, England. He made several collecting expeditions, to Newfoundland and as far south as the Carolinas. His close friend, Thomas Walter of South Carolina, named *Magnolia fraseri* in his honour.

fra'seri Of John Fraser (see *Frasera*).

fraxi'neus, -a, -um Resembling ash or *Fraxinus*.

fraxinifo'lius, -a, -um With leaves resembling ash or *Fraxinus*.

Fra'xinus f. Ash. The classical Latin name.

Free'sia f. Bulbous plants with fragrant flowers named in honour of a German physician, Friedrich Heinrich Theodor Freese (d. 1876), of Kiel, a friend of Ecklon who named the genus.

Fremon'tia f.; **Fremontoden'dron** n. Flannel-bush. Discovered by and named in honour of Major-General John Charles Frémont (1813–1890) who made four hazardous journeys exploring the Far West of the United States between 1842 and 1848. He brought back many notable trees and shrubs to be studied and named by Torrey, Gray, and Darlington and in doing so made great contributions to both botany and horticulture. (See *Darlingtonia, Torreya*.)

fresno'ensis, -is, -e From Fresno county, California. Spanish *fresno*, ash-tree; referring here to *Fraxinus oregona*.

Freycine'tia f. Climbing shrubs named for Admiral Louis Claude De Saulses de Freycinet (1779–1842), French navigator and a founder of the Paris Geographical Society. He made various exploring voyages to Australasia and the Pacific, returning with important collections.

fri'gidus, -a, -um Growing in cold regions.

Fritilla'ria f. Fritillary. From L. *fritillus*, a dicebox; from the spotted markings on the flowers of some species.

Froeli'chia f. American annuals named for Joseph Aloys Froelich (1796–1841), German physician and botanist, author of a work on gentians.

frondo'sus, -a, -um Leafy.

fruc'tifer, fructi'fera, -um Fruit-bearing; fruitful.

frumenta'ceus, -a, -um Pertaining to grain; grain-bearing.

frutes'cens; fru'ticans; frutico'sus, -a, -um Shrubby; bushy. From L. *frutex*, a shrub.

frutici'cola A dweller in bushy places.

fruticulos'us, -a, -um Shrubby and dwarf.

fuca'tus, -a, -um Painted; dyed.

Fu'chsia f. Flowering shrubs and small trees named for Leonhart Fuchs (1501–1566), German physician and herbalist, professor at Tübingen, who published a herbal with unusually beautiful woodcuts of plants.

fuchsioi'des Resembling *Fuchsia*.

fu'gax Withering or falling off quickly; fleeting.

ful'gens; ful'gidus, -a, -um Shining; glistening.

fuligino'sus, -a, -um Dirty-brown; sooty.

fullo'num Pertaining to fullers. Specific epithet of the fuller's teasel (*Dipsacus fullonum*) used in the manufacture of woollen cloth to raise the nap.

fulves'cens Becoming tawny, yellow-brown.

fulvi- Latin prefix meaning tawny. Thus *fulvicaulis*, tawny-stemmed; *fulvinervis*, tawny-nerved.

ful'vus, -a, -um Tawny-orange; fulvous. Thus, *fulvidus*, slightly tawny.

Fuma'ria f. Fumitory. From L. *fumus*, smoke; in allusion to the smoky scent of some species, and to the fact that the juice, like smoke, brings tears to the eyes. Also, the bluish leaves may be described as a smoky green. This old garden plant was once very widely grown as a remedy for scurvy.

fumariifo'lius, -a, -um With leaves resembling fumitory or *Fumaria*.

fu'midus, -a, -um Smoky; dull and greyed in colour.

fumo'sus, -a, -um Smoky.

fu'nebris, -is, -e Funereal; belonging to graveyards.

fungo'sus, -a, -um Relating to or resembling fungus; spongy.

funicula'tus, -a, -um Resembling a thin cord or rope.

Fun'kia Plantain lily. Now named *Hosta*.

fur'cans; furca'tus, -a, -um Forked; furcate.

Furcra'ea f. Succulent plants named for Antoine François Fourcroy (1755–1809), French chemist and naturalist, who played an important part in the French Revolution.

furfura'ceus, -a, -um Mealy; scurfy.

fu'riens Exciting to madness.

furseo'rum In honour of Admiral Paul Furse (b. 1904) and Mrs. (Polly) Furse who have made extensive plant collections in Turkey, Iran and Afghanistan.

fusca'tus, -a, -um Brownish.

fuscifo'lius, -a, -um With dusky-brown leaves.

fusco-ru'ber, -rubera, rum Brownish-red.

fus'cus, -a, -um Brown; dusky.

fusifor'mis, -is, -e Spindle-shaped, that is thickest in the middle and drawn out at each end.

fu'tilis, -is, -e Useless.

G

Ga'gea f. Small bulbous plants named for Sir Thomas Gage (1761–1820), Hengrave Hall, Suffolk, who botanized in Ireland and Portugal. The greengage plum, the English name for the Reine Claude, was named for his grandfather who introduced it from France.

Gaillar'dia f. Blanket-flower. Herbaceous plants named in 1786 by Fougeroux de Bondaroy for a French magistrate and patron of botany, Gaillard de Charentonneau.

galacifo'lius, -a, -um With leaves like *Galax*.

galacti'nus, -a, -um Milky.

Galacti'tes f. Gr. *gala*, milk. The veins of the leaves of these herbs are milky-white.

Galan'thus m. Snowdrop. From Gr. *gala*, milk; *anthos*, flower; an allusion to the colour of the flowers.

galan'thus, -a, -um With milky-white flowers.

Ga'lax f. Wand plant. Gr. *gala*, milk; possibly an allusion to the white flowers of this evergreen perennial.

ga'le English vernacular name cognate with German Gagel adopted as specific epithet for *Myrica gale* by Linnaeus.

Galean'dra f. Gr. *galea*, a helmet; *andros*, a stamen. One stamen in this orchid is helmet-shaped.

galea'tus, -a, -um; galericula'tus, -a, -um Helmet-shaped.

Gale'ga f. Goatsrue. From Gr. *gala*, milk. It was once thought that feeding on these perennials would improve the milk flow.

galegifo'lius With leaves resembling goatsrue or *Galega*.

Galeop'sis f. Gr. *galĕŏpsis*, name for a member of *Labiatae*, from *galē*, weasel; *ŏpsis*, appearance.

Galinso'ga f. Weeds, ironically and punningly named 'Gallant Soldier' in England, commemorating Mariano Martinez Galinsoga, 18th-century Spanish doctor in Madrid, whose botanical accomplishments match the smallness of their flowers.

galioi'des Resembling *Galium* or bedstraw.

Ga'lium n. Bedstraw, cleavers. From Gr. *gala*, milk. Yellow bedstraw (*G. verum*) can be used in cheese-making to curdle the milk.

gal'licus, -a, -um From France.

Galto'nia f. Spire lily, giant summer hyacinth. Named in honour of Sir Francis Galton (1822–1911), anthropologist and geneticist who travelled widely in Africa. He was an early proponent of fingerprinting for identification, laid the foundation of modern eugenic thinking, and wrote a distinguished book on meteorology in which he first stated the anti-cyclonic theory.

Gamo'lepis f. Gr. *gameo*, to marry; *lĕpis*, a scale. In these South African herbs and shrubs the bracts of the urn-shaped involucre are united.

gandaven'sis Relating to Ghent, Belgium.

gange'ticus, -a, -um From the region of the Ganges.

Garci'nia f. Mangosteen. Named in honour of Laurent Garcin (1683–1751), French botanist, who travelled widely in India. The mangosteen (*G. mangostana*) is possibly the most delectable and one of the most local of all East Indian fruits. Native in Malaya, it does not do as well elsewhere. It is about the size of a tangerine, and the leathery, russet skin conceals white- or rose-coloured edible segments of delicious flavour and scent. It is very wholesome and is a fruit that can be eaten harmlessly during bouts of malaria.

Garde'nia f. Evergreen flowering shrubs or small trees named for Dr. Alexander Garden (1730–1791), correspondent of Linnaeus and a native of Aberdeenshire, Scotland: educated at Aberdeen and for many years a physician in Charleston, South Carolina. *G. jasminoides* is the species commonly offered by florists. It is often called Cape jasmine, but it is not a jasmine and it originated in China.

garga'nicus, -a, -um From Monte Gargano in southern Italy.

Gar'rya f. Silk-tassel bush. Named for Nicholas Garry, Secretary of the Hudson's Bay Company who, in about the decade 1820–1830, assisted David Douglas in his explorations of the Pacific North-west.

Gaste′ria f. Gr. *gaster*, a belly. The asymmetrically swollen base of the flower distinguishes this genus from allied *Liliaceae*.

Gastrochi′lus m. From Gr. *gaster*, a belly; *cheilos*, a lip. The lip of the flower on these herbs is swollen.

Gaulnet′tya f. Name for intergeneric hybrids between *Gaultheria* and *Pernettya* antedating × *Gaulthettia*.

Gaulthe′ria f. Ornamental woody evergreen plants named for Jean François Gaultier (*c*. 1708–1756), also called Gaulthier, French physician and botanist of Quebec. Perhaps the best-known member of the family is the wintergreen or checker-berry (*G. procumbens*) of the north-eastern woods of America.

Ga′ura f. Gr. *gauros*, superb; an allusion to the striking flowers of these herbs.

Gaylussa′cia f. Huckleberry. Named in honour of Joseph Louis Gay-Lussac (1778–1850), celebrated French chemist who is best remembered for his law concerning the volume of gases. His work laid the foundation of the food-canning industry. He was very famous for his balloon ascents made in the interests of scientific investigation. In 1804 he ascended to 12,000 feet and then to 22,000 feet.

Gaza′nia f. Treasure flower. Herbaceous plants named in honour of Theodore of Gaza (1398–1478), who translated the botanical works of Theophrastus from Greek into Latin.

Geitonople′sium n. Gr. *geiton*, a neighbour; *plesion*, near; from this twiner's close relationship to another Australian genus, *Eustrephus*.

ge′lidus, -a, -um From icy-cold regions.

Gelse′mium n. Latinized version of gelsomino, the Italian name for jasmine. *G. sempervirens* is called false jasmine.

gemina′tus, -a, -um Paired.

geminiflo′rus, -a, -um Having twin or several flowers. Also *geminispinus*, having twin or many spines.

gemma′tus, -a, -um Jewelled.

gem′mifer, gemmi′fera, -um Bearing buds.

genera′lis, -is, -e Prevailing; normal.

genestieria′nus, -a, -um In honour of A. Genestier (b. 1858), French missionary in China.

geneven′sis, -is, -e From Geneva.

genicula′tus, -a, -um Bent sharply like a knee.

Geni′pa f. Latinized version of the Guyana vernacular name of one species, the Genipap tree, the juicy fruit of which is used to make a preserve locally called genipot.

Genis'ta f. Broom. The Latin name from which the Plantagenet kings and queens of England took their name (*planta genista*). Dyer's greenweed (*G. tinctoria*) is now well established as an adventive wild flower in north-eastern United States. It was first introduced for the yellow dye it yields by Governor John Endicott of Salem, Massachusetts, about 1630.

Genistel'la f. Diminutive of *Genista*.

genistifo'lius, -a, -um With leaves like broom or *Genista*.

Gentia'na f. Gentian. Named for King Gentius of Illyria, *c.* 500 B.C., who was reputed to have discovered the medicinal virtues of the root of the yellow gentian or bitterwort (*G. lutea*) from which a tonic bitters is still made.

geocar'pus, -a, -um With fruit ripening in the earth.

Geogenan'thus m. Seersucker plant. From Gr. *gĕ*, earth; *gĕnĕa*, race, birthplace; *anthos*, flower. In allusion to the flowers arising almost at ground level.

geoi'des *Geum*-like.

geomet'ricus, -a, -um; geomet'rizans With markings arranged in a formal pattern.

Geo'noma f. Gr. *gĕōnŏmŏs*, colonist. These feather palms send out buds from the apex of the stems which eventually grow into trees themselves.

geonomifor'mis, -is, -e In the form of *Geonoma*.

geor'gei In honour of George Forrest (1873–1932); see *forrestii*.

georgia'nus, -a, -um From the state of Georgia, in the United States.

geor'gicus, -a, -um From Georgia, Caucasus, U.S.S.R.

geraniifo'lius, -a, -um Geranium-leaved.

geranioi'des Resembling *Geranium*.

Gera'nium n. Cranesbill, herb-Robert. *Gĕranion* is the classical Greek name, from *gĕranos*, a crane; in allusion to the long beak of the carpels; see *Erodium*.

Gerar'dia f. North American herbs named for John Gerard (1545–1612), garden superintendent to Lord Burghley, Minister to Queen Elizabeth I. He grew many exotic plants in his Holborn garden but he is best known for his *Herball*, first published in 1597 with a much amended edition, the so-called 'immaculate' one cited as *Ger. emac.* by old authors, in 1633. The first edition is very rare.

The *Herball* is to some extent an adaptation and translation from

Latin of the writings of Rembert Dodoens with many entertaining personal additions by Gerard and later by his editor Thomas Johnson. The illustrations were mostly impressions from the wood blocks employed by Tabernaemontanus in his book published in Frankfurt seven years previously. In Gerard's day, however, plagiarism was generally accepted and widely practised; it showed both learning and a proper appreciation of source material. The fact remains that the *Herball* is a great book. It was one of the first botanical books in English rather than Latin.

Gerbe′ra f. Transvaal daisy. Perennial herbs named for Traugott Gerber, a German naturalist who travelled in Russia and who died in 1743.

germa′nicus, -a, -um From Germany.

Gesne′ria f. Herbaceous perennials named for Conrad von Gessner (1516–1565) of Zurich, the most celebrated naturalist of his day, the father of bibliography, and one of the most learned men of all time. Although he died from plague when only forty-nine, he published the vast *Historiae Animalium* (1551–1587), a critical review of zoological knowledge from the time of Aristotle on. His early death prevented him from undertaking a similar botanical study although plants were probably his chief interest throughout his life. His accomplishments were extraordinary since he had to contend with great poverty in his youth and continuous ill health. He is credited with having brought the tulip into repute in Western Europe after seeing a specimen growing in a garden in Augsburg in 1559. Within a dozen years, tulip-fancying had gripped Holland, to culminate in the ruinous tulip mania of the 1630's. (See *Tulipa*.)

Ge′um n. Avens, herb-bennet. The classical Latin name for these perennial herbs.

Gevu′ina f. The native name for the tree known as the Chilean hazel.

gibbero′sus, -a, -um Hunchbacked; humped on one side.

gibbiflo′rus, -a, -um Having flowers with a swelling or hump on one side.

gibbo′sus, -a, -um; gib′bus, -a, -um Swollen on one side.

gib′bsii In honour of the Hon. Vicary Gibbs (1853–1932), English merchant banker, with a fine garden at Aldenham; see *aldenhamensis*.

gibralta′ricus, -a, -um From Gibraltar.

gigante′us, -a, -um Unusually tall or large.

gi′gas Giant. The Gigantes, huge men of incredible strength and terrifying aspect, were the sons of Terra, who stormed the heavens but were killed by the lightning of Jove and the blows of Hercules.

Gilia

Gi'lia f. Herbs, named for Filippo Luigi Gilii (1756–1821), astronomer in Rome, author with Caspar Xuarez of *Osservazioni fitologiche* (1789–92).

Giliber'tia f. Tropical American shrubs named after Jean Emmanuel Gilibert (1741–1814), French physician and botanist, professor in Vilna, then in Lyons.

Gille'nia f. Perennial herbs named for Arnold Gille (Gillenius), 17th-century German botanist, who had a garden at Cassel; this genus is now called *Porteranthus*.

giluus, -a, -um Dull yellow.

Gin'gko f. Maidenhair tree. The puzzling name *Gingko* is a misrendering of the Japanese name *gin* (silver), *kyō* (apricot), used in the 17th century in Japan but now obsolete. This genus, of which many fossil forms exist to indicate a former wide distribution, now consists of one species only which is not known to exist in the wild state. The male and female flowers are on separate trees. The female trees are undesirable because of the foul-smelling fruit; the kernel is said to be edible but it would require a strong stomach to approach it.

Gin'seng See *Panax*.

giraldia'nus, -a, -um; giral'dii In honour of Giuseppe Giraldi, Italian missionary in China who collected plants in Shensi province, 1890 to 1895.

glabel'lus, -a, -um; glabra'tus, -a, -um; glabre'scens; glabrius'-culus, -a, -um Rather or somewhat glabrous.

gla'ber, gla'bra, -um Without hairs; glabrous. Also *glaberrimus*, completely glabrous.

glacia'lis, -is, -e From icy-cold regions, especially the neighbourhood of glaciers.

gladia'tus, -a, -um Swordlike.

Gladi'olus m. Latin for a small sword; in allusion to the shape of the leaves. The plants are also sometimes called sword-lily. People generally use the Latin plural gladioli instead of adding *s* to the singular Latin name, as in antirrhinums, petunias, crocuses.

glandifor'mis, -is, -e In the form of a gland or shaped like one. Also:

glandu'lifer   gland-bearing
glanduliflo'rus   with glandular flowers
glandulo'sus   glandular

glaphy'rus, -a, -um Smoothed; polished.

glauces'cens Having some bloom; somewhat blue or sea-green.

Glauci'dium n. From *Glaucium* in allusion to the form of the few-petalled poppy-like flowers.

glaucifo'lius, -a, -um Having grey-green (glaucous) leaves.

glauciifo'lius, -a, -um Having leaves like *Glaucium*.

Glau'cium n. Horned poppy; sea poppy. From Gr. *glaukos*, greyish-green; from the colour of the leaves.

glaucocar'pus, -a, -um With glaucous fruits.

glaucoi'des Appearing to be coated with bloom.

glaucopep'lus, -a, -um With a greyish or bluish covering.

glaucophy'llus, -a, -um Having grey- or bluish-green leaves; having bloom on the leaves.

glau'cus, -a, -um Having bloom, the fine, whitish, powdery coating which occurs on certain leaves (e.g. cabbage) and fruits (e.g. plum).

Glaux f. Gr. name used by Dioscorides for another plant.

Glecho'ma f. From Gr. *glēchōn*, a kind of mint.

Gledit'sia (also, incorrectly, **Gleditschia**) f. Honey locust. Ornamental trees named for Gottlieb Gleditsch (d. 1786), Director of the Botanical Garden, Berlin.

Gleiche'nia f. Named in honour of Friedrich Wilhelm von Gleichen (1717–1783), German botanist interested in microscopical studies.

Glirici'dia L. *glis*, a dormouse; *caedo*, kill; the name is a Latinization of the Colombian vernacular name *mata-raton* and refers to the use of the seeds and bark to poison rats and mice.

glischrus, -a, -um Gluey; sticky.

Glob'ba f. Name based on an Indonesian plant name, *galoba*.

glo'bifers, globi'fera, -um Bearing globe-shaped or spherical clusters.

globo'sus, -a, -um Round; spherical.

Globula'ria f. Globe daisy. L. *globulus*, a small round ball; from the form of the flower heads of these herbs and shrubs.

globula'ris, -is, -e Pertaining to a small sphere or ball.

globu'lifer, globuli'fera, -um Bearing clusters in the form of small globes or spheres.

globulo'sus, -a, -um Small and globular.

glomera'tus, -a, -um Clustered into more or less rounded heads. Also *glomerulifer*, having small rounded heads or clusters.

glomerula'tus, -a, -um In small clusters or glomerules.

glomeruliflo'rus, -a, -um Having flowers in glomerules. A glomerule

Gloriosa

is an inflorescence consisting of a tightly clustered cyme (a group or cluster of flowers opening from the centre outwards as in sweet cherry).

Glorio'sa f. Glory-lily. From L. *gloriosus*, glorious. Gloriosa daisies are tetraploid hybrids of *Rudbeckia q.v.*

glorio'sus, -a, -um Superb; glorious.

glosso'phorus, -a, -um Tongue-bearing.

Glottiphyl'lum n. Gr. *glōtta*, a tongue; *phyllon*, a leaf. The leaves of these succulent plants are very thick, soft, and fleshy, and shaped like tongues. They are brittle and easily snapped.

Gloxi'nia f. What is usually called *Gloxinia* by florists is, in fact, *Sinningia*. The true *Gloxinia*, much less common, is named for Benjamin Peter Gloxin, a doctor at Colmar, who published *Observationes botanicae* at Strasbourg in 1785.

gloxinioi'des Resembling *Gloxinia*.

gluma'ceus, -a, -um Having glumes, the chaffy bracts enclosing the flowers of grasses or sedges.

glutino'sus, -a, -um Sticky; gluey; glutinous.

Glyce'ria f. Manna grass. From Gr. *glykys*, sweet; with reference to the edible grains of *G. fluitans*.

Glyci'ne f. Soybean. From Gr. *glykys*, sweet; in allusion to the sweetness of the roots and leaves of some species, none of which begins to compare in importance with the soybean. Quick to mature, drought-resistant and easy to grow, this plant has been in cultivation for at least three thousand years. It yields oil, meal, cheese curds, and cake as well as milk, which in some places is delivered daily as we do cow's milk. About two-thirds of the world's crop is grown in Manchuria where it has caused wars as well as political and commercial manoeuvring. It is imported by Western countries for use in soap, paint, synthetic-rubber manufacture, etc., as well as for food, both for human beings and cattle.

glycinioi'des Resembling soybean or *Glycine*.

Glycos'mis f. Gr. *glykys*, sweet; *osme*, smell. Both the flowers and the leaves of these shrubs and small trees are fragrant.

Glycyrrhi'za f. Liquorice. From Gr. *glykys*, sweet; *rhiza*, a root. The root of *G. glabra*, the only species of any interest, provides the liquorice of commerce of which the Catalonian is the most highly regarded.

glyptostroboi'des Resembling the genus *Glyptostrobus*.

Glypto'strobus m. From Gr. *glypto*, to carve; *strŏbilŏs*, a cone; an allusion to the depressions and pits on the cone scales of these deciduous cypress-like trees.

Gmeli'na f. Trees and shrubs named for Johann Gottlieb Gmelin (1709–1755), German traveller and naturalist, who made extensive travels in Siberia and Kamchatka in 1733–1743, then returned to Tübingen, his native town, where he was a professor from 1749–1755.

Gnapha'lium n. Cudweed. From Gr. *gnaphaliŏn*, a downy plant with soft white leaves used for stuffing cushions.

gnaphaloi'des Resembling cudweed or *Gnaphalium*.

Gni'dia f. From Latin geographical name referring to Gnidus (Gr. *Knidos*) in Caria, Asia Minor, where a kind of Daphne grew.

Gode'tia f. Showy annuals named for Charles H. Godet (1797–1879), Swiss botanist, author of a *Flore du Jura* (1858–1869).

Goe'thea f. In honour of the German poet Johann Wolfgang von Goethe (1749–1832).

Goldfus'sia f. In honour of Georg August Goldfuss (1782–1848), professor of zoology and mineralogy in Bonn, Germany.

Gome'sa f. Epiphytic orchids named for Bernardino Gomez, Portuguese naval surgeon, who published a book on the plants of Brazil in 1803.

Gomphocar'pus m. Milkweed. Gr. *gŏmphŏs*, bolt, nail; *karpos*, fruit. Sometimes included in *Asclepias*.

Gompholo'bium n. Gr. *gŏmphŏs*, bolt; *lŏbŏs*, a pod; from the appearance of the pod on these Australian evergreen flowering shrubs.

Gomphre'na f. Globe amaranth. Latin name for a kind of amaranth, possibly this one; usually grown as an 'everlasting'.

Gon'gora f. Epiphytic orchids named for Don Antonio Cabellero y Gongora (1740–1818), Spanish Viceroy of New Granada, now Colombia, and patron of José Celestino Mutis (1732–1818), naturalist and plant explorer, who was responsible for the Bogotà Observatory which became a renowned centre of scholarship and research.

gongylo'des Swollen; roundish.

gonio'calyx With prominently angled calyx.

Goode'nia f. In honour of Samuel Goodenough (1743–1827), Bishop of Carlisle from 1808 onwards, a keen amateur botanist.

Goo'dia f. Shrubs named for Peter Good (d. 1803), Kew gardener who accompanied Robert Brown on Flinders' voyage round Australia.

Goodye'ra f. Rattlesnake plantain. A terrestrial orchid named in honour of John Goodyer (1592–1664) of Mapledurham, Oxfordshire, who assisted Johnson in his edition of Gerard's *Herball*, and translated Dioscorides's *Materia medica* into English.

Gordo'nia f. Ornamental evergreen or deciduous shrubs and trees with white flowers named for James Gordon (d. 1781), correspondent of Linnaeus and nurseryman of Mile End, London. *Franklinia* is often placed in this genus.

Gorma′nia f. Plants often included in *Sedum* named for M. W. Gorman of Portland, Oregon, collector of plants of the Pacific Northwest.

gossy′pinus, -a, -um Resembling cotton or *Gossypium*.

Gossy′pium n. Cotton. L. *gossypion*. The English name derives ultimately from Arabic. Cotton has been cultivated since prehistoric times in the Old World and was found already in use in the New when the Spanish arrived. The most important of vegetable fibres, it used to supply about 90 per cent of the clothing used in the world, but synthetic fibres and wool have much reduced this.

Gourli′ea f. Shrubs and small trees named for Robert Gourlie (d. 1832), Scottish botanist in Chile.

Grabow′skia f. Spiny shrubs named for Henri Emanuel Grabowski (1792–1842), German botanist at Breslau.

gracilen′tus, -a, -um Slender; thin.

gra′cilis, -is, -e Graceful; slender. Also:

graciliflo′rus	With slender or graceful flowers
graci′lior	More graceful
graci′lipes	With a slender stalk
gracilisty′lus	With a slender style
gracil′limus	Most graceful

grae′cus, -a, -um Greek; Grecian.

grami′neus, -a, -um Resembling grass; grassy.

graminifo′lius, -a, -um With leaves resembling grass.

Grammatophyl′lum n. From Gr. *grammata*, letters; *phyllon*, a leaf; an allusion to the markings on the flowers of these orchids.

grammope′talus, -a, -um With striped petals.

granaden′sis, -is, -e; granaten′sis, -is, -e Relating to Granada, Spain or to Colombia, South America.

gran′dis, -is, -e Big; showy. Also:

gran′diceps	large-headed
grandicus′pis	having big points
grandidenta′tus	having big teeth
grandiflo′rus	large flowered
grandifo′lius	with big leaves
grandifor′mis	on a large scale
grandipuncta′tus	having large spots

grani'ticus, -a, -um Growing on or in crevices of granite or other hard quartz rocks.

granula'tus, -a, -um With grain-like tubercles or knobs.

granulo'sus, -a, -um Composed of minute grains or appearing as though covered with them.

Graptope'talum n. Gr. *graptos*, painted, marked with letters, to write; *petalon*, a petal; from the variegated markings on the petals of these glaucous succulents.

Graptophyl'lum n. Gr. *graptos*, painted, marked with letters, to write; *phyllon*, a leaf; from the variegated markings on the leaves of these shrubs.

gratianopolita'nus, -a, -um Of Grenoble, Dauphiné, France.

Gra'tiola f. L. *gratia*, agreeableness or pleasantness; in allusion to the medical uses of these herbs.

gratiss'imus, -a, -um Very pleasing or agreeable.

gra'tus, -a, -um Pleasing.

grave'olens Heavily scented.

Greeno'via f. Succulent plants named in honour of the English geologist George Bellas Greenough (1778–1855).

Grei'gia f. Large showy herbs named for Major-General Samuel Alexeivich Greig (1827–1887), President of the Russian Horticultural Society.

Grevil'lea f. Evergreen trees and shrubs named for Charles Francis Greville (1749–1809), a founder of the Horticultural Society of London and a Vice-President of the Royal Society.

Gr'ewia f. A genus of trees and shrubs named for Nehemiah Grew (1641–1712), English physician, a pioneer microscopist, author of a celebrated *Anatomy of Plants* (1682).

Grey'ia f. A small tree from Natal named for Sir George Grey (1812–1898), notable colonial governor who ultimately became the elected Prime Minister of New Zealand where he initiated great reforms.

griersonia'nus, -a, -um In honour of R. C. Grierson, customs officer at Tengyueh, Yunnan, China, friend of George Forrest.

Griffi'nia f. In honour of William Griffin (d. 1827), English horticulturist. Stove bulbs from South America.

griffithia'nus, -a, -um; griffi'thii In honour of William Griffith (1810–1845), British botanist and medical man who collected in India and Afghanistan but died before he could publish his extensive and detailed observations.

Grinde'lia f. Gumplant. Named for David Hieronymus Grindel (1766–1836), German professor of botany at Riga.

griseba'chii In honour of August Heinrich Rudolf Grisebach (1814–1879), German botanist, professor at Göttingen, noted for his work on plants of the Balkan Peninsula (Rumelia), the West Indies and Argentina, on *Gentianaceae* and plant geography.

Griseli'nia f. Evergreen shrubs and trees named for Francesco Griselini (1717–1783), Venetian naturalist.

gris'eus, -a, -um Grey.

groenlan'dicus, -a, -um From Greenland.

grono'vii Specific epithet in honour of Johan Fredrik Gronovius (1690–1762) of Leyden, a keen naturalist and supporter of Linnaeus during his stay in Holland; he belonged to the celebrated Gronovius family of classical scholars and was the author of a *Flora Virginica* (1739–1743).

grosse-serra'tus, -a, -um With large saw teeth.

Grossula'ria f. Gooseberry. Latinization of the French *groseille*.

gros'sus, -a, -um Very large.

grui'nus, -a, -um Resembling a crane.

Guaia'cum n. From the South American vernacular word guaiac, the name for lignum vitae (*G. officinale*), meaning wood of life, so named because of its high repute in medicine. Guaiacum resin is used for a number of acute and chronic conditions, as well as in certain tests of blood. The wood of these trees is hard, dense, and virtually unsplittable. It is used for such things as mallets, ships' blocks, and formerly for propeller shaft bearings. Lignum vitae was used in the lock-gate hinges on the Erie Canal where they lasted for a century.

guianen'sis, -is, -e Of Guiana, northern South America.

Guiliel'ma f. Palms named for Queen Frederica Guilielma Carolina of Bavaria, d. 1841.

Guizo'tia f. Annuals named for François Pierre Guillaume Guizot (1787–1874), French historian and statesman.

gum'mifer, gummi'fera, -um Producing gum.

gummo'sus, -a, -um Gummy.

Gunne'ra f. Perennial herbs. Named in honour of Ernst Gunnerus (1718–1773), Norwegian bishop and botanist, the author of *Flora Norvegica* (1766–1772). Of all foliage plants one of the most handsome and impressive. The leaves of *G. manicata* may be six feet across and the crown as large as a man's body.

gunnerifo'lius With leaves like *Gunnera*.

gutta'tus, -a, -um Spotted; speckled with small dots.

Guzma'nia f. Herbaceous perennials. Named for an 18th-century Spanish naturalist, Anastasio Guzman.

Gymnade'nia f. Gr. *gymnos*, naked; *adēn*, gland; the viscidia of the stamens of these orchids are not enclosed in a pouch.

Gymnocaly'cium n. Gr. *gymnos*, naked; *kalyx*, a bud; because the flower buds of these succulent plants are naked.

gymnocar'pus, -a, -um Having naked fruit.

Gymno'cladus f. Gr. *gymnos*, naked; *klados*, a branch. The branches are bare of leaves in winter. One species, the Kentucky coffee-tree (*G. dioica*), has seeds which the pioneers used as a coffee substitute.

Gymnogram'ma f. Gr. *gymnos*, naked; *gramma*, a line. In allusion to the naked sori of these ferns being in lines.

Gymno'pteris f. Gr. *gymnos*, naked; *pteris*, fern. In allusion to the naked sori of these ferns.

Gynandro'psis f. Gr. *gynē*, female; *andros*, male; *opsis*, appearance; from the manner in which the stamens appear to be inverted on the top of the ovary on these annuals.

Gyne'rium n. Gr. *gynē*, female; *erion*, wool; a reference to the hairy spikelets of the female plants of these grasses.

Gynu'ra f. Gr. *gynē*, female; *oura*, a tail; from the long, rough stigma of these foliage plants.

Gypso'phila f. Baby's breath. Gr. *gypsos*, gypsum; *philos*, loving. Some species of these small-flowered plants of the stitchwort family are lime-loving.

gy'rans Going around in circles.

H

haagea'nus, -a, -um In honour of J. N. Haage (1826-1878), seed grower of Erfurt, Germany.

Habena'ria f. Fringed orchis, rein orchis. From L. *habēna*, a rein, strap. In some species the spur is long and shaped like a strap; the divisions of the lip are also narrow and strap-like.

Haber'lea f. Tufted herbs named for Carl Constantin Haberle (1764–1832), professor of botany at Budapest.

Habli'tzia f. In honour of Carl Ludwig Hablizl (1752–1821), naturalist and traveller in the Caucasus, Iran and the Crimea, of which he was vice-governor.

Habran'thus m. From Gr. *habros*, graceful, beautiful, pretty; *anthos*, flower.

habro'trichus, -a, -um With beautiful or soft hairs or bristles.

Hacque'tia f. Herbaceous perennial named for Balthasar Hacquet (1740–1815), Austrian writer on Alpine plants.

hadria'ticus, -a, -um From the shores of the Adriatic.

haema'leus, -a, -um Blood-red.

Haeman'thus m. Red Cape tulip, blood-lily. From Gr. *haima*, blood; *anthos*, a flower.

haeman'thus, -a, -um With blood-red flowers. Also:

haema'stomus	with blood-red mouth
haematoca'lyx	with blood-red calyx
haematocar'pus	with blood-red fruits
haematochi'lus	with blood-red lip
haemato'des	blood-red

Haematoxy'lum n. Logwood. From Gr. *haima*, blood; *xylon*, wood. Since the 16th century it has been a very valuable export from Central America and the West Indies for the sake of the brilliant red dye produced from the heart wood which synthetic chemistry has never been able to match. *H. campechianum* is known by several other names, such as blood-tree and campeachy-wood.

Ha'kea f. Australian evergreen shrubs named for Baron Christian Ludwig von Hake (1745–1818), German patron of botany, councillor in Hanover.

hakeoi'des Resembling *Hakea*.

Hale'sia f. Silver bell. Beautiful flowering trees and shrubs named in honour of the Reverend Stephen Hales (1677–1761), curate of Teddington, near London, physiologist, chemist, and inventor whose wide-ranging curiosity resulted in important and far-reaching experiments in plant physiology as well as, possibly, the first measurement of blood pressure. It was performed one Sunday on a white mare, thrown and tied to a gate, whose blood, transferred through a goose's windpipe (an efficient medium before the invention of rubber tubing), rose in a glass tube to a height of 8 feet 3 inches.

He also invented an effective method of ventilating prisons to control jail fever, and ships to prevent dry rot; a 'sea gauge' for taking soundings, and a method of fumigating wheat with brimstone to preserve it from weevils.

halimifo'lius, -a, -um With leaves like *Atriplex halimus* (Gr. *halimŏn*), the tree purslane with silvery ovate and obovate leaves used in seaside planting in Britain.

Halimoden'dron n. Salt tree. From Gr. *halimos*, maritime; *děndrŏn*, a tree. A salt-tolerant shrub from Central Asia.

hal'leri In honour of the many-sided Albrecht von Haller (1708–1777), Swiss anatomist, botanist, bibliographer, administrator and poet, author of the first textbook of physiology and numerous other works.

hallia'nus, -a, -um; ha'llii Plants so named may commemorate Herman Christin van Hall (1801–1874), Dutch botanist, professor in Groningen, if Indonesian; Elihu Hall (1822–1882), American surveyor, or Harvey Monroe Hall (1874–1932), professor of botany, University of California, pioneer in experimental taxonomy, if North American; George Rogers Hall (1820–1899), American medical man, then trader in the China Sea, who introduced many plants from Japan into the United States, if Japanese.

halli'eri In honour of Hans Hallier (1868–1932), German botanist.

halo'philus, -a, -um Salt-loving.

Hamame'lis f. Witch-hazel. The Greek name for a plant with a pear-shaped fruit, possibly the medlar. The bark and twigs of *H. virginiana* supply the witch-hazel of pharmacy. The twigs are also a favourite choice of dowsers or water-diviners.

Hamatocac'tus n. L. *hamatus*, hooked, plus *cactus*; referring to the long central spine of each areole which is usually hooked at its tip.

hama'tus, -a, -um; ham-o-'sus, -a, -um Hooked.

Hame'lia f. Handsome shrubs named in honour of Henri Louis du Hamel du Monceau (1700–1781), celebrated French writer on trees and shrubs.

hancea'nus, -a, -um In honour of Henry Fletcher Hance (1827–1856), British consul in China.

hanco'ckii In honour of William Hancock (1847–1914), Irish customs official in Chinese Imperial Maritime Customs, who collected in China and Taiwan.

Haplopap'pus m. Gr. *haplos*, single; *pappos*, pappus. The down on fruits of Compositae.

Harbou'ria f. Herbaceous plants named for J. P. Harbour, collector of Rocky Mountain plants.

Hardenber'gia f. Australian sarsaparilla. Evergreen twiners named in honour of Franziska, Countess von Hardenberg; she was the sister of Austrian baron Carl A. A. von Hügel (1795–1870), an enthusiastic patron of horticulture, who travelled widely in the Philippines and elsewhere.

Harpephyl'lum n. Gr. *harpē*, sickle; *phyllon*, leaf.

harpophyllus

harpophyl′lus, -a, -um With leaves shaped like sickles.

Harri′sia f. Cacti named for William Harris (1860–1920), Superintendent of the Public Gardens, Jamaica, whose collecting throughout the island vastly increased knowledge of its flora.

harrovia′nus, -a, -um In honour of George Harrow, manager of Messrs Veitch's Coombe Wood nursery where many Chinese plants introduced by E. H. Wilson were first raised.

Hartwe′gia f. Epiphytic orchids named for Carl Theodor Hartweg (1812–1871), German gardener who collected for the Horticultural Society of London in Mexico, 1836–1837, and California, 1846–1847.

hasta′tus, -a, -um Spear-shaped. Also:
has′tifer bearing a spear
hastila′bius with a spear-shaped lip
hasti′lis of a spear
hastula′tus somewhat spear-shaped

Hatio′ra f. Anagram of *Hariota,* a genus of cacti which had to be renamed on account of the earlier use of the name *Hariota* for another genus. Originally named in honour of Thomas Hariot (1560–1621), mathematician and cartographer on the 1585 expedition to Virginia and author of *A briefe and true Report of the Newfound Land of Virginia* (1588).

Hawor′thia f. Succulent plants named for Adrian Hardy Haworth (1768–1833), English authority on succulent plants and *Lepidoptera,* the author of a monograph on *Mesembryanthemum* (1794), etc.

He′be f. Australasian and South American evergreen shrubs named in honour of the goddess of youth, cup-bearer to the gods, who later married Hercules when he ascended to Olympus. *Hebe* differs from *Veronica* primarily in the form of the capsule.

hebecar′pus, -a, -um With fruit covered in down; from Gr. *hēbē,* youth, down of puberty; *karpos,* fruit.

Hebenstrei′tia f. A genus of shrubs, sub-shrubs, and annuals named for Johann Ernst Hebenstreit (1702–1757), professor of medicine at Leipzig.

hebephyl′lus, -a, -um Downy-leaved.

Hech′tia f. Mexican succulents named for J. H. G. Hecht (d. 1837), a counsellor to the King of Prussia.

hecto′rii In honour of Sir James Hector (1834–1907), New Zealand geologist and botanist of Scottish origin, who took a very active part in educational affairs.

164

Hedeo′ma f. American pennyroyal. Gr. *hēdys*, sweet; *ŏsmē*, scent; from the aromatic foliage.

He′dera f. Ivy. The Latin name. Ivy was held sacred to Bacchus, the god of wine, and was intimately connected with his revels.

hedera′ceus, -a, -um Resembling ivy.

hederifo′lius Ivy-leaved.

Hedy′chium n. Ginger-lily. From Gr. *hēdys*, sweet; *chion*, snow. The flower of one species is white and fragrant.

Hedy′sarum n. Greek name *hēdysarŏn*, from *hēdys*, sweet. This genus of ornamental herbs or sub-shrubs includes the handsome and fragrant French-honeysuckle (*H. coronarium*).

Hedy′scepe f. Umbrella palm. From Gr. *hēdys*, sweet; *scĕpē*, a covering; from the manner in which the flowers of this feather palm are produced in a dense cluster from the leaf crown.

Hei′mia f. Shrubs named for Dr. Heim (d. 1834), of Berlin.

Hele′nium n. Sneezeweed. From the Greek name *hĕlĕniŏn*, for another plant named for Helen of Troy.

Heliam′phora f. Gr. *hēlios*, the sun; *amphora*, a jar. From the sun-loving habit and leaf-shape of these pitcher plants.

Helianthel′la f. Diminutive of *Helianthus* to which it is closely related.

Helian′themum n. Sunrose. From Gr. *hēlios*, the sun; *anthemon*, a flower.

helianthoi′des Resembling *Helianthus*.

Helian′thus m. Sunflower. From Gr. *hēlios*, the sun; *anthos*, a flower. *H. tuberosus* is the 'Jerusalem artichoke'.

Helichry′sum m. Everlasting. Immortelle. From Gr. *hēlios*, the sun; *chrysos*, golden.

Helicodi′ceros m. Dragon's mouth. Twist-arum. From Gr. *hĕlix*, anything spirally twisted; *dis*, twice; *kĕras*, a horn. The basal divisions of the leaves are twisted and stand erect like horns.

Helico′nia f. After Mount Helicon in Greece where it was supposed that the Muses lived, so named from its relationship to *Musa*.

Helioce′reus m. Gr. *hēlios*, the sun; *cereus*, cactus; because of the desert habitat of these procumbent Mexican plants.

helio′lepis With sun-like (i.e. bright or glittering) scales.

Helio′phila f. Gr. *hēlios*, the sun; *philein*, to love. From the habitat of these South African plants.

Heliop′sis f. Gr. *hēlios*, the sun; *ŏpsis*, resembling. In allusion to the rayed yellow flower heads of these American perennials.

Heliotropium

Heliotro'pium n. Heliotrope, cherry-pie. Gr. *hēliotrōpiŏn*, from *hēlios*, the sun; *trŏpē*, a turning; in allusion to an old disproved idea that the flower heads turned with the sun. The leaves and flowers of many plants do this and are known as heliotropic.

Helip'terum n. Everlasting. From Gr. *hēlios*, the sun; *pteron*, a wing; a technical reference to the plumed pappus.

hel'ix Greek for anything of spiral shape, hence applied to twining plants.

Helle'borus m. Gr. *hĕllĕbŏros*, name of the Lent hellebore (*H. orientalis*). The roots of some species yield a violent poison once used in the treatment of insanity.

helle'nicus, -a, -um Of Greece, Grecian.

helo'des Of bogs.

helodox'a Glory of the marsh.

Helo'nias f. Stud flower, swamp pink. From Gr. *hĕlos*, a marsh; the natural habitat of these plants.

Helonio'psis f. *Helonias*; *opsis*, resemblance. A genus of liliaceous plants from Japan and Formosa.

helve'ticus, -a, -um Of Switzerland.

hel'volus, -a, -um Pale brownish yellow.

hel'vus, -a, -um Dim-coloured, light bay.

Helwin'gia f. Deciduous shrubs named after Georg Andreas Helwing (1666–1748), German pastor who wrote on *Pulsatilla* and the flora of Prussia.

Helxi'ne f. Baby's tears. A creeping herb with a classical name formerly applied to a related plant (pellitory or *Parietaria*); see *Soleirolia*.

Hemerocal'lis f. Day-lily. Gr. *hēmera*, day; *kallos*, beauty. Each flower lasts but one day.

Hemian'dra f. Gr. *hēmĭ*, half; *andros*, male. This genus of labiates, which has only one anther cell per stamen, was discovered by Robert Brown, a most able botanist, who at Sir Joseph Banks' suggestion visited Australia with Flinders, from 1801–1805.

hemidar'tus, -a, -um Half-flayed, i.e. with patchy hair-covering.

Hemigra'phis f. Gr. *hēmĭ*, half; *graphis*, brush; i.e. for one half of a brush, referring to the dense hair-covering on the filaments of the outer stamens.

Hemioni'tis f. Gr. *hēmiŏnītis*, from *hēmiŏnŏs*, a mule; because these ferns were once regarded as promoting sterility and hence worn by women as a charm against pregnancy.

Hemipte′lea f. Gr. *hēmĭ*, half; *ptĕlĕa*, elm; in allusion to the fruits of this spiny tree, which have a wing only on the upper half, whereas in the elm (*Ulmus*) the fruit is encircled by a wing.

hemisphae′ricus, -a, -um Hemispherical; in the shape of half a ball.

Hemite′lia f. Gr. *hēmĭ*, half; *tēlia*, cover.

hemitricho′tus, -a, -um Hairy on one side but not the other.

hemsle′yi; hemsleya′nus, -a, -um In honour of William Botting Hemsley (1843–1924), from 1899 to 1908, Keeper of the Kew Herbarium; a leading authority on the plants of Central America and China.

hen′rici In honour of Prince Henri d'Orléans (1867-1901), French explorer, who collected in western China.

hen′ryi In honour of Augustine Henry (1857–1930), Irish medical man, plant-collector and dendrologist, who made vast collections in central China and Formosa and later became a professor of forestry.

Hepa′tica f. Liverwort. From Gr. *hēpar*, the liver. This is a double allusion to the colour and shape of the leaves, whence the belief, according to the Doctrine of Signatures, that they would be good for complaints of the liver.

hepaticifo′lius, -a, -um With leaves like *Hepatica*.

hepta- In compound words signifying seven. Thus *heptaphyllus*, seven-leaved.

heracleifo′lius, -a, -um With leaves like *Heracleum* or cow-parsnip.

Heracle′um n. Cow-parsnip. Variant of Gr. *hēraklĕia* or *panakĕs hēraklĕiŏn* in honour of Hercules (Gr. *Hērakles*).

herba′ceus, -a, -um Not woody; herbaceous.

Herman′nia f. In honour of Paul Hermann (1646–1695), professor of botany in Leiden, Holland, who spent some years in Ceylon.

Hermodac′tylus m. Variant of old Gr. name for a *Colchicum*, from *Hĕrmēs*, Mercury; *daktylos*, finger; referring to the tubers.

Hernia′ria f. Herniary, rupturewort, burstwort. From *hernia*, a rupture; for which the plant was once supposed to be effective.

Hespera′loe f. Gr. *hĕspĕrŏs*, western. The plants are New World but in habit like the Old World aloes.

Hesperan′tha f. Gr. *hĕspĕra*, evening; *anthos*, flower. The flowers open late in the day.

Hes′peris f. Sweet rocket, dame's violet. From Gr. *hĕspĕra*, the evening. The plants have a marked fragrance at sunset and later.

Hesperoyuc′ca f. From Gr. *hĕspĕrŏs*, western, and *Yucca*. Native to California.

hes'perus, -a, -um Of the West.

heter-, hetero- In Greek compound words signifying various, diverse.
Thus:

heteracan'thus	diversely spined
heteran'thus	diversely flowered
heterocar'pus	diversely fruited
hete'rodon	diversely toothed
heterodo'xus	differing from the type of the genus
heteroglos'sus	diversely tongued
hetero'lepis	diversely scaled
heteromor'phus	differing in form
heterope'talus	diversely petalled
heterophyl'lus	diversely leaved
hetero'podus	diversely stalked

Heteranthe'ra f. From Gr. *hĕtĕros*, varying, different; *anthera*, anther. In allusion to one anther being larger and different in shape from the other two.

Heterocen'tron n. Gr. *hĕtĕros*, different; *kĕntrŏn*, a spur, sharp point. The longer anthers have two bristle-like appendages, the shorter anthers merely two basal swellings.

Heterome'les f. Toyon, Christmas-berry. From Gr. *hĕtĕros*, different; *mēlŏn*, apple.

Heterospa'the f. Gr. *hĕtĕros*, differing. The spathes of this palm are of unequal length.

Heu'chera f. Alum-root. In honour of Johann Heinrich von Heucher (1677–1747), professor of medicine at Wittenberg. Pronunciation can be either *hew-ke-ra* or *hoy-sher-ra*.

Heucherel'la f. Name for intergeneric hybrids between *Heuchera* and *Tiarella*.

Heur'nia See *Huernia*.

Heve'a f. Latinized form of the Guyanan name *hevé* for *H. guianensis*. The related Para rubber tree (*H. brasiliensis*) is the most important source of natural rubber. After the discovery of the secret of vulcanization (Hancock in England, 1834; Goodyear in the United States, 1839) the demand for rubber soon exceeded the Amazon supplies. The Director of Kew, Sir Joseph Hooker, determined in the 1870's to obtain seed. Difficulties in transporting it in fertile condition were solved by H. A. Wickham who in 1870 chartered a special steamship to carry his prepared baskets of 70,000 seeds. Hothouses at Kew were summarily emptied and within two weeks there were over 2,000 seedlings, most of which went to Ceylon. Thus was started the great rubber-growing industry in south-east Asia which, thanks largely to

Ridley's work on tapping the latex economically, by 1910 eclipsed the rubber-collecting industry of the Amazon basin, largely based on the barbarous exploitation of native people.

hexa- In compound words signifying six. Thus:

hexagono'pterus six-angled wings
hexago'nus six-angled
hexan'drus with six stamens
hexape'talus six-petalled
hexaphyl'lus six-leaved

Hexaglot'tis f. From Gr. *hex*, six; *glotta*, tongue. The six narrow divisions of the stigma suggesting the tongues of small birds.

hi'ans Gaping.

Hibber'tia f. Evergreen shrubs named in honour of George Hibbert (1757–1837) who had a private botanic garden at Clapham, near London, England and employed a collector, James Niven, to collect plants in South Africa.

hiberna'lis, -is, -e Pertaining to winter.

hiber'nicus, -a, -um Irish.

hiber'nus, -a, -um Winter-flowering or winter-green.

hibiscifo'lius, -a, -um With leaves resembling *Hibiscus*.

Hibis'cus m. Rose-mallow. The Greek name for mallow. *H. rosa-sinensis*, generally known simply as *Hibiscus*, is a magnificent shrub widely grown in frost-free areas and in greenhouses. It is said that the blooms are worn by the ladies of Tahiti and other Pacific islands in conventional fashions to indicate freedom or otherwise from commitments to the opposite sex. In Jamaica it has been known as 'shoe flower' or 'shoe black' because the mucilaginous petals can be used to polish shoes, 'a use which seems little consistent with their elegance and beauty', as a Jamaican writer said. The genus also includes rose-mallow (*H. moscheutos*), rose-of-Sharon (*H. syriacus*), the confederate rose (*H. mutabilis*) and roselle (*H. sabdariffa*).

Hicke'nia f. In honour of Cristobál Maria Hicken (1875–1933), professor of botany at Buenos Aires. A genus of cactus.

Hidalgo'a f. Climbing dahlia. In honour of a Mexican, Miguel Hidalgo (*fl.* 1824).

hiema'lis, -is, -e Of the winter; winter-flowering.

Hiera'cium n. Hawkweed. The classical name *hiĕrakion* for various yellow-flowered Mediterranean composites, supposedly from *hierax*, falcon.

Hiero'chloë f. Gr. *hiĕrŏs*, sacred; *chloë*, grass. In some parts of northern

Europe these fragrant grasses are strewn before church doors on saints' days.

hierochun'ticus, -a, -um Specific epithet of rose-of-Jericho or resurrection plant (*Anastatica*). From the classical name of the town of Jericho.

himala'icus, -a, -um; himalayen'sis, -is, -e Of the Himalaya.

Hind'sia f. In honour of Richard Brinsley Hinds (d. before 1861), naval surgeon in Royal Navy, naturalist on voyage of H.M.S. *Sulphur*, 1836–1842.

Hippeas'trum n. Evidently from Gr. *hippŏs*, horse; *hippeus*, rider; apparently suggested by the epithet *equestre*, belonging to a horseman, of an included species; as stated in *Bot. Mag.* 9: t. 305 (1795), 'the spatha is composed of two leaves, which standing up at a certain period of the plant's flowering like ears give to the whole flower a fancied resemblance to a horse's head'. These bulbous plants are popularly known as *Amaryllis*.

Hippo'broma f. From Gr. *hippŏs*, horse; *brŏmŏs*, rage, fury. In allusion to its poisonous nature, driving horses mad.

hippocas'tanum Latin name for the horse-chestnut, for which see *Aesculus*. There is a clearly marked horseshoe under the leaf axils.

Hippocre'pis f. Horseshoe vetch. Gr. *hippŏs*, a horse; *krēpis*, a shoe; from the shape of the pod of this trailing herb.

hippoma'nicus, -a, -um From Gr. *hippŏmanēs*, used for herbs which drive horses mad or are poisonous to them or of which they are madly fond.

Hippo'phaë f. Sea-buckthorn. The classical Greek name for another plant, probably prickly spurge.

hippophaeoi'des Resembling *Hippophae*.

hirci'nus, -a, -um Smelling of goat or goatlike.

hirsu'tus, -a, -um Hairy. Thus:

hirsutiss'imus	very hairy
hirsu'tulus	somewhat hairy
hirtel'lus	rather hairy
hirtiflo'rus	hairy-flowered
hir'tipes	hairy-footed
hir'tus	hairy

hispa'nicus, -a, -um Spanish.

his'pidus, -a, -um Bristly.

hodgso'nii In honour of Bryan Houghton Hodgson (1800–1894), British Resident in Nepal, amateur naturalist especially interested in Himalayan birds and mammals.

Hoffman'nia f. Herbs or shrubs named in honour of George Franz Hoffmann (1761–1826), professor of botany at Göttingen, 1792–1804, and Moscow, 1804–1826.

Hohe'ria f. Latinized version of the Maori name for these New Zealand trees and shrubs.

Holboel'lia f. In honour of Frederik Ludvig Holbøll (1765–1829), of the Copenhagen botanic garden; an evergreen twining plant.

Hol'cus m. Greek name for Sorghum.

hollan'dicus, -a, -um Of Holland, Netherlands.

Holmskiol'dia f. Shrubs named in honour of Theodor Holmskiold (1732–1794). Danish botanist. Name best pronounced as 'Holmshol-dia'.

holo- In compound words signifying completely. Thus:
holocar'pus   whole-fruited (that is not lobed or split)
holochry'sus   completely golden
hololeu'cus   wholly white
holope'talus   with entire (undivided) petals
holoseri'ceus   covered all over with silky hairs

Holodis'cus m. Gr. *hŏlŏs*, entire; *diskos*, a disc. An allusion to the disc of the flower of these ornamental shrubs.

holos'tea Specific epithet for *Stellaria*, starwort. From the Greek for a chickweed-like plant.

Homalan'thus m. Gr. *hŏmalos*, flat; *anthos*, a flower; from the appearance of the flower of these tropical trees.

Homaloce'phala f. Gr. *hŏmalos*, flat; *kĕphalē*, a head; referring to the flat top of the flower of these plants of the Cactus family.

Homalome'na f. Name coined by Rumphius and adopted by Schott, from Gr. *homalos*, flat; *mēnē*, moon; as a translation of a Malayan vernacular name, which, according to Backer, was misrendered. Herbaceous and shrubby plants from Asia and tropical Africa.

Home'ria f. Not in honour of the Greek poet Homer but from Gr. *hŏmĕrĕo*, to meet; in allusion to the joining of the filament of the stamens into a tube.

homo'lepis Having structurally similar scales.

hondoen'sis, -is, -e Of Hondo, Japan.

hookeri In honour of Sir William Jackson Hooker (1785–1865), unsuccessful brewer and extremely successful professor of botany, from 1841 to 1865 Director of the Royal Botanic Gardens, Kew; or his son Sir Joseph Dalton Hooker (1817–1911), much-travelled botanist and author of many important works, from 1865 to 1885 Director of Royal Botanic Gardens, Kew.

Hor'deum n. Wild barley. Latin name for barley (*H. vulgare*). Squirrel-tail grass is *H. jubatum.*

horizonta'lis, -is, -e Flat to the ground; horizontal.

Hormi'num n. Greek name for sage to which this genus is related.

hormo'phorus, -a, -um Bearing a necklace; from Gr. *hŏrmŏs*, chain, necklace, collar, wreath.

hor'ridus, -a, -um Very prickly.

horten'sis, -is, -e; hor'torum; hortula'nus, -a, -um Of or pertaining to gardens.

hortula'norum Of gardeners.

Hosa'ckia f. Herbaceous plants named in honour of David Hosack (1769–1835), professor of botany and materia medica at Columbia College. In 1801 he purchased land, part of which is now occupied by the Rockefeller Centre, to form the Elgin Botanic Garden which he maintained at his own cost. Nine years later the Garden was transferred to the state of New York under the control of the College of Physicians and Surgeons which, in 1814, became part of Columbia University.

Hos'ta f. Plantain lily. Named for Nicholaus Tomas Host (1761–1834), physician to the Emperor of Austria.

Hotto'nia f. Water-violet. In honour of Peter Hotton (1648–1709), Dutch physician and botanist, professor at Leyden.

Houlle'tia f. Epiphytic orchids. Named for M. Houllet (1811–1890), head gardener at the Jardin des Plantes, Paris.

Housto'nia f. Bluet. Named for Dr. William Houstoun (1695–1733), Scottish surgeon, who collected plants in Central America and the West Indies, making many drawings and notes which are now in the British Museum (Natural History), London.

Houttuy'nia f. An aquatic perennial named for Martin Houttuyn (1720–1794), Dutch naturalist.

Ho'vea f. Evergreen flowering shrubs named in honour of Anton Pantaleon Hove, 18th-century Polish botanist and collector for the Kew botanic garden and Sir Joseph Banks in West Africa, India and the Crimea.

Hove'nia f. Japanese raisin tree. Named for David ten Hove (1724–1787), Dutch senator, who helped to finance Thunberg's collecting expedition in 1772–1776 to South Africa, Java and Japan.

Ho'wea f. From Lord Howe Island, east of Australia, where alone this genus of palms is found.

howel'lii In honour of Thomas Howell (1842–1912), pioneer investigator and industrious collector of the flora of Oregon and Washington, western North America.

Hoy'a f. Wax-flower. Climbing flowering evergreen plants named for Thomas Hoy (*fl.* 1788–1809), 18th-century head gardener to the Duke of Northumberland at Syon House, Isleworth, by the Thames opposite the Royal Botanic Gardens, Kew.

Hudso'nia f. Beach heather. Named for William Hudson (1730–1793), London apothecary and Praefectus of the Chelsea Physic Garden, author of *Flora Anglica*.

Huer'nia f. Succulent dwarf perennials named in honour of Justin Heurnius (1587–1652), Dutch missionary and the first collector of Cape of Good Hope plants. The genus was named by Robert Brown who sailed to Australia with Flinders and who originated the misspelling.

hugo'nis Specific epithet of *Rosa hugonis* in honour of Father Hugh Scallon, missionary in West China, from 1890–1900.

Huma'ta f. L. *humatus*, covered with earth, from L. *humus*, soil; apparently in allusion to the terrestrial rather than epiphytic habit of these ferns.

Hu'mea f. Herbs and shrubs named for Lady Amelia Hume (1751–1809), wife of Sir Abraham Hume of Wormleybury, Hertfordshire, and a pupil of J. E. Smith. *Humea elegans* is a synonym of *Calomeria amaranthoides*, which had been published a month or so earlier.

humifu'sus, -a, -um Sprawling on the ground.

hu'milis, -is, -e Low-growing; more dwarfish than most of its kindred.

humulifo'lius, -a, -um Hop-leaved.

Hu'mulus m. Hop. Apparently Latinized from a Low German name for the hop. There are two species widely grown commercially—*H. americanus* and *H. lupulus*, the European hop (*lupulus*, meaning a small wolf, an allusion to the plant's habit of smothering the trees over which it grows). The principal use is in brewing, the ripe flower heads or cones being dried and used to supply aroma and the bitter flavour as well as to assist in the technical processes of brewing. Dried hops can be used to raise bread and are particularly useful when yeast is not available.

hunga'ricus, -a, -um Hungarian.

Hunneman'nia f. A Mexican perennial named for John Hunneman (d. 1839), English bookseller, agent for sale of herbarium specimens and introducer of plants.

hunnewellia'nus, -a, -um For the Hunnewell family of Wellesley, Massachusetts, founders of the Hunnewell Arboretum.

Huntley'a f. In honour of the Rev. J. T. Huntley (*fl.* 1837), an English clergyman and orchid enthusiast.

hupehen'sis, -is, -e From Hupeh, China.

Hu'ra f. Sand-box tree, monkey-dinnerbell. Latin version of the South American name for these curious and ornamental trees.

Hutchin'sia f. Low herbs named in honour of Ellen Hutchins (1785–1815) of Bantry, Ireland, accomplished in cryptogamic (flowerless plant) botany, who made many contributions to *English Botany*.

hyacin'thinus, -a, -um; hyacin'thus, -a, -um Dark purplish-blue; hyacinthine; resembling a hyacinth.

hyacinthiodo'rus, -a, -um Scented like *Hyacinthus odorus*.

Hyacin'thus m. Greek name for hyacinth, a word derived from an earlier non-Greek language (Thraco-pelasgian) and apparently connected with the blue colour of water but, according to mythology, in honour of Hyakinthos, a youth of great beauty, who, it is said, was accidentally killed by Apollo when teaching him to throw the discus. From his blood sprang a plant named in his memory which is thought not to have been our hyacinth but a gladiolus, probably *Gladiolus italicus*, the lower petals of which have markings which might be taken to represent the mournful exclamation 'AI AI' (alas, alas).

hya'linus, -a, -um Transparent or nearly so.

hy'bridus, -a, -um Mixed; hybrid.

Hydrange'a f. Gr. *hydōr*, water; *aggos*, a jar. The fruit of these shrubs is cup-shaped.

hydrangeoi'des Resembling *Hydrangea*.

Hydras'tis f. Golden seal, orange root. Gr. *hydōr*, water. Name suggested by the leaf of *Hydrophyllum canadense*, water-leaf, with which *Hydrastis* was confused.

Hydriaste'le f. Gr. *hydōr*, water; *stēlē*, a column. In the wild the trees are often near springs.

Hydril'la f. Said to be probably an ill-formed diminutive of *hydra*, water serpent. A widely distributed aquatic plant.

Hydro'charis f. Gr. *hydōr*, water; *charis*, grace. They are pretty water plants.

Hydro'cleys f. Water poppy. Gr. *hydōr*, water; *kleis*, a key. There is no clear reason for the allusion.

Hydroco'tyle f. Water pennywort. Gr. *hydōr*, water; *kŏtylē*, a small cup; from the form of the leaves in *H. umbilicus*.

Hydro'lea f. Gr. *hydōr*, water; L. *olea* (Gr. *elaia*), olive; from its habitat near water and some resemblance of the leaves of *H. spinosa* to those of the olive.

Hydromy'stria f. From Gr. *hydōr*, water; *mystriŏn*, a little spoon. In allusion to the leaves of this South American water plant.

Hydrophyl'lum n. Water-leaf. Gr. *hydōr*, water; *phyllon*, leaf. The stems and leaves of the original species are very watery.

Hydros'me f. Gr. *hydra*, water snake; *osmē*, smell; in cryptic allusion to the evil smell of these plants when in flower.

hyema'lis, -is, -e Of winter; flowering in winter.

hylae'us, -a, -um Belonging to woods.

Hyloce'reus m. Gr. *hyle*, a wood; *cereus*, cactus. These epiphytes prefer some shade.

Hylome'con f. Gr. *hyle*, a wood; *mecon*, poppy.

hylo'philus, -a, -um Loving woods.

hymen- In compound words signifying membranous. Thus:
    hymenan'thus   with flowers bearing a membrane
    hymeno'des   resembling a membrane
    hymenorrhi'zus   membranous rooted
    hymenose'palus   with membranous sepals

Hymenae'a f. After Hymen, the Greek god of marriage, in allusion to the twin leaflets on these tropical evergreen trees.

Hymenanthe'ra f. Gr. *hymēn*, a membrane; the anthers of these shrubs or small trees are terminated by a membrane.

Hymenocal'lis f. Spider lily. Gr. *hymēn*, a membrane; *kallos*, beauty; in allusion to the membrane uniting the stamens.

Hymenophyl'lum From Gr. *hymēn*, a membrane; *phyllum*, leaf; in allusion to the membranous fronds of these ferns.

Hymenospo'rum n. Gr. *hymēn*, a membrane; *spora*, a seed. The seeds of this ornamental Australian evergreen have a membranous wing.

Hyophor'be f. Pignut palm. Gr. *hys*, a pig; *phorbe*, food. The fleshy fruits are eaten by pigs.

Hyoscy'amus m. Henbane. The Greek name meaning pigbean, the epithet 'pig' being used here purely in a derogatory sense, as this plant is highly poisonous and yields hyoscyamine which has certain medical applications as a narcotic.

hyperbo'reus, -a, -um Far-northern.

hypericifo'lius, -a, -um With leaves like *Hypericum*.

hypericoi′des Resembling *Hypericum.*

Hyperi′cum n. St. John's wort. The Greek name *hypereikon* is from *hyper*, above; *eikon*, picture; the plant was supposed to keep evil spirits at bay, in revenge for which the Devil pierced the leaves with a needle. The flowers of some species were placed above images to ward off evil at the ancient midsummer festival Walpurgisnacht, which later became the feast of St. John (24 June), when they are in flower.

Hyphae′ne f. Gr. *hyphaino*, to entwine; in allusion to the fibres of the fruit on these palms.

hypnoi′des Mosslike.

hypo- In compound words signifying under. Thus:

hypoglau′cus	glaucous beneath
hypoleu′cus	white beneath
hypophae′us	dusky below
hypophyl′lus	under the leaf

Hypocalym′ma n. Gr. *hypo*, under; *kalymma*, a veil. The calyx falls like a veil or cape from these evergreen shrubs.

Hypochoe′ris f. Cat's ear. Perennial herbs with dandelion-like flowers. A name used by Theophrastus for this or a related genus.

hypochondri′acus, -a, -um Of melancholy appearance; with sombre-coloured flowers.

Hypocyr′ta f. From Gr. *hypo*, under; *kyrtos*, bulging, swelling. In allusion to the swollen lower side of the corolla.

hypoga′eus, -a, -um Underground; developing underground.

Hypo′lepis f. Gr. *hypo*, under; *lĕpis*, a scale; from the position of the sori on these ferns.

Hypo′xis f. Star grass, yellow star grass. An old Greek plant name from Gr. *hypo*, under; *oxys*, sharp; transferred to these plants to which it has no particular relevance.

Hypse′la f. From Gr. *hypsēlos*, high; apparently in allusion to the habitat of the type-species high in the Andes.

hyrca′nus, -a, -um From the Caspian area.

hyssopifo′lius, -a, -um With leaves like hyssop.

Hysso′pus m. Hyssop. The classical name for this sweet herb adapted from a Semitic plant name *ēzōb*. The plant to which it is now applied is almost certainly not the hyssop of the Bible, for *Hyssopus* is not indigenous in Palestine. The biblical hyssop is generally thought to be *Origanum maru.*

hys′trix Bristly; porcupine-like.

I

ian'thinus, -a, -um Violet-blue.

iber'icus, -a, -um Of Iberia, i.e. Spain and Portugal, or Georgia, Caucasus, U.S.S.R.

iberi'deus, -a, -um Resembling *Iberis*.

iberidifo'lius, -a, -um *Iberis*-leaved.

Ibe'ris f. Candytuft. Gr. *ibēris*, indicating a plant from Iberia.

Ibicel'la f. Diminutive of *Ibex*, in allusion to the fruit with two long slender upcurved horns.

ibo'ta Japanese name of *Ligustrum ibota*.

Ibo'za f. Latin version of the Kaffir name for this showy shrub.

ica'co Spanish name for the cocaplum (*Chrysobalanus icaco*).

icosan'drus, -a, -um Having twenty stamens (*icos-* in compound words signifies twenty).

idae'us, -a, -um Of Mount Ida of ancient Greece.

ida-mai'a Specific epithet of *Brevoortia* (*q.v.*) after Ida May Burke, daughter of a California stage-coach driver, for whom the plant was named in 1867.

Ide'sia f. A flowering tree with heart-shaped leaves, named for Eberhard (or Evert) Ides, German or Dutch explorer of northern Asia in the service of the Russian Tzar Peter the Great between 1691 and 1695.

ido'neus, -a, -um Suitable; worthy.

ignes'cens; ig'neus, -a, -um Fiery red.

ika'riae From the island of Nikaria (Ikaria), Aegean Sea.

I'lex f. Holly. From the Latin name for the holm oak (*Quercus ilex*).

ilicifo'lius, -a, -um *Ilex-* or holly-leaved.

illecebro'sus, -a, -um Enticing, charming.

Illi'cium n. Allurement; from the enticing aromatic scent of these small trees and shrubs.

illini'tus, -a, -um Smeared; smirched.

illustra'tus, -a, -um Pictured.

illus'tris, -is, -e Brilliant, lustrous.

illy'ricus, -a, -um Of Illyria, or Illyricum, region corresponding roughly to western Jugoslavia and comprising Liburnia and Dalmatia.

imbecil'lis, -is, -e; imbecil'lus, -a, -um, weak, feeble.

imber'bis, -is, -e Without a beard.

imbrica'tus, -a, -um Overlapping in regular order like tiles, e.g. scales.

immacula'tus, -a, -um Spotless.

immer'sus, -a, -um Immersed, growing under water.

Impa'tiens f. Touch-me-not, balsam. L. impatient; in allusion to the violent discharge of the seeds from the pods when ripe.

impa'tiens Impatient; with seed capsules dehiscing as soon as ripe.

impedi'tus, -a, -um Hindered, difficult of passage, tangled.

Impera'ta f. After Ferrante Imperato (1550–1625), an apothecary of Naples. Another name for *miscanthus, q.v.*

impera'tor L. for leader, chief, emperor, used as the epithet for an outstanding species.

imperatri'cis Of the empress, i.e. of the Empress Joséphine (1763–1814) who, as the wife of Napoleon, established an outstanding garden at Malmaison, with rare plants brought from Belgium, Holland and England for its adornment, and employed the celebrated flower-painter P. J. Redouté as her botanical artist. The genera *Josephinia* and *Lapageria* (based on her earlier name Joséphine de la Pagerie) were named in her honour; see p. 277.

imperia'lis, -is, -e Imperial; showy.

imple'xus, -a, -um Tangled.

impres'sus, -a, -um Sunken or impressed, as veins may be.

impu'dicus, -a, -um Shameless; lewd.

inaequa'lis, -is, -e Unequal; in compound words, *inaequi-*, e.g. *inaequidens*, with unequal teeth.

inaper'tus, -a, -um Not open; closed.

inca'nus, -a, -um Hoary; quite grey.

incarna'tus, -a, -um Flesh-coloured.

Incarvil'lea f. Showy herbs named after Fr. Pierre d'Incarville (1706–1757), French missionary in China and botanical correspondent of the great botanist Bernard de Jussieu, by whom various north Chinese plants, including *Lycium barbarum*, were introduced into European gardens.

incer'tus, -a, -um Doubtful; uncertain.

incisifo'lius, -a, -um With cut leaves.

inci'sus, -a, -um Incised; deeply and irregularly cut.

inclina'tus, -a, -um Bent down.

incompara'bilis, -is, -e Incomparable.

incomp'tus, -a, -um Unadorned.

inconspi'cuus, -a, -um Inconspicuous.

incrassa'tus, -a, -um Thickened.

incurva'tus, -a, -um; incur'vus, -a, -um Bent inward, as in the rays of some chrysanthemums.

in'dicus, -a, -um Literally of India but also applied to plants originating throughout the East Indies and from as far away as China. *Rosa indica*, for instance, came from near Canton. It seems as if any plant that came home in an Indiaman might be given this specific epithet without further ado—to the confusion of future generations.

Indigo'fera f. Indigo. From indigo; plus L. *fero*, to bear. A widely distributed genus in tropical and subtropical countries, from which the dye indigo is obtained. There are some handsome flowering plants belonging to this genus.

indivi'sus, -a, -um Undivided.

indura'tus, -a, -um Hard, hardened.

ine'brians Intoxicating.

iner'mis, -is, -e Unarmed; without prickles.

infaus'tus, -a, -um Unfortunate.

infecto'rius, -a, -um Dyed; coloured.

infes'tus, -a, -um Dangerous; troublesome.

infla'tus, -a, -um Swollen; inflated.

infle'xus, -a, -um Bent inwards.

infortuna'tus, -a, -um Unfortunate, hence applied to poisonous plants.

infrac'tus, -a, -um Curving inward.

infundibulifor'mis, -is, -e Funnel- or trumpet-shaped.

In'ga f. The Latinized West Indian name for these acacia-like trees and shrubs.

in'gens Enormous.

ingramii In honour of Douglas C. Ingram (1882–1929), American forester, killed in a forest fire in Washington, U.S.A., or Collingwood Ingram (b. 1880), English traveller, ornithologist and horticulturist, author of *Birds of the Riviera* (1926), *Isles of the Seven Seas* (1936), *Ornamental Cherries* (1948) and *Garden of Memories* (1970).

innomina'tus, -a, -um Unnamed.

inno′xius, -a, -um Harmless, e.g. without prickles.

inodo′rus, -a, -um Unscented.

inopi′nus, -a, -um Unexpected.

inorna′tus, -a, -um Without ornament; unadorned.

in′quinans Stained; flecked.

inscrip′tus, -a, -um Inscribed; marked with what might seem to be letters.

insig′nis, -is, -e Distinguished; remarkable.

insi′pidus, -a, -um Tasteless; insipid.

insiti′tius, -a, -um Grafted.

insub′ricus, -a, -um Relating to southern Switzerland (Ticino) and adjacent north Italy.

insula′nus, -a, -um; insula′ris, -is, -e Pertaining to islands, growing on an island.

intac′tus, -a, -um Untouched; intact.

in′teger, inte′gra, -um Entire.

integrifo′lius, -a, -um With entire or uncut leaves.

interme′dius, -a, -um Intermediate in colour, form, or habit.

interrup′tus, -a, -um Interrupted; not continuous, as with scattered leaves or flowers.

intertex′tus, -a, -um Intertwined.

intor′tus, -a, -um Twisted.

intrica′tus, -a, -um Tangled.

intror′sus, -a, -um Turned inward toward the axis.

intumes′cens Swollen.

intyba′ceus, -a, -um Chicory-like.

in′tybus Specific epithet for chicory; from the Latin name *intubus* for wild chicory.

I′nula f. The Latin name for these herbaceous perennials. *I. helenium* is elecampane.

inunda′tus, -a, -um Flooded, hence growing in places liable to flooding.

inu′tilis, -is, -e Useless.

inver′sus, -a, -um Turned over.

invi′sus, -a, -um Unseen; not visible.

involucra′tus, -a, -um Provided with an involucre, a ring of bracts surrounding several flowers.

involu'tus, -a, -um Rolled inward, as leaves in a bud.

Iochro'ma n. Gr. *iŏn*, violet; *chrōma*, colour; from the colour of the flowers of these shrubs or small trees.

io'des Violet-like, from Gr. *iŏn*, violet. Rust-like, rust-coloured, from Gr. *iŏs*, rust.

ioen'sis, -is, -e From Iowa, U.S.A.

ionan'drus, -a, -um With violet-coloured stamens.

ionan'thus, -a, -um With violet-coloured flowers.

Ioni'dium n. Gr. *iŏn*, violet; *eidos*, a resemblance. The flowers resemble violets.

Ionopsid'ium n. Gr. *iŏn*, violet; *opsis*, like. These annuals somewhat resemble tufted violets.

iono'pterus, -a, -um Violet-winged.

Iphei'on n. Derivation obscure. *I. uniflorum* is the spring starflower often called *Brodiaea* (or *Triteleia*) *uniflora*.

Ipomoe'a f. Morning-glory. From Gr. *ips*, worm; *hŏmoios*, similar to.

Iresi'ne f. Apparently a contraction of Gr. *ĕirĕsiōnē*, a branch wound round with wool, from Gr. *ĕiros, ĕirĕos*, wool. These herbaceous plants have brilliantly coloured leaves and woolly flowers.

Iriar'tea f. South American palms commemorating Bernardo de Yriarte, 18th-century Spanish patron of learning.

i'ricus, -a, -um Irish.

irides'cens Iridescent.

iridiflor'us, -a, -um Iris-flowered. Also:
    iridifo'lius, -a, -um   iris-leaved
    iridoi'des   resembling iris

I'ris f. Rhizomatous and bulbous perennials named for the Greek goddess of the rainbow.

irregula'ris, -is, -e With parts of dissimilar size.

irri'guus, -a, -um Watered, soaked.

ir'ritans Causing discomfort.

irrora'tus, -a, -um Bedewed, dew-sprinkled, hence minutely spotted.

isabelli'nus, -a, -um Yellowish, tawny yellow.

isan'drus, -a, -um With equal stamens.

I'satis f. Woad. The classical Greek name. *I. tinctoria* provided the blue dye used by ancient Britons to stain their bodies. Until recently it was cultivated on a small scale in Britain as one of the best of the blue vegetable dyes. It is a handsome garden perennial, both in flower and seed.

Iser'tia f. South American shrubs and trees named for Paul Erdmann Isert (1757–1789), a German surgeon.

islan'dicus, -a, -um Of Iceland.

Isolo'ma n. Gr. *isos*, equal; *lōma*, a border. The lobes of the corolla are equal on these tropical herbaceous plants.

isophyl'lus, -a, -um With equal-sized leaves.

Isople'xis f. Gr. *isos*, equal; *plĕkō*, to plait, devise, contrive. Application to these shrubs related to foxgloves is not clear, but apparently it refers to the almost equal upper and lower lips of the corolla being 'equally contrived'.

Isopo'gon m. Gr. *isos*, equal; *pōgōn*, a beard. The inflorescence of these Australian evergreen shrubs has a beardlike fringe.

Isopy'rum n. Gr. *isos*, equal; *pyros*, wheat. The ancient Greek name of a *Fumaria*, probably transferred to these low-growing perennial herbs on account of similar foliage.

Iso'toma f. Gr. *isos*, equal; *tŏma*, a section. The segments of the corolla are equal in these *Lobelia*-like herbs.

Isotre'ma n. Gr. *isos*, equal; *trēma*, hole; in allusion to the almost regular calyx-limb around the open mouth of the flower, which distinguishes this group from *Aristolochia* proper.

istri'acus, -a, -um Of Istria.

ita'licus, -a, -um Of Italy.

I'tea f. Greek name for the willow, applied to these trees and shrubs of the saxifrage family.

iteophyl'lus Willow-leaved.

I'xia f. African corn lily. Gr. *ixia*, bird-lime; from the sticky sap of these South African plants.

ixioi'des *Ixia*-like.

Ixioli'rion n. *Ixia* plus *leirion*, a lily; in allusion to the similarity to *Ixia*.

ixocar'pus, -a, -um With sticky or glutinous fruit.

Ixo'ra f. Portuguese rendering of Sanskrit *Israra*, lord, referring to the god Siva.

J

Jabosa f. Latinized form of Malay name *jambu* for the rose-apple of Indo-Malaya.

Jacaran'da f. The Latinized Brazilian name for these flowering trees and shrubs.

jack'ii Specific epithet honouring John George Jack (1861–1949), Canadian dendrologist at the Arnold Arboretum, Boston, Massachusetts.

jackma'nii Specific epithet of *Clematis*. Named after G. Jackman of Woking, Surrey, c. 1865.

jacobae'us, -a, -um Specific epithet applied to various plants, in honour of St. James (Jacobus), whose saint's day is 25 July, or referring to the island of St. Iago, Cape Verde Islands.

Jacobi'nia f. Tubular-flowered plants possibly named for the town of Jacobina near Bahia, Brazil.

Jacquemon'tia f. Herbs or sub-shrubs related to *Convolvulus* named for Victor Jacquemont (1801–1832), French naturalist and traveller in the West Indies and in India, where he died.

jacqui'nii In honour of Nicolaus Joseph von Jacquin (1727–1817), Austrian botanist of Dutch origin, who travelled extensively in the Caribbean region as a young man and later published many sumptuous folio botanical works when a professor at Vienna.

jala'pa Of Jalapa (once Xalapan) Veracruz, Mexico. Specific epithet for four-o'clock (*Mirabilis jalapa*) from which the drug jalap was once supposed to be derived. It comes, in fact, from *Ipomoea purga*.

jamaicen'sis, -is, -e From Jamaica.

Jame'sia f. A white-flowering shrub named in honour of Dr. Edwin James (1797–1861), American botanical explorer of the Rocky Mountains area.

Jankae'a f. Balkan plant related to *Ramonda* commemorating the Hungarian botanist and cavalry officer Victor de Janka (1837–1890).

japo'nicus, -a, -um Japanese. Japonica is a common name for both Camellia and Japanese quince.

Jasi'one f. Sheep's-bit scabious. Greek name for another plant.

jasmi'neus, -a, -um; jasminoi'des Resembling jasmine. Also *jasminiflorus*—jasmine-flowered.

Jasmi'num n. Jasmine (sometimes called jessamine). The Latin version of the Persian name *yasmin* for these sweet-scented shrubs and climbers.

Jat′ropha f. Gr. *iatros,* a physician; *trophe,* food. Some species of these herbs, shrubs, or trees have medicinal value.

java′nicus, -a, -um From Java.

Jefferso′nia f. Twin leaf. Attractive small herbs named in honour of Thomas Jefferson (1743–1826), President of the United States, 1801–1809. A truly great man of many parts, he was an outstanding plantsman and patron of botany with deep interest in horticulture and farming.

jonquil′la Variant of Spanish vernacular name *junquillo,* from *Juncus,* rush, referring to the rush-like leaves.

Jovella′na f. Herbs or sub-shrubs named for Gaspar Melchor de Jovellanos, 18th-century patron of botany in Peru.

Jua′nia f. A palm. From Juan Fernandez, a rocky archipelago some 400 miles west of Valparaiso, Chile.

Jubae′a f. A palm named for King Juba of Numidia who committed suicide in 46 B.C. when his ancient kingdom in North Africa was absorbed as a province by the Romans.

juba′tus, -a, -um Crested, hence with long awns.

jucun′dus, -a, -um Agreeable; pleasing.

juga′lis, -is, -e; jugo′sus, -a, -um Yoked.

Ju′glans f. Walnut. The Latin name. From *jovis,* of Jupiter; *glans,* an acorn.

ju′liae In honour of Julia Mlokosewitsch, who discovered *Primula juliae* in the Caucasus in 1900.

julia′nae In honour of Juliana Schneider, first wife of C. K. Schneider.

jun′ceus, -a, -um Rushlike.

Jun′cus, m. Rush. The classical Latin name for the bulrush.

juniperifo′lius, -a, -um Juniper-leaved.

juniperi′nus, -a, -um Juniper-like; bluish-brown like juniper berries.

Juni′perus f. Latin name for the juniper. Apart from ornament, a principal use of juniper is as the source of the berries which flavour gin.

Juri′nea f. In honour of André Jurine (1780–1804), son of the Geneva professor Louis Jurine, student of medicine and botany.

Jussiae′a f. Primrose willow. Tropical herbs or shrubs named for Bernard de Jussieu (1699–1777), French botanist, who laid the foundation for the natural system of plant classification. This genus is now included in *Ludwigia*.

Justi'cia f. Herbs, some of them showy, named for James Justice (*fl.* 1730–1763), celebrated 18th-century Scottish horticulturist.

Juttadin'tera f. In honour of Jutta Dinter, wife of Professor K. Dinter (see *Dinteranthus*). This was once included in *Mesembryanthemum*.

K

Kadsu'ra f. The Japanese name for this evergreen twining shrub.

Kaempfe'ria f. Tuberous fleshy-rooted plants named for Engelbert Kaempfer (1651–1716), a very learned and adventurous German physician, who travelled widely throughout the East, lived for two years in Japan, and wrote a book *Amoenitates exoticae*, which includes accurate descriptions of hitherto unknown Japanese plants, as well as a posthumously published history of Japan.

kaki Japanese name for *Diospyros kaki*, in full, *kaki-no-ki*; in no way connected with 'khaki'.

Kalan'choe f. From the Chinese name of one species of these succulent herbs or sub-shrubs.

Kal'mia f. Laurel—mountain, sheep, swamp, etc. Calico bush. Named in honour of Pehr Kalm (1715–1779), Finnish pupil of Linnaeus, who was sent by the Swedish government in 1748 to report on the natural resources of North America. His descriptions of the domestic economy and natural history are both interesting and trustworthy.

kalmiiflo'rus, -a, -um With flowers like *Kalmia*.

Kalmio'psis f. Western American shrub named from its resemblance (*opsis*) to *Kalmia*.

Kalo'panax m. Gr. *kalos*, beautiful; *Panax*, name of a related genus.

kamtscha'ticus, -a, -um From Kamchatka.

kansuen'sis, -is, -e Of Kansu, the north-western province of China whence Reginald Farrer introduced some good hardy plants.

karatavien'sis, -is, -e Of the Kara Tau (black mountain) range in Kazakhstan, Central Asia, U.S.S.R.

kashmiria'nus, -a, -um Of Kashmir.

keiskea'nus, -a, -um; keiskei In honour of Keisuke Ito (1803–1901), celebrated Japanese physician and botanist, who as a young man studied under von Siebold at Nagasaki, promoted western medicine in

Japan and became professor of botany at Tokyo, honoured in his 98th year as one of the 'Twelve Heroes of Modern Japan'.

kele′ticus, -a, -um Gr. *kēlētikos*, charming.

kellog′gii In honour of Albert Kellogg (1813–1887), American physician and pioneer Californian botanist.

Kel′seya f. A small shrub with densely imbricated leaves named for Harlan P. Kelsey (1872–1958), nurseryman of Boxford, Massachusetts, U.S.A.

Kenne′dia f. Trailing or climbing perennials named for John Kennedy (1775–1842), the original partner of Lee and Kennedy, nurserymen of Hammersmith, London.

kermesi′nus, -a, -um Carmine.

Ker′nera f. Perennial herbs named for Johann Simon von Kerner (1755–1830), professor of botany at Stuttgart.

ker′neri In honour of Anton Joseph Kerner von Marilaun (1831–1898), professor of botany at Vienna, author of a monumental *Pflanzenleben* or *Natural History of Plants* which remains a treasure-house of botanical information.

Ker′ria f. Jew's mallow. A showy flowering shrub named for William Kerr (d. 1814), Kew gardener and collector, who collected in China, Java and the Philippines.

Ketelee′ria f. Chinese evergreen trees named in honour of Jean Baptiste Keteeler (1813–1903), horticulturist of Brussels.

kewen′sis, -is, -e Of the Royal Botanic Gardens, Kew, England; mostly applied to interspecific hybrids which originated there.

khasia′nus, -a, -um From the Khas; hills of Assam, India.

Kick′xia f. Annual creeping herbs named for Jean Kickx (1775–1831), apothecary of Brussels and writer on cryptogamic (flowerless) plants.

Kige′lia f. Sausage-tree. Latinized from the Mozambique vernacular name *kigeli-keia*.

Kingdon-wardia f. In honour of F. Kingdon-Ward. See *wardii*.

Kirengesho′ma f. Japanese name for a tall yellow-flowered perennial from *ki*, yellow; *renge*, lotus blossom; *shoma*, hat; *Rengeshoma* being the Japanese vernacular name for *Anemonopsis macrophylla*.

kir′kii Specific epithet in honour of Thomas Kirk (1828–1898), author of *The Forest Flora of New Zealand*, etc. Born in Coventry, England, he worked in a timber mill before migrating to New Zealand where his outstanding botanical work led to his gaining university teaching posts, and finally to his appointment as Chief Commissioner of State Forests. Also in honour of his son Harry Bower Kirk (d.

1948), professor of biology in the University of New Zealand (1903–1944), outstanding naturalist, humanist and teacher. Also used in remembrance of Sir John Kirk (1832–1922), for many years British Consul at Zanzibar, ardent botanist and father of Colonel J. W. C. Kirk, author of *A British Garden Flora.*

Kitaibe'lia f. In honour of Paul Kitaibel (1757–1817), Hungarian botanist, professor of botany at Pest.

kiusia'nus, -a, -um Of Kyushu, the large southern island of Japan.

Klei'nia f. In honour of Jacob Theodor Klein (1685–1759), German botanist in East Prussia.

Klu'gia f. In honour of J. Chr. Fr. Klug (1774–1856), German entomologist, and his daughter.

Knau'tia f. In honour of Christoph Knaut (1638–1694), German doctor and botanist.

Knigh'tia f. Trees and shrubs named for Thomas Andrew Knight (1759–1838), a president of the Horticultural Society of London and a well-known pomologist.

Knipho'fia f. Red-hot poker. Named for Johann Hieronymus Kniphof (1704–1763), professor of medicine at Erfurt and author of a folio of nature-printed illustrations of plants (1747).

Ko'chia f. Summer cypress. A large genus of herbs and sub-shrubs which includes Belvedere, an annual grown for its autumn foliage; named for Wilhelm Daniel Josef Koch (1771–1849), professor of botany at Erlangen.

Koellike'ria f. In honour of Rudolf Albrecht Koelliker (1817–1905), Swiss anatomist, professor at Würzburg.

Koellikohle'ria f. Hybrids of *Koellikeria* and *Kohleria.*

Koelreute'ria f. Deciduous flowering trees named for Joseph Gottlieb Koelreuter (1733–1806), professor of natural history at Karlsruhe, a pioneer experimental investigator of plant hybridization.

Kohle'ria f. In honour of Michael Kohler, 19th-century teacher of natural history at Zürich, Switzerland.

koka'nicus, -a, -um Of Khokand, Turkistan, U.S.S.R.

Kolkwit'zia f. Beauty bush. A deciduous flowering shrub named for Richard Kolkwitz (b. 1873), professor of botany in Berlin *c.* 1900.

kolomik'ta Vernacular name in the Amur region, eastern Asia, for *Actinidia kolomikta.*

kongboen'sis, -is, -e From Kongbo, south-eastern Tibet.

korea'nus, -a, -um (also **korianus**); **koriaen'sis, -is, -e** Of Korea.

Korthal'sia f. Climbing palms named for Peter Willem Korthals (1807–1892), Dutch botanist, from 1831–1836 in Java, Sumatra and Borneo.

Kosteletz'kya f. Perennial herbs or shrubs related to *Hibiscus*, named in honour of Vincenz Franz Kosteletzky (1801–1887) of Prague, writer on medical botany.

kot'schyi; kotschya'nus, -a, -um In honour of Theodor Kotschy (1813–1866), Austrian botanist, plant-collector and traveller in North Africa and western Asia.

kou'sa Japanese name for *Cornus kousa*.

kouytchen'sis, -is, -e Of Kweichow province, central China.

Krame'ria f. Woody plants and perennial herbs named for Johann Georg Heinrich Kramer, Austrian botanist and military doctor in the first half of the 18th century.

Kri'gia f. Dwarf dandelion. Named for David Krieg, 18th-century German physician who collected in Maryland.

Kuh'nia f. Small perennial herbs named for Adam Kuhn (1741–1817), American botanist, professor at Philadelphia, a student of Linnaeus's at Uppsala, 1761–1765.

Kun'zea f. Evergreen shrubs or small trees named for Gustav Kunze (1793–1851), professor of botany at Leipzig.

kur'dicus, -a, -um Referring to the mountainous region of western Asia inhabited by the Kurds.

kur'roo Vernacular name, better rendered as *karu*, of *Gentiana kurroo* and *Picrrorhiza kurrooa*, used in Himalaya as a bitter tonic.

L

labia'tus, -a, -um; labio'sus, -a, -um Lipped, i.e. with well-developed lip.

Labi'chea f. In honour of a French naval officer Labiche (d. 1818).

la'bilis, -is, -e Slippery; unstable.

laburnifo'lius, -a, -um *Laburnum*-leaved.

Laburnocy'tisus m. Compounded name for the graft hybrid between *Cytisus* and *Laburnum*.

Labur'num n. Latin name for these trees, all parts of which are poisonous.

lac Milk. L. *lac*, milk. Thus:

lactes'cens becoming milky

lac′teus milk-white
lacti′color milk-white
lac′tifer producing milky sap
lactiflo′rus milk-white flowers

la′cer, la′cera, -um Torn or cut into fringelike segments.

Lachena′lia f. Cape cowslip. Named for Werner de la Chenal (1736–1800), professor of botany at Basle.

Lachnan′thes f. Gr. *lachne*, wool, down; *anthos*, a flower; an allusion to the woolly flowers of this eastern United States swamp-loving perennial.

lachno′podus, -a, -um With downy pedicels.

lacinia′tus, -a, -um; lacinio′sus, -a, -um Slashed or torn into narrow divisions.

lac′rimans Weeping.

lac′ryma-jo′bi Job's tears. Specific epithet of a plant in the genus *Coix*. The edible seeds form a dirty-white beadlike structure. ('My face is foul with weeping, and on my eyelids is the shadow of death.' Job 16:16.)

lact- See above under *lac*.

Lactu′ca f. Lettuce. L. *lac*, milk; from the milky sap.

lactucifo′lius, -a, -um Lettuce-leaved.

lacuno′sus, -a, -um With deep holes or pits.

lacus′tris, -is, -e Pertaining to lakes.

lada′nifer, ladani′fera, -um Bearing ladanum (occasionally labdanum), a fragrant resinous juice used in medicine, collected from *cistus*. This is the myrrh of the Bible.

Lae′lia f. A genus of epiphytic orchids named after one of the Vestal Virgins.

Laeliocattley′a f. *Laelia* plus *Cattleya*; a collective designation for bigeneric hybrids derived from these genera.

lae′tus, -a, -um Bright; vivid. Also: *laetiflorus*, bright flowered; *laetivirens*, vivid green.

laevicau′lis, -is, -e Smooth-stemmed.

laeviga′tus, -a, -um; lae′vis, -is, -e Smooth.

Lagarosi′phon m. From Gr. *lagaros*, thin, narrow; *siphōn*, tube.

Lagenan′dra f. From Gr. *lagēnos*, flask; *anēr*, man, male organ. In allusion to the peculiar anther-cells prolonged upwards into a slender tube like the neck of a flask.

Lagena′ria f. Bottle-gourd. Gr. *lāgēnos*, a flask.

Lagerstroe'mia f. Showy flowering trees and shrubs named for the Swedish merchant Magnus von Lagerström (1691–1759) of Göteborg by his friend Linnaeus.

lag'geri In honour of Franz Josef Lagger (1802–1870), Swiss physician and botanist.

lagodechia'nus, -a, -um From Lagodechi in the Caucasus.

Lago'tis f. From Gr. *lagos*, hare; *ŏus, ōtŏs*, ear.

Laguna'ria f. A handsome flowering tree named for Andréa Laguna (1494–1560), Spanish botanist.

Lagu'rus m. Hare's-tail grass. From Gr. *lagos*, a hare; *oura*, a tail, referring to the hairy inflorescence.

Lalleman'tia f. In honour of Julius Leopold Avé-Lallemant (1803–1867), of Lübeck, German botanist, for a time an official of the botanic garden, Leningrad, who travelled in South America.

Lamar'ckia f. Ornamental annual grass. Named for the Chevalier Jean Baptiste de Monet Lamarck (1744–1829), distinguished French naturalist and author of the botanical part of the *Encyclopédie méthodique* and *Flore française*; his work on evolution was an important forerunner of Darwin's.

Lamber'tia f. Australian shrubs named for Aylmer Bourke Lambert (1761–1842), English botanist, author of *The Genus Pinus* (1804).

lamella'tus, -a, -um Layered, i.e. made of or possessing thin layers of tissue.

La'mium n. Dead-nettle. The Latin name.

Lampran'thus m. From Gr. *lampros*, shining, glossy; *anthos*, flower. A genus of succulent plants, formerly included in *Mesembryanthemum*.

lana'tus, -a, -um Woolly.

lanceola'tus, -a, -um Lanceolate, spear-shaped, i.e. of narrow shape with curved sides tapering to a point.

lan'ceus, -a, -um Spear-shaped. Thus: *lancifolius*, lance-leaved.

lan'iger, lani'gera, -um; lano'sus, -a, -um; lanugino'sus, -a, -um Woolly.

lannesia'nus, -a, -um In honour of Lannes de Montebello, a Frenchman who sent plants from Japan to France about 1870.

Lanta'na f. Late Latin name for *Viburnum*, transferred to this genus. Mostly Western Hemisphere shrubs. Since introduction into India and elsewhere as ornamentals they have established themselves as difficult weeds in tea gardens and other cultivated areas.

Lapage'ria f. An evergreen twining shrub with bell-shaped flowers named after Joséphine de La Pagerie, the Empress Joséphine who as a girl lived on the La Pagerie estate in Martinique.

Lapeirou'sia f. Cormous flowering plants named after Baron Philippe Picot de la Peyrouse (1744–1818) of Toulouse, French botanist, author of a flora of the Pyrenees. *Davallia lapeyrousii* is, however, named after Jean François de Galaup, Comte de la Pérouse (1741–1788), a French admiral and explorer. After active service in the Seven Years and American Revolutionary wars, he sailed in 1785 with two ships to continue Cook's exploration of the Pacific, and, following the example of Cook and Banks, he carried with him a complement of scientists. He disappeared in 1788, the wrecks of his ships being found in 1826 off the Santa Cruz Islands.

lapha'mii In honour of Increase Allen Lapham (1811–1875), American naturalist.

Lapor'tea f. Named after an obscure Frenchman, possibly a companion of Gaudichaud on the voyage of the corvette *L'Uranie*, 1817–1820.

lappa Latin name for a bur.

lappa'ceus, -a, -um Bur-like.

lappo'nicus, -a, -um Of Lapland.

Lardizaba'la f. Evergreen flowering climbers named for Miguel Lardizábal, Spanish patron of botany.

larici'nus, -a, -um Resembling larch. Also *laricifolius*, larch-leaved.

La'rix f. Larch. The classical name for the tree highly valued for centuries for its tough, strong wood, particularly useful in underwater work.

Lar'rea f. Creosote bush. Evergreen shrubs named for Juan Antonio de Larrea, Spanish patron of science, *c.* 1810.

Laserpi'tium n. The classical Latin name for these perennial herbs.

lasi- In compound words signifying woolly. Thus:

lasiacan'thus	woolly-spined
lasian'drus	woolly-stamened
lasiocar'pus	woolly-fruited
lasiodon'tus	woolly-toothed
lasioglos'sus	having a rough or hairy tongue
lasiope'talus	woolly-petalled
lasio'podus	woolly-footed

Lasthe'nia f. Yellow-flowered glabrous herbs from California, given the name of the girl Lasthĕnïa who, in order to attend Plato's classes, dressed as a boy.

Las'trea f. In honour of Charles Jean Louis Delastre (*c.* 1792–1859), author of a flora of the Département de la Vienne, France.

Lata'nia f. From the Mauritius vernacular name for these palms.

latebro'sus, -a, -um Pertaining to dark or shady places.

latera'lis, -is, -e At the side. In compound words *lateri-*, as in *lateripes* (with stalk on the side).

lateri'tius, -a, -um Brick-red.

la'thyris Greek name of a kind of spurge (*Euphorbia*).

La'thyrus m. Sweet pea, beach pea, everlasting pea. The Greek name *lathyros* for pea or pulse. Sweet pea is *L. odoratus.* The edible garden pea comes under *Pisum* (*P. sativum*).

la'tus Broad; wide. In compound words *lati-*. Thus:

latiflo'rus	broad-flowered
latifo'lius	broad-leaved
lati'frons	with broad fronds
latilab'rus	wide-lipped
latilo'bus	wide-lobed
latimacula'tus	with broad spots
la'tipes	with a broad stalk
latispi'nus	with wide thorns
latisqua'mus	with wide scales

laudan'dus, -a, -um; lauda'tus, -a, -um Praiseworthy.

Laure'lia f. Bay-tree. Latinization of Spanish vernacular name *laurel*, from L. *laurus*, transferred to this South American plant on account of the fragrance of the leaves.

laurifo'lius, -a, -um Bay-leaved.

lauri'nus, -a, -um Resembling *Laurus*, true laurel, or bay-tree.

Lauroce'rasus f. Cherry-laurel. From L. *laurus*, laurel; *cerasus*, a cherry.

Lau'rus f. Latin name for the laurel or bay. The laurel crown was of bay leaves. The sweet bay provides the leaves so essential in good cookery.

Lavan'dula f. Lavender. From L. *lavo*, to wash; from its use in soaps and toiletries of various kinds.

lavandula'ceus, -a, -um Lavender-violet.

lavandulifo'lius, -a, -um With lavender-like leaves.

Lavate'ra f. Tree-mallow. Named for J. R. Lavater, 16th-century physician and naturalist in Zürich.

lavateroi'des Resembling *Lavatera*.

Lawson'ia f. Henna. Named by Linnaeus to honour his friend and patron Dr. Isaac Lawson (d. *c.* 1747), Scottish army doctor who, while in Leyden, helped to pay for the publication of Linnaeus's *Systema*

Naturae (1735). Despite statements by authors who should have known better, he was not the brother of John Lawson, surveyor-general of North Carolina and author of *The History of North Carolina.*

lawsonia′nus, -a, -um For Charles Lawson (1794–1873), author of *Pinetum Britannicum,* to whose nursery in Edinburgh *Cupressus lawsoniana* was introduced from western North America in 1854.

la′xus, -a, -um Loose or open. Thus, *laxiflorus, laxifolius,* loose-flowered and loose-leaved.

La′yia f. Annual herbs with yellow or white flowers, named for George Tradescant Lay (d. *c.*1845), naturalist on the voyage of exploration made in 1825–1828 by Frederick William Beechey to the Pacific North-west and the Bering Straits.

la′zicus, -a, -um Epithet for an iris found in Lazistan on the shores of the Black Sea; closely related to *Iris unguicularis,* it flowers in winter and grows well in damp, peaty conditions.

ledifo′lius, -a, -um With leaves resembling *Ledum.*

Le′dum n. Labrador tea, wild rosemary, etc. From Gr. *lēdon,* cistus.

Le′ea f. Tropical small trees or shrubs named for James Lee (1715–1795), nurseryman of Hammersmith, London.

leian′thus, -a, -um Smooth-flowered. Also:
leiocar′pus smooth-fruited
leio′gynus with a smooth pistil
leiophyl′lus with smooth leaves

leichtli′nii For Max Leichtlin (1831–1910) of Baden-Baden, who introduced many plants into cultivation, notably from the Near East.

Leiophyl′lum n. Sand-myrtle. From Gr. *leios,* smooth; *phyllon,* a leaf; from the glossy foliage.

Leitne′ria f. Corkwood. Named in memory of Edward F. Leitner (d. 1838), German physician and naturalist killed by Indians when collecting in Florida. *L. floridana,* a shrub, is the only member of the family *Leitneriaceae.*

Lemaireoce′reus m. Named for Charles Lemaire (1801–1871), French specialist in cacti.

Lem′na f. Duckweed. The Greek name for some water weed.

lemoi′nei For Victor Lemoine (1823–1911), and his son Emile Lemoine (1862–1942) of Nancy, France, nurserymen and plant-breeders, who introduced many new lilacs, peonies and other hybrids, including the bigeneric × *Heucherella tiarelloides.*

Lens f. Lentil. The classical name for this ancient pulse. Esau sold Jacob his birthright for a red-lentil pottage.

lenticula'ris, -is, -e; lentifor'mis, -is, -e Shaped like a lens.

lentigino'sus, -a, -um Freckled.

len'tus, -a, -um Tough but pliant.

leo'nis Coloured or toothed like a lion.

Leono'tis f. Gr. *lĕōn*, a lion; *ŏus, ōtis*, an ear. The corolla of these annual and perennial herbs might be imagined to look like a lion's ear.

Leon'todon m. From Gr. *Lĕōn*, lion; *ŏdŏus*, tooth. In allusion to the toothed leaves of the dandelion (*dent de lion*) to which the name was also applied.

leontoglos'sus, -a, -um With a throat or tongue like a lion.

Leontopo'dium n. Edelweiss. The classical name *lĕŏntŏpŏdĭŏn*, meaning lion's foot, for another plant transferred to this genus.

Leonu'rus m. Motherwort or lion's-tail. From Gr. *lĕōn*, a lion; *ouros*, a tail.

leonu'rus, -a, -um Like a lion's tail.

leopardi'nus, -a, -um Conspicuously spotted like a leopard.

Le'pachys f. Cone-flower. From Gr. *lepis*, scale; *pachys*, thick; from the thickened tips of the involucral scales. This name has now been discarded in favour of the earlier *Ratibida*, the meaning of which is obscure.

Lepi'dium m. Garden cress, pepper-grass. The classical name.

lepidosty'lus, -a, -um With a scaly style.

lepido'tus, -a, -um Scaly; covered with small scales.

le'pidus, -a, -um Graceful; elegant.

Lepis'mium n. From Gr. *lepis*, scale. A genus of *Cactaceae*, separated from *Rhipsalis*, with small scales surrounding the areoles.

lept- In Greek compound words, signifies thin or slender. Thus:

leptan'thus	slender-flowered
leptocau'lis	thin-stemmed
lepto'cladus	thin-branched
lepto'lepis	thin-scaled
leptope'talus	thin-petalled
leptophyl'lus	thin-leaved
lep'topus	thin-stalked
leptose'palus	thin-sepalled
lepto'stachys	slender-spiked
leptoth'rius, -a, -um	with thin fine leaves
leptu'rus	slender-tailed

Leptoder'mis f. Gr. *lĕptos*, thin; *derma*, skin; from the membranous fused bracteoles forming a sheath or skin covering the calyx.

Lepto'pteris f. Gr. *lĕptos*, slender; *pteris*, a fern.

Lepto'pyrum n. Gr. *lĕptos*, slender; *pyros*, wheat; from the form of the fruits of these perennial herbs.

Leptosper'mum n. Gr. *lĕptos*, slender; *sperma*, seed; referring to the narrow seeds of these evergreen aromatic shrubs.

Lepto'syne f. Gr. for delicacy, slenderness.

Lepto'tes f. Gr. *lĕptŏtēs*, delicateness.

lepturoi'des Like a hare's tail.

Leschenaul'tia f. Ornamental heathlike Australian plants named for Louis Théodore Leschenault de la Tour (1773–1826), French botanist and traveller.

Lespede'za f. Bush clover. Named by the French botanist Michaux (who collected widely in North America) to honour the Spanish governor of Florida, Vincente Manuel de Céspedes, *c.* 1790.

Lesquerel'la f. A genus of herbs named in honour of Leo Lesquereux (1805–1889), foremost authority on American fossil botany in the latter part of the 19th century.

Lettso'mia f. Climbing shrubs named for John Coakley Lettsom (1744–1815), London physician, botanist and philanthropist, an original member of the Horticultural Society of London. He became involved in the dispute surrounding Forsyth's plaister (see *Forsythia*). An advocate of prison reform, he is stated to have accomplished more in twelve months than his predecessors had in thirty years: 'no proposal for the public benefit was ever offered to his consideration in vain'. He was notable for his generosity to the poor and came to the aid of Curtis in 1778 with £500, a very large sum then, towards the publication of the *Flora Londinensis*.

leuc- In compound words signifies white. Thus:

leucan'thus	white-flowered
leucocau'lis	white-stemmed
leucoce'phalus	white-headed
leucochi'lus	white-lipped
leucoder'mis	white-skinned
leuconeu'rus	white-nerved
leucope'talus	white-petalled
leucopha'eus	dusky-white
leucophyl'lus	white-leaved
leucorrhi'zus	white-rooted
leuco'stachys	white-spiked
leuco'trichus	white-haired
leucoxan'thus	whitish-yellow
leuco'xylon	white-wooded

Leucaden'dron n. Cape silver tree. From Gr. *leukos,* white; *dĕndron,* a tree. An allusion to the foliage.

Leucae'na f. Gr. *leukos,* white; in allusion to the flowers of these *Acacia*-like trees and shrubs.

leucanthemifo'lius, -a, -um With leaves like *Leucanthemum,* a group often placed in the genus *Chrysanthemum,* the best known being *C. leucanthemum,* the ox-eye daisy.

leucas'pis With a white shield.

Leuce'ria (incorrectly, **Leucheria**) f. Gr. *leucheres,* white. Some species of these South American herbs are white and woolly.

Leuchtenber'gia f. A genus of *Cactaceae* with one species (*L. principis*) apparently named in honour of Prince Maximilian E. Y. N. von Beauharnais (1817–1852), Duke of Leuchtenberg, Oberpfaz, Germany.

Leucoco'ryne f. From Gr. *leukos,* white; *kŏrynē,* club.

Leuco'crinum m. Sand lily. Gr. *leukos,* white; *crinon,* a lily. A small early-blooming bulbous plant.

Leuco'jum n. Snowflake, St. Agnes' flower. The Greek name for various scented white-flowered plants, from *leukŏn iŏn* (white violet) later transferred to the snowflake.

Leuco'thoe f. Ornamental shrubs named in honour of Leucothoë, one of the many loves of Apollo.

Leu'zia f. Herbaceous perennials named for a friend of de Candolle, Joseph Philippe François Deleuze (1753–1835) of Avignon and Paris.

Levis'ticum n. Lovage. Said to be a corruption of *Ligusticum.*

Lewi'sia f. Bitterwort. Small showy herbaceous perennials from the Rocky Mountains. In honour of Captain Meriwether Lewis (1774–1809), senior leader of the Lewis and Clark expedition, the first coast-to-coast crossing of the American continent.

Leyceste'ria f. Deciduous shrubs with whorled flowers. Named for William Leycester, Chief Justice of Bengal *c.* 1820.

Li'atris f. Unknown derivation. This genus of North American perennial herbs includes plants under many English names: prairie-button, snakeroot, Kansas gay-feather, blue blazing star, etc.

li'bani; libano'ticus, -a, -um Of Mount Lebanon.

libe'ricus, -a, -um From Liberia.

Liber'tia f. Perennial herbs in the iris family, named for Marie A. Libert (1782–1863), Belgian botanist who wrote on liverworts.

Libo'cedrus f. Incense cedar. From Gr. *libos,* tear, drop; *cedrus,* cedar; presumably in allusion to the tear-like gummy exudations and suggested by the name *Dacrydium* for an allied genus.

libur′nicus, -a, -um From Croatia (Liburnia) on the shores of the Adriatic.

liby′cus, -a, -um Referring to Libya, North Africa.

Licua′la f. From the native Moluccan name *leko wala* for these palms.

ligno′sus, -a, -um Woody.

Ligula′ria L. *ligula*, a strap; from the strap-like ray florets of these handsome herbaceous perennials of the daisy family.

ligula′ris, -is, -e; ligula′tus, -a, -um Straplike.

ligusticifo′lius, -a, -um Lovage-leaved.

Ligus′ticum n. Lovage. Named for Liguria, the Italian province in which is the city of Genoa, where the plant abounds.

ligus′ticus, -a, -um Of Liguria (see *Ligusticum*).

ligustrifo′lius, -a, -um Privet-leaved.

ligustri′nus, -a, -um Privetlike.

Ligus′trum n. Latin for privet.

likiangen′sis, -is, -e From the Lichiang range, Yunnan, Western China, a favourite collecting area of George Forrest.

lila′cinus, -a, -um Lilac in colour.

lili- In compound words signifying lily. Thus:

lilia′ceus lilylike
liliflo′rus lily-flowered
lilifo′lius lily-leaved

Li′lium n. Lily. The Latin name akin to Gr. *leirion*, the madonna-lily.

lilliputia′nus, -a, -um Of very small and low growth, suitable for the land of Lilliput.

limba′tus, -a, -um Bordered.

limen′sis, -is, -e From Lima, Peru.

Limnan′thes f. Meadow foam. Gr. *limne*, a marsh; *anthos*, a flower; from the habitat of these American annuals.

Limno′bium n. Gr. *limne*, a marsh; *biŏs*, life.

Limno′charis f. Gr. *limne*, a marsh; *charis*, beauty. From the habitat and beauty of the flowers of these aquatic perennials.

limno′philus, -a, -um Swamp-loving.

limoniifo′lius, -a, -um With leaves like sea-lavender.

Limo′nium n. Sea-lavender, sea-pink. From Gr. *leimon*, a meadow; in allusion to the common habitat in salt meadows.

limo′sus, -a, -um Of marshy or muddy places.

Linanthus

Linan'thus m. Gr. *linon*, flax; *anthos*, a flower. Now often classified under *Gilia*.

Lina'ria f. Toadflax, butter-and-eggs. From Gr. *linon*, flax; from the flaxlike leaves.

linariifo'lius, -a, -um *Linaria*-leaved.

lindaue'anus, -a, -um; lindavia'nus, -a, -um In honour of Gustav Lindau (1866–1923), German botanist who specialized in the study of *Fungi* and *Acanthaceae*.

Lindelo'fia Herbaceous perennials named for Friedrich von Lindelof, of Darmstadt, patron of botany *c.* 1850.

Linde'ra f. Feverbush, Benjamin bush, benzoin. Deciduous and evergreen shrubs and trees with aromatic bark named after Johann Linder, (1676–1723), a Swedish botanist and physician.

Lindhei'mera f. In honour of Ferdinand Jacob Lindheimer (1801–1879), German political exile who made large botanical collections in Texas.

lindleyi In honour of John Lindley (1799–1865), one of the most industrious of British botanists, long associated with the Horticultural Society of London, first professor of botany in London University, a specialist in the study of *orchidaceae*, author and editor of numerous comprehensive botanical and horticultural publications.

Lindsa'ea f. A genus of about 75 species of ferns, named in honour of John Lindsay (*fl.* 1785–1803), surgeon in Jamaica.

linearifo'lius, -a, -um Linear-leaved.

linea'ris, -is, -e Narrow, with sides nearly parallel.

line'atus, -a, -um With lines or stripes.

lin'gua A tongue or tonguelike.

linguifor'mis, -is, -e Tongue-shaped.

lingula'tus, -a, -um Tongue-shaped.

liniflo'rus, -a, -um Flax-flowered.

linifo'lius, -a, -um Flax-leaved.

Linna'ea f. Twin-flower. Named for and by Carl Linnaeus, ennobled as Carl von Linné (1707–1778), and described by him as 'a plant of Lapland, lowly, insignificant, disregarded, flowering but for a brief space—from Linnaeus who resembles it'. Born in Sweden, Linnaeus trained as a physician but soon concentrated on natural history. His *Systema Naturae* (1735) began a series of encyclopaedic works dealing mostly with plants but also with animals, minerals and materia medica, seeking in a very methodical and systematic manner to classify and name the whole organic world, thereby providing a basis for further and wider research.

He was a teacher without peer. His pupils, such as Pehr Kalm (see *Kalmia*), became botanic leaders in their own right. He was a prodigious writer and his published works total more than 180 titles. His garden at Uppsala still attracts pilgrims from all over the world. It was inevitable that his so-called 'sexual system' of classifying plants into major artificial groups based primarily on the number of stamens and pistils should become modified and then replaced by more natural systems based on more characters, but this does not in any way detract from his essential genius. His most permanent contribution to biology was his invention of a consistent binomial system of naming species by means of a single-word generic name followed by a single-word specific epithet. (See pp. 7, 350.)

His 'herborizing lectures' while he walked through lanes and fields at the head of two or three hundred students were social events of the Swedish summer. Attended by a band of trumpets and French horns, his pupils were divided into detached companies to be called together by music when he wished to discourse on any particular subject.

He made a close study of what he called 'floral nuptials', his metaphorical descriptions being taken too literally by his opponents. The petals of flowers he regarded as 'bridal beds which the Creator has so gloriously arranged, adorned with such noble bed-curtains and perfumed with so many sweet scents that the bride-groom may celebrate his nuptials with his bride with all the greater solemnity.' The great Linnaean collections are now at Burlington House in London, having been purchased by Sir James Smith for one thousand guineas in 1784 and later bought by the Linnean Society of London.

The epitaph on Linnaeus's monument in the cathedral at Uppsala describes him simply as 'Princeps Botanicorum'.

linnaea′nus, -a, -um; linnaei In honour of Linnaeus.

Li′num n. Flax. The Latin name. Until cotton came into supply in the 18th century, flax and hemp were the most important vegetable fibre at least as far as Western man was concerned. Indeed, for Western prehistoric man it was the only fibre and, as far as can be determined, the principles of manufacture have remained unchanged. Flax is also the source of linseed.

Lipa′ris f. Twayblade (in America). From Gr. *liparŏs*, oily or smooth; in allusion to the glossy surface of the leaves.

Lip′pia f. Lemon verbena. Named for Augustin Lippi (1678–1701), Italian naturalist and botanist. He was killed in Abyssinia.

Liquidam′bar f. Sweetgum. From L. *liquidus*, liquid; *ambar*, amber.

Liriodendron

A fragrant resin called liquid storax is prepared from the inner bark of one species (*L. orientalis*).

Lirioden'dron n. Tulip tree. From Gr. *leiriŏn*, a lily; *dĕndron*, a tree.

Liri'ope f. A perennial herb named in honour of a Greek woodland nymph, Līriŏpē, the mother of Narcissus.

Lissochi'lus m. From Gr. *lissos*, smooth; *cheilos*, lip.

Listera f. Small terrestrial orchids named for Martin Lister (*c.* 1638–1712), British zoologist, naturalist, physician to Queen Anne. This is sometimes called twayblade in Britain, but more often adder's tongue or sweethearts in the United States.

litangen'sis, -is, -e From Litang, Szechwan, China.

Li'tchi f. From the Chinese name for the handsome tree producing the well-known fruits which are popularly called nuts and spelled in any one of a half-dozen different ways.

Lithocar'pus m. Tanbark or chestnut-oak. From Gr. *lithos*, a stone; *karpos*, fruit. In allusion to the hard acorns.

litho'philus, -a, -um Rock-loving.

Lithophrag'ma n. Californian perennial herbs. From Gr. *lithos*, a stone; *phragma*, a fence.

Li'thops f. Pebble plants. From Gr. *lithos*, a stone; *ops*, like; from the resemblance of these South African succulents to small stones.

Lithosper'mum n. Gromwell. The classical name from Gr. *lithos*, stone; *sperma*, seed; in allusion to the hard nutlets. The plant has been reported as having been used by North American Indians as an oral contraceptive.

lithosper'mus, -a, -um With very hard stonelike seeds.

Lithrae'a f. From the Chilean vernacular name for these evergreen shrubs or small trees.

litien'sis, -is, -e From Litiping, Yunnan, China.

Litse'a f. Latinization of a Japanese name for these widely distributed trees and shrubs from tropical and sub-tropical regions.

Litto'nia f. In honour of Samuel Litton (1779–1847), professor of botany at Dublin.

littora'lis, -is, -e; littore'us, -a, -um Of the seashore.

li'vidus, -a, -um Lead-coloured; bluish-grey.

Livisto'na f. A genus of tall palms, named by Robert Brown for Patrick Murray, Baron of Livingston, near Edinburgh, who before 1680 had a well-stocked garden; his plants later became the foundation of the Edinburgh Botanic Garden.

Lloydia f. In honour of Edward Lloyd (1660–1709), Welsh antiquary and botanist, for many years keeper of the Ashmolean Museum, Oxford.

Loa'sa f. Origin obscure. The Chilean name for these plants with showy flowers and stinging foliage is usually *ortiga* (nettle).

loba'tus, -a, -um Lobed. Also:
lobocar'pus   with lobed fruits
lobophyl'lus   with lobed leaves
lobula'tus   with small lobes

Lobe'lia f. A large genus of herbs and sub-shrubs. Named for Mathias de l'Obel (1538–1616), Flemish botanist and physician to James I of England. Cardinal flower is *L. cardinalis*.

lobelioi'des Resembling *Lobelia*.

Lobi'via f. Anagram of Bolivia where these cacti are found.

Lobula'ria f. Sweet alyssum. From L. *lobulus*, small pod.

loch'mius, -a, -um Of thickets.

Lockhar'tia f. Epiphytic orchids named for David Lockhart (d. 1846), Kew gardener, first superintendent of the Trinidad Botanic Garden.

Lodoi'cea f. Double coconut. Coco-de-mer. Named for Louis XV of France (1710–1774). Louis can be Latinized as Ludovicus or Lodoicus.

Loese'lia f. Central American herbs named for Johann Loesel (1607–1657), German botanical writer, professor at Königsberg.

Loga'nia f. New Zealand and Australian herbs and sub-shrubs named by Robert Brown in honour of James Logan (1674–1751), born in Ireland, Governor of Pennsylvania and a botanical writer, contemporary and friend of John Bartram.

loganobac'cus Loganberry, commemorating Judge J. H. Logan of California, *c*. 1880.

Loiseleu'ria f. A trailing evergreen flowering shrub named for Loiseleur, whose full name was Jean Louis Auguste Loiseleur-Deslongchamps (1774–1849), one of the great company of botanist-physicians.

lok'jae In honour of Hugo Lokja (1845–1887), Hungarian teacher and lichenologist, who made extensive collections of Alpine plants in the Caucasus in 1885.

lolia'ceus, -a, -um Like *Lolium*.

Lo'lium n. Rye-grass. The classical name for one species.

Loma'ria f. From Gr. *lōma*, border, edge; in allusion to the sori on the edge of the fronds in these ferns.

lomariifo'lius, -a, -um With leaves like the fern genus *Lomaria*.

Loma′tia f. Gr. *lōma*, border, edge; a reference to the winged edges of the seeds of these evergreen shrubs or trees.

Lo′nas f. African daisy. The derivation of this generic name is unknown.

Lonchocar′pus m. Gr. *lonche*, a lance; *karpos*, fruit. An allusion to the shape of the pods of these trees or climbing shrubs.

lon′gus, -a, -um Long. In compound words signifying long. Thus:

longebractea′tus	with long bracts
longepeduncula′tus	with a long peduncle
longicau′lis	long-stalked
longi′comus	long-haired
longicus′pis	long-pointed
longiflo′rus	long-flowered
longifo′lius	long-leaved
longihama′tus	with long hooks
longi′labris	long-lipped
longi′lobus	with long lobes
longimucrona′tus	ending in a long sharp point
lon′gipes	long-stalked
longipe′talus	with long petals
longipinna′tus	having leaflets arranged on each side of a long common stalk; long-pinnate
longiracemo′sus	with long racemes
longiros′tris	long-beaked
longisca′pus	with a long scape
longise′palus	with long sepals
longi′spathus	with long spathes
longispi′nus	long-thorned
longisquama′tus	with long scales
longi′stylus	with a long style
Also: **longis′simus**	very long

Lonice′ra f. Honeysuckle. Named for Adam Lonitzer (1528–1586), German botanist, the author of a herbal (*Kreuterbuch*), many times reprinted between 1557 and 1783.

Lope′zia f. Mexican herbs and sub-shrubs named for Tomás Lopez, Spanish botanist, who wrote on the plants of South America *c.* 1540.

lophan′thus, -a, -um Having crested flowers.

lopho′gynus, -a, -um With a crested ovary.

Lopho′phora f. Gr. *lŏphŏs*, a crest; *phoreo*, to bear; from the tufts of hairs borne on the equivalent of leaf axils of these cacti.

lorifo′lius, -a, -um With strap-shaped leaves.

Loroglos′sum From Gr. *lōrŏn*, strap, thong; *glōssa*, tongue; in reference to the strap-shaped lip of these orchids.

Lorope'talum n. Gr. *lōrŏn*, strap, thong; *petalon*, petal. These evergreen shrubs have flowers with long narrow petals.

lotifo'lius, -a, -um *Lotus*-leaved.

Lo'tus f. The classical Greek name *lōtos*, applied to a diversity of plants, e.g. clover, birdsfoot trefoil, fenugreek, melilot, etc.

louisia'nus, -a, -um From Louisiana.

Loury'a f. In honour of M. Loury, head gardener at the Muséum d'Histoire Naturelle, Paris, in 1889.

lownde'sii In honour of Colonel George Lowndes (1899–1956) of the Royal Garhwal Rifles, who collected in Nepal in 1950.

lu'cens; lu'cidus, -a, -um Bright; shining; clear.

luci'liae In honour of Lucile Boissier (1822–1849), who died while accompanying her husband, the celebrated Genevese botanist Edmond Boissier, on an expedition in Spain.

Lucu'lia f. From the Nepalese vernacular name *lukuli swa* for *Luculia gratissima*.

ludlowii Epithet of a beautiful tree peony and some other Himalayan and Tibetan plants named in honour of Frank Ludlow (b. 1885), political officer, ornithologist and botanist, who explored south-eastern Tibet (1933–1949) in company with Major George Sheriff.

ludovicia'nus, -a, -um Of Louisiana.

Ludwi'gia f. In honour of Christian G. Ludwig (1709–1773), professor at Leipzig.

Luet'kea f. In honour of Feodor Lütke (1797–1882), Russian admiral.

Lu'ffa f. Dish-cloth gourd. Vegetable sponge. From the Arabic name. *L. cylindrica* produces the curious fruit, the dried fibrous interior of which is sold as a bath sponge.

Lui'sia f. In honour of Luis de Torres, 19th-century Spanish botanist.

lukiangen'sis, -is, -e From Lukiang, Yunnan, China.

Luna'ria f. Honesty. Satin-flower. L. *luna*, the moon; from the form of the seed vessel.

luna'tus, -a, -um; lunula'tus, -a, -um Shaped like a crescent moon.

Lupi'nus m. Lupin or lupine. The classical name. Supposed to be derived from *lupus*, a wolf, because of the superstition that these plants destroyed the fertility of the soil. Today they are a favourite crop for ploughing in as green manure, since, like other leguminous plants, they are able to fix nitrogen and thus increase the fertility of soil.

lupuli'nus, -a, -um Hoplike.

lu′pulus Specific name of hop (*Humulus lupulus*). Literally, lupulus means a small wolf, the vine once having been locally called willow-wolf (*lupus salictarius*) from its persistent habit of climbing over willow trees.

lu′ridus, -a, -um Smoky yellow; sallow; wan.

lusita′nicus, -a, -um From Portugal (Lusitania).

lute′olus, -a, -um Yellowish.

lutes′cens Yellowish, becoming yellow.

lutetia′nus, -a, -um Of Paris (Lutetia).

lu′teus, -a, -um Yellow from *lutum*, dyer's greenweed, the source of a yellow dye. Also in compound words *luteo-*.

luxu′rians Luxuriant.

Luzula f. From Italian *lucciola*.

Lycas′te f. Presumably in honour of Lycaste, a daughter of Priam.

lychnidifo′lius, -a, -um *Lychnis*-leaved.

Lych′nis f. Catchfly, Maltese cross, evening campion. The classical name, said to be derived from Gr. *lychnōs*, a lamp; possibly referring to the brilliant flowers of some species, or to the ancient use of leaves of a woolly species for wicks.

Ly′cium n. Matrimony vine or box-thorn. From the Greek name for a thorny tree, with juice and roots used medicinally, from Lycia, a region of Asia Minor.

ly′cius, -a, -um Referring to Lycia, south-west Asia Minor.

lycoc′tonum Specific epithet for *Aconitum lycoctonum* or wolf's-bane, one of the more poisonous species of monkshood, from Gr. *lykŏs*, wolf; *ktŏnŏs*, murder.

Lycoper′sicon n. Tomato. Gr. *lykŏs*, a wolf; *persicon*, a peach; originally the name of an Egyptian plant, later transferred to these American plants.

lycopodioi′des Resembling *Lycopodium* or club-moss.

Lycopo′dium n. Ground pine. Club-moss. From Gr. *lykŏs*, a wolf; *pŏdiŏn*, a foot; from some fancied resemblance to a wolf's foot.

Ly′copus m. Water hoarhound. Gr. *lykŏs*, wolf; *pŏus*, a foot; from some fancied resemblance of these herbaceous perennials to a wolf's foot.

Lyco′ris f. Golden spider lily. Named in honour of a Roman beauty, the mistress of Mark Antony, famed for intrigues.

ly′dius, -a, -um From Lydia in Asia Minor.

Lygo′dium n. Climbing-fern. From Gr. *lygōdēs*, like a willow; in allusion to the supple twisting shoots.

Lyo'nia f. Ornamental shrubs named for John Lyon, a Scottish gardener and botanist of great skill who introduced many North American plants into European gardens. He was associated with Frederick Pursh in developing William Hamilton's beautiful garden at Woodlands, Philadelphia. He achieved enthusiastic notice from Captain Mayne Reid and died exploring in North Carolina sometime before 1818.

Lyonotham'nus m. An ornamental woody plant named for W. S. Lyon (1851–1916), an American collector who sent plants to Asa Gray. The second half of the name is from Gr. *thamnos*, a shrub.

lyra'tus, -a, -um Of lyrate form, usually referring to leaves with a broad rounded tip and sinuate sides diminishing in width towards the base.

Lysichi'ton m. From Gr. *lysis*, a loosening, releasing; *chitōn*, a cloak. In these almost stemless swamp herbs the spathe is cast off.

Lysima'chia f. Loosestrife. Gr. *lȳsimacheion*, said to be named after King Lysimachos of Thrace, from Gr. *lȳsimachōs*, ending strife, whence the English name loosestrife.

lysimachioi'des *Lysimachia*-like.

Lysiono'tus m. From Gr. *lysis*, a loosening, releasing; *nōtŏs*, the back; in allusion to the capsule opening away from the dorsal (back) suture.

lysiste'mon With a loose stamen or stamens, i.e. free from the other stamens.

lyso'lepis With loose scales.

Ly'thrum n. Gr. *lythron*, blood; from the colour of the flowers on these herbs. Purple loosestrife is *Lythrum salicaria*.

M

Maa'ckia f. Flowering trees and shrubs named for Richard Maack (1825–1886), Russian naturalist, explorer of the Ussuri region, eastern Asia, whence he introduced many plants into cultivation.

Ma'ba f. Vernacular name in the Tonga Islands for these ebony-like trees and shrubs.

Macada'mia f. In honour of John Macadam (1827–1865), chemist, medical man and lecturer in Melbourne, Australia.

macedo'nicus, -a, -um From Macedonia.

Machaeroce'reus m. Gr. *machaira*, a dagger, short sword; *cereus*, cactus. A reference to the spines.

macilen'tus, -a, -um Thin; lean.

Mackaya

Mackay'a f. In honour of James Townsend Mackay (*c.* 1775–1862), Scottish botanist and gardener, who emigrated to Ireland in 1804 and founded the botanic garden of Trinity College, Dublin in 1806.

mackli'niae Specific epithet of a lily discovered by Kingdon-Ward and named after his second wife, Jean, *née* Macklin.

Maclea'nia f. In honour of John Maclean, Scottish merchant at Lima, Peru, between 1832 and 1834, who sent many plants from there to Britain.

Macleay'a f. Handsome herbaceous perennials named for Alexander Macleay (1767–1848), Colonial Secretary for New South Wales and once Secretary of the Linnean Society of London. Plume poppy is *Macleaya cordata*, syn. *Bocconia cordata*.

Maclu'ra f. Osage orange. Named for William Maclure (1763–1840), American geologist.

Macrade'nia f. Gr. *makros*, long; *adēn*, a gland; an allusion to the length of the stalk of the pollen bundles on these epiphytic orchids.

macr- In compound words signifying either long or big. Thus:

macracan'thus	with large spines
macrade'nus	with large glands
macran'drus	with large anthers
macranthe'rus	with large anthers
macran'thus	with large flowers
macrobot'rys	with large grapelike clusters
macrocar'pus	large-fruited
macroce'phalus	large-headed
macrodon'tus	with large teeth
macro'meris	with large parts
macrophyl'lus	with long or large leaves
macro'podus	stout-stalked
macrorrhi'zus	with large roots or root stocks
macrosper'mus	with large seeds
macrosta'chyus	with long or large spikes
macru'rus	long-tailed

Macroza'mia f. Gr. *makros*, large; plus *Zamia*, a related genus. These Australian plants are highly poisonous to sheep.

macula'tus, -a, -um; macu'lifer, maculi'fera, -um; maculo'sus, -a, -um Spotted.

madagascarien'sis, -is, -e From Madagascar.

Madde'nia f. Trees and shrubs, natives of India and China, named for Colonel Edward Madden (d. 1856), who collected in India.

maderaspaten'sis, -is, -e From Madras, southern India.

Madhu'ca f. Latinized form of Indian name.

Ma'dia f. Tarweed. From the native Chilean name of these annuals.

Mae'sa f. From the Arabic name *maas* for these shrubs.

magella'nicus, -a, -um From the area of the Straits of Magellan, South America.

magni'ficus, -a, -um Great, splendid, magnificent.

Magno'lia f. Sweet bay. Deciduous and evergreen flowering trees and shrubs named for Pierre Magnol (1638–1751), professor of botany and director of the botanic garden, Montpellier, France.

mag'nus, -a, -um Great; big.

Maher'nia f. Honey-bell. Anagram of *Hermannia*, a genus related to these herbs and evergreen sub-shrubs.

Mahober'beris f. Hybrids of *Berberis* and *Mahonia*.

Maho'nia f. Handsome evergreen flowering shrubs named for Bernard M'Mahon (1775–1816), American horticulturist and author of *The American Gardener's Calendar* (1807).

Maian'themum n. False or wild lily-of-the-valley in North America, May Lily in Britain. From Gr. *Maios*, May; *anthemon*, blossom. A scented woodland herb of north temperate regions.

mai'rei In honour of Edouard Ernest Maire (b. 1848), Roman Catholic missionary in Yunnan, west China, who made large collections of herbarium specimens between 1905 and 1916, if a Chinese plant; in honour of René C. J. E. Maire (1878–1949), for many years professor of botany in Algeria, if a North African plant.

maja'lis, -is, -e May-flowering.

majes'ticus, -a, -um Majestic.

ma'jor; ma'jus, -a, -um Bigger; larger.

Majora'na f. Sweet marjoram. From the Arabian *marjamie*.

maki'noi In honour of Tomitaro Makino (1863–1957), celebrated Japanese botanist.

malaba'ricus, -a, -um From the Malabar coast, southern India.

Malach'ra f. Variant of Gr. *malache*, malva, now applied to these hairy tropical and sub-tropical herbs.

Malacocar'pus m. Gr. *malakos*, soft; *karpos*, fruit. From the fleshy fruit of these South American cacti.

malacoi'des Soft; mucilaginous; like mallow.

malacosper'mus, -a, -um With soft seeds.

Malco'mia f. Virginia-stock. Named by Robert Brown for William

Malcolm (d. 1820), London nurseryman, who published a catalogue of greenhouse plants in 1778 and for his son, also William (1769–1835).

malifor'mis, -is, -e Apple-shaped.

Mallo'tus m. Kamila tree. Gr. *mallōtŏs*, woolly. The fruit in some species has white spines.

Ma'lope f. Name apparently of Greek origin used by Pliny for a kind of mallow to which family these annual herbs belong.

Malpi'ghia f. Barbados-cherry, named for Marcello Malpighi (1628–1694), famous Italian anatomist and professor at Bologna.

Ma'lus f. Latin name for apple. For derivations of some named varieties of *Malus pumila*, the cultivated apple, see *Apple*.

Mal'va f. Mallow.

malva'ceus, -a, -um Mallow-like.

Malvas'trum n. False mallow. L. *malva*; mallow; plus *aster*; indicating incomplete resemblance. From a similarity to *Malva*.

Malvavis'cus m. L. *malva*, mallow; and possibly *viscus*, glue, then referring to the pulp around the seeds.

malviflo'rus, -a, -um Mallow-flowered.

malvi'nus, -a, -um Mauve.

Mam'mea f. Mammee-apple. From the West Indian name.

mammifor'mis, -is, -e Nipplelike.

Mammilla'ria f. L. *mammilla*, a nipple; from the form of this cactus.

mammilla'tus, -a, -um; mammilla'ris, -is, -e; mammo'sus, -a, -um Furnished with nipples or breasts.

Mandevil'la f. Climbing milky-juiced shrubs named for Henry John Mandeville, British Minister in Buenos Aires *c.* 1837.

Mandra'gora f. Mandrake. The Greek name. Long known for its poisonous and narcotic properties and used in classical times to deaden pain in surgical operations, it has also been much esteemed as an aphrodisiac and in love potions, presumably because of the shape of the root. It was supposed that the plant shrieked when uprooted ('And shrieks like mandrake torn out of the earth that mortals, hearing them, run mad,' *Romeo and Juliet*), a notion probably encouraged by the herbalists and root-gatherers of antiquity in order to protect the plant from undue exploitation. In the East the plant was supposed to facilitate pregnancy. The Greek *mandragoras* is apparently a corruption of Assyrian *nam tar ira* (male drug of the plague god Namtar).

mandshu'ricus, -a, -um Of Manchuria.

manesca'vii Of Manescau (d. 1875), merchant at Pau, who botanized in the Pyrenees.

Manet'tia f. Ornamental twining plants named for Saverio Manetti (1723–1785), Florentine medical man and prefect of the Florence botanic garden.

Mangi'fera f. Mango. The Indian vernacular name plus L. *fero*, to bear. The Bombay Alfonso, at its best, is probably the most delectable of the many varieties.

manica'tus, -a, -um With long sleeves.

Ma'nihot f. Cassava or tapioca. From the Brazilian name, *manioc*. The tuberous roots in their natural state are very poisonous and become edible only after steeping in water.

Manilka'ra f. Malabar (South Indian) vernacular name cited by von Rheede in 1683.

manipuliflo'rus, -a, -um With few-flowered clusters; from *manipulus*, a handful.

manipuren'sis, -is, -e From Manipur State, India.

mantegazzia'nus, -a, -um In honour of Paolo Mantegazzi (1831–1910), Italian anthropologist and traveller.

Manti'sia f. Named from a resemblance of the fantastically shaped flowers to the Mantis insect.

Manu'lea f. From L. *manus*, a hand; from the fingerlike divisions of the corolla of these South African herbs and sub-shrubs.

manzanil'la Specific epithet of *Arctostaphylos*. From the Spanish word meaning a small apple.

Maran'ta f. Arrowroot. Named for Bartolommeo Maranti, Venetian botanist *c.* 1559.

Marat'tia f. Evergreen ferns named for Giovanni Francesco Maratti (d. 1777), Italian botanist, professor in Rome.

margarita'ceus, -a, -um; margari'tus, -a, -um Pertaining to pearls, pearly.

margari'tifer, margariti'fera, -um Pearl-bearing.

margina'lis, -is, -e; margina'tus, -a, -um Margined.

marginel'lus, -a, -um With a narrow margin.

Margyricar'pus m. Gr. *margarites*, a pearl; *karpos*, fruit; from the white fruit of these South American sub-shrubs.

maria'nus, -a, -um Specific epithet of various plants with white-mottled leaves, notably Our Lady's or blessed thistle (*Silybum marianum*); the spots were supposed to have resulted from drops of her milk falling on the leaves. It is also applied to plants from Maryland (*Terra Mariana*).

Mari'ca f. Name of a nymph in Roman mythology worshipped as a goddess. See *Neomarica*.

marie'sii In honour of Charles Maries (d. 1902) who collected for Messrs. Veitch in Japan and China, 1877 to 1879.

marilan'dicus, -a, -um (also **marylandicus**) Of Maryland.

mari'timus, -a, -um Pertaining to the sea.

marlo'thii In honour of Hermann W. R. Marloth (1855–1931), South African pharmacist and botanist of German origin, author of a profusely illustrated *Flora of South Africa*.

marmora'tus, -a, -um; marmo'reus, -a, -um Marbled; mottled.

marocca'nus, -a, -um Of Morocco.

Marru'bium n. Hoarhound. The classical name for this familiar cough remedy.

Marsde'nia f. Tropical twining shrubs named for William Marsden (1754–1836), secretary to the Admiralty, orientalist and traveller, author of a *History of Sumatra* (1783).

Marshal'lia f. Perennial North American herbs named for Humphry Marshall (1722–1801), a cousin of John Bartram, and author of *Arbustum Americanum* (1785).

Marsi'lea f. Aquatic flowerless plants named for Count Luigi Ferdinando Marsigli (1656–1730) of Bologna, Italian botanist.

mar'tagon n. Origin obscure, possibly alchemical.

Martine'zia f. Ornamental palms named for Baltasar Jacobo Martinez, archbishop of Santa Fé de Bogota, a helper of Ruiz and Pavon.

Marty'nia f. In honour of John Martyn (1699-1768) London physician, professor of botany at Cambridge, England, although for many years of his professorship he neither resided nor lectured there. His son Thomas Martyn (1735–1825) succeeded him as professor.

mas; mas'culus, -a, -um Male; masculine; used by early writers in a metaphorical sense to distinguish robust species from more delicate ones, red-flowered species from blue-flowered, etc.

Mascarenha'sia f. Small trees or shrubs named for the Mascarene Islands (Réunion, Mauritius and Rodriguez) in the south-west Indian Ocean, which take their name from the 16th-century Portuguese navigator Mascarenhas.

Masdeval'lia f. Tufted epiphytic orchids named for José Masdevall (d. 1801), Spanish botanist and physician.

massilien'sis, -is, e of Marseilles, France.

massonia'nus, -a, -um; masso'nii In honour of Francis Masson (1741–c. 1805), Scottish gardener and traveller in South Africa, West Indies and North America, the first plant collector sent out from Kew.

Matrica'ria f. Mayweed. Medieval name possibly from L. *matrix*, womb, because of its one-time medical use in affections of the uterus.

matriten'sis, -is, -e Relating to Madrid, Spain.

matrona'lis, -is, -e Pertaining to March 1st which was the Roman festival of the matrons.

Matteu'cia f. Ostrich fern. In honour of Carlo Matteuci, 19th-century Italian physicist.

Matthio'la f. Stock. Named for Pierandrea Mattioli (1500–1577), Italian physician and botanist, author of a celebrated commentary on Dioscorides, which exists in Latin, Italian, German, French and Czech versions.

Mauran'dia f. Showy perennial herbs, some climbing. Named for Catalina Pancratia Maurandy, Spanish botanist, wife of the director of the Cartagena botanic garden, A. Juan y Poveda, late 18th-century.

maurita'nicus, -a, -um From North Africa, particularly Morocco.

Mauri'tia f. Latinized from Surinam name 'Mauritii-boom' for this palm.

mauritia'nus, -a, -um Relating to Mauritius, Indian Ocean.

Maxilla'ria f. L. *maxilla*, the jaw. The flowers of some of these epiphytic orchids resemble the jaws of an insect.

maxilla'ris, -is, -e Pertaining to the jaws.

Maximilia'na f. Palms named for Maximilian A. P. zu Wied-Neuwied (1782–1867), German prince who travelled and collected in Brazil, 1813–1817.

maximowi'czii In honour of Carl Ivanovich Maximowicz (1827–1891) pronounced 'Maksimovich', Russian botanist, author of many scholarly publications on plants of eastern Asia.

ma'ximus, -a, -um Largest.

Maya'ca f. French Guiana vernacular name.

mays From Mexican vernacular name for maize (Indian corn).

May'tenus m. From the Chilean name *maitén* (Mapuche, *mañtün*) for *M. boaria*.

Ma'zus m. Gr. *mazos*, a teat; from the tubercles closing the mouth of the corolla of these low-growing herbs.

mea'dia Specific epithet for shooting star (*Dodecatheon meadia*). Named for Richard Mead (1673–1754), English physician.

Mecono'psis f. Indian poppy, blue poppy. Gr. *mēkōn*, a poppy; *opsis*, like.

Mede'ola f. Indian cucumber root. Named for the sorceress Medea, who helped Jason to get the golden fleece, because of its reputed medicinal properties.

Medica'go f. Alfalfa, lucerne. The classical name for some grass.

me'dicus, -a, -um Medicinal.

Medinil'la f. Handsome tropical shrubs named for José de Medinilla of Pineda, Governor of the Marianna Islands, *c.* 1820.

Mediocac'tus m. L. *medius*, middle; plus *Cactus*. This genus is intermediate between two others.

mediopic'tus, -a, -um Striped or coloured down the middle.

mediterra'neus, -a, -um Mediterranean, i.e. belonging to the Mediterranean region or, on the contrary, in the middle of the land, far inland.

me'dium From Italian vernacular name *erba media*.

me'dius, -a, -um Intermediate; middle.

medulla'ris, -is, -e; medul'lus, -a, -um Pithy.

meebol'dii In honour of Alfred Karl Meebold (1863–1952), botanical collector in many countries, novelist, poet and essayist.

mega- In Greek compound words signifies big. Thus:

megaca'lyx	with large calyx
megacan'thus	with big spines
megacar'pus	big-fruited
megalan'thus	big-flowered
megalophyl'lus	with big leaves
megapota'micus	of the big river (generally the Amazon)
megarrhi'zus	big-rooted
megasper'mus	with big seeds
megasta'chyus	with a big spike
megasti'gmus	with a big stigma

Megacli'nium n. From Gr. *mĕga*, big, wide; *klinion*, a little bed. In allusion to the flattened inflorescence of these orchids.

megaseifo'lius, -a, -um With leaves like *Megasea* (large-leaved saxifrages now called *Bergenia*).

Megaskepas'ma n. Gr. *mĕga*, large; *skĕpasma*, covering. With reference to the conspicuous bracts.

mege'ratus, -a, -um Gr. *mĕgĕrătŏs*, passing lovely, much contended for.

meiacan'thus, -a, -um Small or few-spined.

mekongen'sis, -is, -e From a region by the Mekong River, usually where it flows through Yunnan, China.

Melaleu'ca f. Bottlebrush. From Gr. *mĕlas*, black; *leukos*, white. Often the trees have a black trunk and white branches.

melancho'licus, -a, -um Drooping; sad-looking.

Melan'thium n. Bunchflower. From Gr. *mĕlas*, black; *anthos*, a flower. The flower segments are persistent and become dark after flowering. The name *mĕlanthion* was originally applied to *Nigella*.

Melasphae'rula f. Gr. *mĕlas*, black; *sphaerula*, a small ball. In allusion to the small black bulbs of this pretty South African plant.

Melas'toma n. Gr. *mĕlas*, black; *stŏma*, a mouth. The berries of these tropical shrubs and trees stain the mouth black.

mele'agris, -is, -e Spotted like the guinea fowl.

Me'lia f. Bead-tree, chinaberry. Greek name *mĕlia* for the ash tree in allusion to the similarity of the leaves.

Melian'thus m. Honeybush. From Gr. *mĕli*, honey; *anthos*, a flower. The flowers produce abundant nectar.

Me'lica f. Melic grass. From Greek name *mĕlikē* for a sweet grass.

Melicoc'ca f. Honeyberry. From Gr. *mĕli*, honey; *kŏkkŏs*, a berry. An allusion to the sweetness of the fruit.

Melicy'tus m. From Gr. *mĕli*, honey; *kytŏs*, a hollow container, jar; in allusion to the hollow staminal nectaries of the flowers of these trees and shrubs, found in New Zealand and a few other Pacific islands. The genus belongs to the *Violaceae*, and the flowers of *M. ramiflorus* are heavily scented.

Melilo'tus f. Sweet clove. Gr. *mĕlilōtos*, from *mĕli*, honey; *lotos*, lotus. The foliage is fragrant, bees enjoy the flowers. The genus is related to *Lotus*.

meliodo'rus, -a, -um Honey-scented.

Melios'ma f. Gr. *mĕli*, honey; *ŏsmē*, fragrance. The flowers of these trees and shrubs are honey-scented.

Melis'sa f. Balm. From Gr. *mĕlissa*, a honeybee, also the name of a Cretan nymph who first discovered how to get honey. The flowers of these herbaceous perennials are liked by bees.

meliten'sis, -is, -e Of Malta.

Melit′tis f. Bastard balm. From Gr. *mĕlissa*, a honeybee. Bees enjoy the flowers.

mel′leus, -a, -um Pertaining to honey.

melli′tus, -a, -um Honey-sweet. Also: mellifer, honey-bearing; melliodorus, honey-scented.

Melocac′tus m. From Latin *mēlo*, a shortened form of *mēlŏpĕpo*, an apple-shaped melon, and *cactus*, with reference to the rounded form of these cacti.

melofor′mis, -is, -e Melon-shaped.

Melo′thria f. From the Greek name *mēlōthron* for another plant, probably *Bryonia*. These are climbing herbs of the cucumber family.

membrana′ceus, -a, -um Skinlike; membranous.

meniscifo′lius, -a, -um With crescent-shaped leaves.

Menisper′mum n. Moonseed. Gr. *mēnē*, the crescent moon; *sperma*, a seed; from the shape of the seed.

Men′tha f. Mint. The Latin name.

Mentze′lia f. Prairie lily, blazing star. A genus of showy-flowered plants named for Christian Mentzel (1622–1701), German botanist, personal physician to the Elector of Brandenburg.

Menyan′thes f. Bogbean, buckbean. Variant of Gr. *mēnanthos*, moonflower, a name for *Nymphoides peltata*.

Menzie′sia f. Low-growing flowering shrubs named for Archibald Menzies (1754–1842), Scottish naval surgeon and botanist who accompanied Vancouver on his voyage of North-west Pacific exploration, 1790–1795.

menzie′sii Specific epithet of *Arbutus*. See *Menziesia*.

Mercuria′lis f. Herb-Mercury. Originally *herba mercurialis* named in honour of the messenger of the gods. The source of a blue dye.

Merende′ra f. From Spanish *Quitameriendas*, ultimately from L. *merenda*, midday meal; the flowering of *Merendera* in autumn warns the shepherds in the mountains that they must leave the pastures there.

meridia′nus, -a, -um; meridiona′lis, -is, -e Of noonday; blooming at noontime.

Merrem′ia f. In honour of Blasius Merrem (1761–1824), professor of natural sciences, Marburg, Germany.

Merten′sia f. Bluebell (in the United States), Virginia cowslip. Named for Franz Carl Mertens (1764–1831), professor of botany at Bremen.

Mery'ta f. Gr. *merytos*, glomerate; in allusion to the male flowers being crowded together in a head.

Mesembryan'themum n. Fig marigold. The etymological tangle of this name began in 1684 when Breyne published it as *Mĕsēmbrian-thĕmum*, derived from Gr. *mĕsēmbria*, midday; *anthemon*, flower; in allusion to the fact that the only species then known bloomed at noon. When night-flowering species became known, and this name accordingly seemed inappropriate, Dillenius in 1719 ingeniously renamed the genus *Mĕsēmbryanthĕmum*; by changing the i to y he altered the derivation, to Gr. *mesos*, middle; *embryon*, embryo; *anthemon*, flower, with reference to the position of the ovary. The group has now been divided into numerous smaller genera based on habit of growth and fruit-characters.

mes'nyi In honour of William Mesny (1842-1919), a Channel Islander from Jersey who became a general in the Chinese Imperial Army and travelled widely in China, occasionally collecting plants, among them the large-flowered yellow jasmine.

Mes'pilus f. Medlar. The Latin name for this fruit which is pleasant when frosted or rotten and which goes well with wine.

metal'licus, -a, -um Metallic.

Metasequo'ia f. Dawn redwood. From Gr. *meta*, with, after, sharing, changed in nature; *Sequoia* (which see), to which it is related and to which fossil specimens were first referred.

Metroside'ros f. Iron tree. From Gr. *mētra*, core, heart-wood; *sidērŏs*, iron; from the hardness of the heart-wood of these handsome trees and climbers; the noble *M. tomentosa* discovered by Banks and Solander in 1769 is the crimson-flowered New Zealand Christmas tree or pohutukawa.

metterni'chii In honour of Prince Metternich (1773-1859), Austrian statesman.

Me'um n. Baldmoney. The Greek name for this perennial aromatic herb.

mexica'nus, -a, -um Mexican.

meyenia'nus, -a, -um; meye'nii In honour of Franz Julius Ferdinand Meyen (1804-1840), German physician and naturalist, who collected many plants on a voyage round the world, 1830-1832. He became a professor of botany in Berlin.

me'yeri In honour of Carl Anton Meyer (1795-1855), German botanist in Russian employ, director of St. Petersburg botanic garden; or Ernst H. F. Meyer (1791-1858), German botanist chiefly celebrated for his very scholarly but never completed history of botany; or

Frank N. Meyer (d. 1918), Dutch-born agricultural explorer and collector in China, Manchuria, Korea, Mongolia and Turkistan for U.S. Department of Agriculture.

mezer'eum Latinized from a Persian name.

mi'cans Glittering.

Michau'xia f. Biennial herbs named for André Michaux (1746–1803), French collector, explorer, and plantsman who collected this, among many others, on an expedition to Persia, 1782–1785, and, in company with his son, François André Michaux (1770–1855), carried out remarkable journeys in North America (1785–1796) and sent back an extensive herbarium. When he himself returned to France, he took with him forty boxes of seeds in addition to the 60,000 living plants he had at various times already dispatched, most of which were lost by shipwreck. He died on an expedition to Madagascar. Much of his *Flora Boreali-Americana* (1803) was written by L. C. M. Richard (1754–1821).

michauxioi'des Resembling *Michauxia*.

Miche'lia f. Evergreen flowering trees and shrubs named for Pietro Antonio Micheli (1679–1737), Florentine botanist, particularly celebrated for his remarkable pioneer researches on cryptogams, including fungi.

Mico'nia f. Tropical American trees and shrubs with showy foliage, named for Francisco Mico (b. 1528), Spanish physician and botanist.

Micran'themum n. Gr. *mikros*, small; *anthemon*, flower.

micro- In compound words signifying small. Thus:

micracan'thus	small thorns
micran'thus	small-flowered
microcar'pus	small-fruited
microce'phalus	with a small head
microchi'lus	small-lipped
micro'dasys	small and shaggy
mic'rodon	small-toothed
microglos'sus	small-tongued
micro'gynus	with small ovary
microle'pis	small-scaled
micro'meris	with a small number of parts or small parts
micrope'talus	small-petalled
microphyl'lus	small-leaved
micro'pterus	small-winged
microse'palus	with small sepals
microsper'mus	with small seeds

Microcit'rus m. Finger lime. Gr. *mikros*, small; *Kitron*, lemon, etc.

Microcoe'lum n. Gr. *mikros*, small; *koilos*, hollow, hollowed. In allusion to a small hollow in the endosperm of the fruit of these palms.

Microglos'sa f. Gr. *mikros*, small; *glossa*, a tongue; in reference to the short ray florets of these Asian and African shrubs.

Microle'pia f. Gr. *mikros*, small; *lepis*, a scale; a reference to the small indusia on these ferns.

Microme'ria f. Yerba buena. Gr. *mikros*, small; *meris*, a part. The flowers of these low-growing herbs and sub-shrubs are very small.

Micros'tylis f. Gr. *mikros*, small; *stylos*, a column. These orchids have small columns.

middendor'ffii In honour of Alexander Theodor von Middendorff (1815–1894), zoologist who travelled and collected extensively in Siberia.

Mika'nia f. Climbing hempweed. Shrubs and twining herbs, named in honour of Joseph Gottfried Mikan (1743–1814), professor of botany at Prague, father of Johann Christian Mikan (1769–1844), likewise a professor at Prague, who collected in Brazil.

milia'ceus, -a, -um Pertaining to millet.

milita'ris, -is, -e Pertaining to soldiers; like a soldier.

Mil'la f. A cormous plant with scented waxy white blossoms named for Julian Milla, 18th-century gardener to the king of Spain.

millefolia'tus, -a, -um; millefo'lius, -a, -um Many-leaved; literally, with a thousand leaves.

mil'leri In honour of Philip Miller (1691–1771), English gardener and botanist, curator of the Chelsea Physic Garden 1722–1770, author of the *Gardener's Dictionary* (1731; 8th ed. 1768).

Milto'nia f. Epiphytic orchids named for Viscount Milton, later Lord Fitzwilliam (1786–1851), patron of gardening.

mim'etes Gr. *mimētēs*, an imitator.

Mimo'sa f. Gr. *mimos*, a mimic. A reference to the sensitive collapse, when touched, of the leaves of some species of these feathery-foliaged trees, shrubs, and herbs.

mimosifo'lius, -a, -um With leaves like *Mimosa*.

mimosoi'des Like *Mimosa*.

Mi'mulus m. Monkey-flower. Latin diminutive of *mimus*, a mimic. The corolla looks like the face of a monkey.

Mimu'sops f. L. *mimus*, a mimic; Gr. *ōps*, eye, face, appearance.

Mi'na f. In honour of Joseph X. F. X. Mina, a Mexican (fl. 1824).

mi′nax Threatening; forbidding.

minia′tus, -a, -um Cinnabar-red.

mi′nimus, -a, -um Smallest.

mi′nor; mi′nus, -a, -um Smaller.

Minuar′tia f. In honour of Juan Minuart (1693–1768) of Barcelona.

minu′tus, -a, -um Very small. Also:
  minutiflo′rus   minute-flowered
  minutifo′lius   minute-leaved
  minutis′simus   most minute

miquelia′nus, -a, -um; mique′lii In honour of Friedrich Anton Wilhelm Miquel (1811–1871), professor of botany at Utrecht and Leyden, author of works on flora of Japan, Indonesia, etc.

Mira′bilis f. Four-o'clock. Tuberous-rooted perennials. Latin word meaning wonderful.

mira′bilis, -is, -e Wonderful; remarkable.

Miscan′thus m. Gr. *miskos*, a stem; *anthos*, a flower; an allusion to the stalked spikelets of these handsome perennial grasses.

mishimien′sis, -is, -e From the Mishmi Hills, Assam, India.

missourien′sis, -is, -e Referring to Missouri, U.S.A.

Mitchel′la f. Partridge berry. An evergreen trailing herb named for Dr. John Mitchell (1711–1768), physician of Virginia, born in Lancaster county, who was a correspondent of Linnaeus.

Mitel′la f. Bishop's cap, mitrewort. Diminutive of Gr. *mitra*, a cap; in allusion to the form of the young fruit of this perennial herb.

mi′tis, -is, -e Mild; gentle; without spines.

Mitra′ria f. Gr. *mitra*, a mitre or cap; from the shape of the seed pod of this evergreen flowering shrub.

mitra′tus, -a, -um Turbaned; mitred.

mitrifor′mis, -is, -e Caplike.

mix′tus, -a, -um Mixed.

modes′tus, -a, -um Modest.

moesia′cus, -a, -um From *Moesia*, classical name of northern Balkan region.

Mogha′nia f. Latinized form of an Indian name. This genus is also known as *Flemingia*.

Moh′ria f. In honour of Daniel Matthias Heinrich Mohr (d. 1808) of Kiel.

molda′vicus, -a, -um From the Danube basin.

Moli′na f. Perennial grasses named for Juan Ignacio Molina (1740–1829), writer on the natural history of Chile.

mol′le Peruvian vernacular name.

molli′comus, -a, -um Soft-haired.

mol′lis, -is, -e Soft; with soft hairs.

mollis′simus, -a, -um Very soft.

mollu′go Specific name, derived from the genus *Mollugo* which has no garden interest except as a pernicious carpetweed.

Moloposper′mum n. Gr. *mōlōps*, a stripe; *sperma*, a seed. The fruit of this handsomely foliaged perennial appears to be striped.

Mol′tkia f. Perennial herbs named for Count Joachim Gadske Moltke (1746–1818), Danish statesman.

molucca′nus, -a, -um From the Moluccas or Spice Islands of Indonesia.

Moluccel′la f. Shell-flower. Derivation of the name of these Mediterranean herbs is uncertain, but is apparently a diminutive of *Molucca*.

mo′ly Gr. *mōly*, name of a magic herb of uncertain identity.

Momor′dica f. L. *mordeo*, to bite, possibly because the seeds of these tropical climbers appear bitten.

monacan′thus, -a, -um One-spined.

monadel′phus, -a, -um From Gr. *mŏnŏs*, one; *adĕlphŏs*, brother; the filaments of the stamens being joined into one brotherhood.

monan′drus, -a, -um With one stamen.

Monan′thes f. Gr. *mŏnŏs*, one; *anthos*, a flower. The type-species was named from a solitary-flowered specimen but it can bear up to eight flowers on a stem.

Monar′da f. Bee-balm, bergamot, horse-mint. Named for Nicholas Monardes (1493–1588), physician and botanist of Seville who published in 1569, 1571 and 1574 a book on American products, translated into English in 1577 with the title *Joyfull Newes out of the newe founde Worlde*. The genus includes bee-balm, lemon mint, wild bergamot.

Monardel′la f. Diminutive of *Monarda* which this genus of aromatic herbs resembles.

mon′do Japanese name used as a specific epithet.

Mo′neses f. One-flowered pyrola. Gr. *mŏnŏs*, single; *ĕsis*, a sending forth. These dwarf perennial herbs have solitary flowers.

mongo′licus, -a, -um Mongolian.

moni′lifer, monili′fera, -um Bearing a necklace.

219

monilifor'mis, -is, -e In the form of a necklace with alternate swellings and constrictions.

mono- In compound words signifying single. Thus:

monoce'phalus	with one head
mono'gynus	with one pistil
monope'talus	single-petalled
monophyl'lus	one-leafed
mono'pterus	with a single wing
monosema'tus	with one mark or blotch
monose'palus	one-sepalled
monosper'mus	one-seeded
monosta'chyus	one-spiked

Monot'ropa f. Indian pipe. Gr. *mŏnŏs*, single; *trŏpŏs*, a turn. The top of the stem of these saprophytic herbs is turned to one side.

monregalen'sis, -is, -e From Mondovi, Piedmont, Italy; or from Monreale. Both are known in Latin as Mons Regalis.

Monso'nia f. In honour of Lady Ann Monson (*c.* 1714–1776) who collected at the Cape of Good Hope and in Bengal.

monspelien'sis, -is, -e; monspessulan'us, -a, -um Of Montpellier in southern France.

Monste'ra f. Derivation obscure, possibly contracted from *monstrifer*, monster-bearing, in allusion to the perforated leaves of these tropical evergreen climbers.

monstro'sus, -a, -um Abnormal.

Monta'noa f. Shrubs named for Luis Montana, supposed to have been a mid-19th-century Mexican politician.

monta'nus, -a, -um Pertaining to mountains.

Montbre'tia f. Cormous plants named for Antoine François Ernest Conquebert de Montbret (1781–1801), one of the botanists accompanying Napoleon's invasion of Egypt (1798), where he died. Some plants formerly placed in *Montbretia* belong to *Tritonia* or *Crocosmia*.

Mon'tia f. In honour of Giuseppe Monti (1682–1760), professor of botany at Bologna.

monti'cola Growing on mountains.

monti'genus, -a, -um Mountain-born.

Monui'llea f. In honour of Monville, French cactaeae specialist in first half of 19th century.

Morae'a (Morea) f. Butterfly iris. Named by Miller in honour of Robert More (1703–1780) of Shrewsbury, England, apparently a keen amateur botanist.

Morican'dia f. In honour of Moise Etienne Moricand (1780–1854), Swiss botanist.

morifo'lius, -a, -um Mulberry-leaved.

Mori'na f. Whorl flower. Perennial herbs named for Louis Pierre Morin (1635–1715), French botanist.

Morin'da f. Indian mulberry. L. *morus*, a mulberry; *indicus*, Indian.

Morin'ga f. Latinized version of the Sinhalese name *murunga*. The seeds of these trees yield ben-oil used in perfumery and for lubricating watches and other fine mechanisms.

Mori'sia f. In honour of Giuseppe Giacinto Moris (1796–1869), professor of botany at Turin, author of a flora of Sardinia.

Moriso'nia f. In honour of Robert Morison (1620–1683), professor of botany at Oxford.

Mormo'des f. From Gr. *Mŏrmō*, phantom; in allusion to the strange flowers of these orchids.

morrisonen'sis, -is, -e Of Mount Morrison, Formosa (Taiwan).

Mo'rus f. Mulberry. The Latin name. The tree is widely cultivated as food for silk-worms, most species being useful for this purpose. The best and sweetest fruit is borne by *M. nigra*.

mosa'icus, -a, -um Coloured in a mosaic fashion; see *musaicus*.

Moscha'ria f. Gr. *mŏschŏs*, musk in allusion to the fragrance. A Chilean annual.

moscha'tus Musky. (See under *muscari*.)

Moschos'ma f. From Gr. *mŏschŏs*, musk; *ŏsmē*, smell, fragrance.

moulmainen'sis, -is, -e From Moulmein, Burma.

moupinensis, -is, -e Of Mupin, western China, a principality in which Armand David made large collections of plants; it is now called Pooksing.

moure'tii In honour of Lieutenant Marcellin Mouret (1881–1915), French soldier and amateur botanist who collected in North Africa.

moy'esii In honour of the Rev. J. Moyes of the China Inland Mission, *c.* 1906.

muco'sus, -a, -um Slimy.

mucrona'tus, -a, -um Pointed; mucronate.

mucronula'tus, -a, -um Having a short hard point.

Mucu'na f. From the Brazilian name for these vines, the pods of which bear irritant hairs.

Muehlenbe'ckia f. Semi-woody shrubs named for Henri Gustave

Muehlenbeck (1798–1845), physician at Mulhouse, France, investigator of the flora of Alsace.

Muhlenber′gia f. Grasses named in honour of Gotthilf Henry Ernest Muhlenberg (1753–1815), Lutheran pastor in Pennsylvania, U.S.A., distinguished amateur botanist.

Mulge′dium n. From *mulgeo*, to milk, by analogy with *Lactuca* to which this genus is closely related.

mulien′sis, -is, -e From the Muli region of south-western Szechwan, China.

multi- In compound words signifies many. Thus:

multibractea′tus	with many bracts
multicau′lis	many-stemmed
multica′vus	with many hollows
mul′ticeps	with many heads
multi′color	many-coloured
multicosta′tus	many-ribbed
multi′fidus	many times divided
multiflo′rus	many-flowered
multi′jugus	many yoked together, generally referring to leaves with many pairs of leaflets
multilinea′tus	many-lined
multiner′vis	many-nerved
mul′tiplex	much-folded, hence doubled
multiradia′tus	with many rays
multiscapoi′deus	with many scapes, that is leafless flower-stems
multisec′tus	much-cut

mu′me Variant of the Japanese name *ume* for a species of *Prunus*.

mun′dulus, -a, -um Trim; neat.

mura′lis, -is, -e Growing on walls.

murica′tus, -a, -um Roughened, with hard points, like the shell of the mollusc (*Murex*) from which the celebrated purple dye of antiquity was obtained.

murie′lae In honour of Muriel Wilson (Mrs George L. Slate), daughter of E. H. Wilson.

Murray′a (also incorrectly **Mur′raea**) f. Curry-leaf, orange jessamine, satinwood tree. Named for Johann Andreas Murray (1740–1791), Swedish pupil of Linnaeus and professor of medicine and botany, Göttingen.

Mu′sa f. Banana. From Arabic *mouz* or *moz*, but accepted by Linnaeus because it could have been named for Antonius Musa (63–14 B.C.), physician to the first Roman emperor, Octavius Augustus. The Arabs regarded it as the 'Tree of Paradise', hence the Linnaean epithet *paradisiaca*. The genus is of Old World origin but quite early was established in the New. Next to the coconut it is probably the most important tropical plant for food, fibre, and many other purposes. *M. paradisiaca*, in one of its many variants, provides most of the edible fruits.

musa′ceus, -a, -um *Musa*-like.

musa′icus, -a, -um Coloured or marked in a mosaic fashion; also spelled *mosaicus*.

Musca′ri n. Grape-hyacinth. Turkish name recorded by Clusius in 1601, the bulbs of *Muscari muscarimi* (*M. moschatum*) being received from Constantinople under the names *Muscari, Muschoromi* or *Muscurimi*, meaning musk of the Romans (i.e. Greeks), or *Muschio greco* (Greek musk), referring to the sweet aromatic scent of the flowers, hence from Persian *mushk*, Sanskrit *mushka*, testicle. The source of musk is a scent gland or 'pod' of the male musk-deer (*Moschus moschifer*).

musci′pulus, -a, -um Fly-catching.

muscito′xicus, -a, -um Poisonous to flies.

musci′vorus, -a, -um Fly-eating.

muscoi′des Fly-like.

musco′sus, -a, -um Resembling moss.

musifo′lius, -a, -um *Musa*-leaved.

Mussaen′da f. Latinization of Sinhalese vernacular name *mussenda*.

mussi′nii Named for the Russian Count Apollos Mussin-Puschkin (d. 1815), who made an expedition to the Caucasus in 1802.

muta′bilis, -is, -e Changeable, especially as to colour.

muta′tus, -a, -um Changed.

mu′ticus, -a, -um Blunt; without a point.

mutila′tus, -a, -um Divided as though torn, as with some leaves.

Muti′sia f. Flowering shrubs, many of them climbers, named for José Celestino Mutis (1732–1809), physician of Cadiz, who went to South America in 1760 and settled at Bogotá, whence he sent specimens to Linnaeus and his son. A team of artists worked under his direction making superb illustrations of plants, which were taken to Spain after his death and that of his assistant Caldas, and of which some were first published in 1955.

Myce′lis f. Old name of obscure origin.

myco′ni, Myconel′la f., **myco′nis** In honour of Francisco Mico

(b. 1528), Spanish botanist and physician, not 'the mythical bear Mycon'. See *Miconia*.

myiag'rus, -a, -um Fly-catching. In allusion to stickiness.

myoporoi'des Like *Myoporum*.

Myo'porum n. Gr. *myō*, to shut, close; *pŏrŏs*, opening, pore; referring to the transparent spots on the leaves of these evergreen trees and shrubs filled (and thus closed) with a pellucid substance.

myosotidiflo'rus, -a, -um With flowers like *Myosotis*.

Myosoti'dium n. Chatham Island lily. This handsome plant, confined to the Chatham Islands, is not a lily but related to *Myosotis*, hence the name.

Myoso'tis f. Forget-me-not. The classical Greek name, *myŏsōtis*, from *mus*, mouse; *ŏus*, *ōtŏs*, ear; applied to plants with short pointed leaves, later transferred to this genus.

Myosu'rus m. Mouse-tail. From Gr. *mus*, mouse; *oura*, tail; with reference to the slender elongated inflorescence.

myri- In compound words signifying very many. Thus:
myriacan'thus   with very many thorns
myriocar'pus   many-fruited
myriophyl'lus   many-leaved

Myri'ca f. Gale, bog myrtle, bayberry, wax myrtle. Derived from the Greek name *myrikē* for tamarisk. The greasy covering of the fruit of these shrubs provides the aromatic tallow from which bayberry candles are made.

Myrica'ria f. False tamarisk. From Gr. *myrikē*, tamarisk, to which this genus is related.

Myrioce'phalus m. Gr. *myrios*, many; *kĕphale*, a head. The flower heads of these Australian herbs are very numerous.

myriophylloi'des Resembling *Myriophyllum*.

Myriophyl'lum n. Water milfoil. Gr. *myrios*, many; *phyllon*, a leaf; with reference to the many divisions of the submerged leaves of these aquatic plants.

Myris'tica f. Nutmeg. Gr. *myristikos*, fit for anointing. A large genus of tropical trees. The fruit of *M. fragrans* is the source of nutmeg.

myrmeco'philus, -a, -um Literally, ant-loving—an epithet applied to plants much frequented by ants as providing them with either food or shelter, as in the case of certain plants with hollow pseudo-bulbs, and many others.

Myroxy'lon n. Gr. *myrŏn*, a sweet-smelling oil; *xylon*, wood. The heartwood of these trees is resinous and fragrant.

Myr'rhis f. Myrrh, sweet cicely. The Greek name for a plant often identified as *M. odorata* as well as for the true myrrh. *M. odorata* is an herbaceous plant of the carrot family useful as a salad plant for its aniseed flavour and aroma. Biblical myrrh is a fragrant gum resin once very highly esteemed for ceremonial, religious, and embalming purposes. It comes from *Commiphora myrrha*, a small East African and Arabian tree.

Myr'sine f. From a Greek name for myrtle; not these trees and shrubs.

myrsinifo'lius, -a, -um *Myrsine*-leaved.

myrsinoi'des Resembling *Myrsine*.

myrtifo'lius, -a, -um Myrtle-leaved.

Myrtillocac'tus m. L. *myrtillus*, a small myrtle. The allusion is to the small myrtle-like fruits, not unlike blueberries, of these cacti.

Myr'tus f. Myrtle. The Greek name. The Greeks held the plant sacred to Aphrodite (the Roman Venus) and in Christian religious painting it became one of the plants associated with the Virgin. In ancient Rome there were two old myrtles, one called the Patrician and the other the Plebeian. The political fortunes of the nobles as against those of the Plebeians could be judged as each plant flourished or languished.

mysoren'sis, -is, -e Of Mysore, southern India.

Mystaci'dium n. From Gr. *mystax*, moustache.

N

Naege'lia f. Herbaceous perennials with velvety leaves named for Karl Wilhelm Naegeli (1817–1891), Swiss botanist, successively a professor at Freiburg, Zürich and Munich, noted for his cytological research and speculations on heredity.

Na'jas f. Water nymph.

nakotil'tus, -a, -um Having the wool plucked off.

Nandi'na f. Latinized form of the Japanese name for these shrubs.

nanel'lus, -a, -um Very dwarf.

nankinen'sis, -is, -e From Nanking, Fukien, China.

nannophyl'lus, -a, -um Small-leaved.

Nan'norhops f. A small tufted palm. From Gr. *nannŏs*, a dwarf; *rhōps*, a bush or low shrub.

na'nus, -a, -um Dwarf.

Napel'lus m. Diminutive of *napus*, meaning a little turnip, in allusion to the tubers.

napifor'mis, -is, -e Turnip-shaped.

napobras'sica Specific epithet for swede or ruta-baga—literally, 'turnip-cabbage'.

narbonen'sis, -is, -e Of Narbonne, southern France.

narcissiflo'rus, -a, -um Narcissus-flowered, i.e. with flowers in an umbel resembling some species of *Narcissus*.

narcissifo'lius, -a, -um Narcissus-leaved.

Narcis'sus m. Daffodil. Classical Greek name in honour of a beautiful youth who became so entranced with his own reflection that the gods turned him into a flower. The word is possibly derived from an ancient Iranian language.

Nardos'mia f. From Gr. *nardŏs*, spikenard; *ŏsmē*, smell.

narino'sus, -a, -um Broad-nosed.

Narthe'cium n. A diminutive of *narthex*, rod.

nastur'tium From L. *nasus tortus*, a twisted nose, due to the plant's pungent qualities. Specific epithet of water-cress (*Rorippa nasturtium-aquaticum*). The plant familiarly known as nasturtium is *Tropaeolum* (*q.v*).

nasu'tus, -a, -um Large-nosed.

natalen'sis, -is, -e From Natal, South Africa.

na'tans Floating.

Nauc'lea f. Said to be a variant of L. *naucula*, a little ship.

nauseo'sus, -a, -um Nauseating.

Nautiloca'lyx m. L. *nautilus*, the nautilus mollusc; *calyx*, calyx.

navicula'ris, -is, -e Boat-shaped.

neapolita'nus, -a, -um From Naples.

nebroden'sis, -is, -e From the Monti Nebrodi, north-eastern Sicily.

nebulo'sus, -a, -um Cloudlike.

neglec'tus, -a, -um Hitherto overlooked.

negun'do From Sanskrit and Bengali *nirgundi*, rendered as *negundo* by Acosta, name for *Vitex negundo* and transferred on account of leaf resemblance to *Acer negundo*.

neilgherren'sis, -is, -e From the Nilgiri Hills, southern India; also rendered *nilagiricus* and *nilgiricus*.

Neil'lia f. Nine bark. Ornamental shrubs named for Patrick Neill

(1776–1851), printer, of Edinburgh, Secretary of the Caledonian Horticultural Society.

nelumbifo'lius, -a, -um With leaves like *Nelumbo* or lotus.

Nelum'bo f. Lotus. Sinhalese name for this large and beautiful aquatic plant.

Nema'stylis f. Gr. *nēma*, a thread; *stylos*, a column; from the slender style of these bulbous plants.

Nematan'thus m. Gr. *nēma*, thread; *anthos*, flower. In allusion to the long thread-like flower-stalks.

Neme'sia f. From the Greek name *něměsiŏn* for a plant similar to these attractive herbs.

Nemopan'thus m. Mountain holly. Contracted from Gr. *nēma*, a thread; *pŏus*, foot; *anthos*, a flower; with reference to the long slender peduncles of this North American shrub.

Nemo'phila f. Gr. *němŏs*, wooded pasture, glade; *phileo*, to love; in allusion to the habitat of some species of these spreading North American annuals.

nemora'lis, -is, -e; nemoro'sus, -a, -um Growing in groves or woods.

Neobentha'mia f. *Neo*-new; *Benthamia*, in honour of George Bentham; see *benthamii*. The name *Benthamia* having been used earlier for another genus, this one had to be named anew.

Neobessey'a f. Cactus. Named for Charles Bessey (1845–1915), professor of botany, University of Nebraska, author of an influential system of classification.

Neoglazio'via f. Named for Auguste F. M. Glaziou (1828–1906), French plant-collector in Brazil.

Neolloy'dia f. Cactus. Named for Francis Ernest Lloyd (1868–1947), American plant physiologist, author of *The Carnivorous Plants*, for many years professor of botany at Montreal. The suffix *neo-*, new, was used owing to the existence of the name *Lloydia*.

Neoma'rica f. From Gr. *neo-*, new; *Marica*, a genus of *Iridaceae*; presumably named after the river nymph Marica.

Neoporte'ria In honour of Carlos Porter, Chilean entomologist.

Neorege'lia f. In honour of E. A. von Regel; see *Aregelia*.

nepalen'sis, -is, -e From Nepal, central Himalayas; also rendered *napaulensis* and *nepaulensis*.

Nepen'thes f. Pitcherplants. From the name of a plant in Greek literature which could assuage grief and, classically, any plant capable of producing euphoria. Used here in connexion with its supposed medical qualities.

Ne'peta f. Ground-ivy. Latin name for these herbaceous plants which include catnip.

nepetoi'des Like *Nepeta*.

Nephro'dium n. From Gr. *něphrodēs*, kidney-like, referring to the indusium of these ferns.

Nephro'lepis f. Sword fern. From Gr. *něphrōs*, a kidney; *lěpis*, a scale; an allusion to the form of the indusium.

nephro'lepis, -is, -e With kidney-shaped scales.

Nephthy'tis f. From Nephthys, in Egyptian mythology, mother of Anubis and wife of Typhon.

Neptu'nia f. From L. *Neptunus*, Neptune, god of the sea, rivers and fountains. In allusion to its watery habitat.

neriifo'lius, -a, -um Oleander-leaved.

Neri'ne f. Bulbous plants named for the water nymph of that name, doubtlessly in allusion to the legend that bulbs of the type-species, the Guernsey lily (*Nerine sarniensis*), were cast on to the shore of Guernsey after a shipwreck: it appears, however, that bulbs brought from South Africa were given to John de Saumarez, Dean of Guernsey, about 1655 by mariners whose ship was cast ashore there.

Ne'rium n. Oleander, rose bay. The classical Greek name. Although oleanders are very poisonous, these flowering shrubs are very popular where the climate is suitable. Men have died from drinking tea stirred with oleander twigs.

neriiflo'rus, -a, -um Oleander-flowered.

Ner'tera f. Gr. *nerteros*, lowly; from the habit of growth of these creeping herbs.

nerva'tus, -a, -um; ner'vis, -is, -e With nerves, i.e. fairly strong unbranched veins or ribs.

nervo'sus, -a, -um With well-developed or conspicuous nerves.

neurocar'pus, -a, -um With prominently nerved or veined fruits.

nevaden'sis, -is, -e Of the Sierra Nevada, Spain or California or the state of Nevada, U.S.A.

Neviu'sia f. Snow wreath. An ornamental shrub named for Ruben Denton Nevius (1827–1913), of Alabama, who discovered it there.

nicae'ensis, -is, -e Of Nice (formerly Nicaea Maritima), southern France, or Iznik (formerly Nicaea), Turkey.

Nican'dra f. Apple-of-Peru. An annual herb named for Nikander of

Colophon, a poet who wrote of plants and their medical uses, *c.* 100 B.C.

Nicode'mia f. In honour of Gaetano Nicodemo (d. 1803), Italian botanist from Naples, curator of the Lyons botanic garden from 1799 to his death by murder or suicide in 1803.

Nicola'ia f. Torch ginger. In honour of the Czar Nicholas I (1796–1855) of Russia.

Nicotia'na f. Tobacco. Named for Jean Nicot (1530–1600), French ambassador to Lisbon who introduced tobacco into France.

nic'titans Blinking; moving.

Nidula'rium n. L. *nidus*, a nest. A reference to the position and appearance of the flower heads of these epiphytes.

ni'dus A nest.

Nierember'gia f. Cup-flower. Perennial herbs named for Juan Eusebio Nieremberg (1595–1658), Spanish Jesuit and author of a book on the marvels of nature published in 1635.

Nigel'la f. Love-in-a-mist. Fennel-flower. From the diminutive of L. *niger*, black; in allusion to the colour of the seeds of these annuals.

ni'ger, ni'gra, -um Black. Often applied to roots, as in the case of the Christmas rose, *Helleborus niger*, as well as to flowers.

ni'gratus, -a, -um; nigres'cens Blackish.

ni'gricans Blackish.

nigropuncta'tus, -a, -um With small black spots or dots.

nilo'ticus, -a, -um From the valley of the Nile.

Ni'pa f. From the Moluccan name for this palm.

Niphae'a f. From Gr. *niphas*, snow. In allusion to the white unspotted flowers of the type-species; *Phinaea* is an anagram of this name.

nippo'nicus, -a, -um Japanese.

ni'tens; nitidus, -a, -um Shining.

niva'lis, -is, -e; niv'eus, -a, -um; nivo'sus, -a, -um Snow-white; growing near snow.

Nive'nia f. In honour of James Niven (*c.* 1774–1826), Scottish gardener and collector in South Africa, 1798–1803.

no'bilis, -is, -e Notable, famous, renowned, excellent.

noctiflor'us, -a, -um; noctur'nus, -a, -um Night-flowering.

nodiflor'us, -a, -um Flowering at the nodes.

nodo'sus, -a, -um Having conspicuous joints or nodes.

nodulo'sus, -a, -um With small nodes.

Nolana

Nola'na f. Chilean bellflower. L. *nola*, a small bell; from the form of the corolla of these prostrate herbs.

Noli'na f. Bear-grass. Short-trunked desert plants named for P. C. Nolin, French agricultural writer *c.* 1755.

noli-tan'gere Touch-me-not. Specific name of *Impatiens*.

Nol'tea f. Soap-bush. South African shrubs named for Ernst Ferdinand Nolte (b. 1791), early 19th-century professor of natural history at Kiel.

Noma'phila f. From Gr. *nŏmŏs*, meadow; *philos*, friend.

Nomocha'ris f. From Gr. *nŏmŏs*, meadow; *charis*, grace, loveliness.

non-scrip'tus, -a, -um Without markings.

nootkaten'sis, -is, -e From Nootka Sound, British Columbia.

Nopa'lea f. From the Mexican name for these cacti, upon a species of which the cochineal insect is cultured.

Nopalxo'chia f. Reputedly from the Aztec name for these epiphytic cacti.

nor'thiae In honour of Marianne North (1830–1890), botanical painter and traveller whose paintings are exhibited in the North Gallery at Kew.

norve'gicus, -a, -um Norwegian.

nota'tus, -a, -um Spotted, marked.

Nothofa'gus f. Southern beech. From Gr. *nŏthŏs*, false; L. *fagus*, beech. Closely related to beech and from the southern hemisphere.

Notholae'na f. From Gr. *nŏthŏs*, false; *chlaina*, cloak; in allusion to the incomplete indosium.

Notholi'rion n. From Gr. *nŏthŏs*, false; *lĕiriŏn*, lily. This group of lily-like bulbous plants has been included both in *Lilium* and *Fritillaria* but is now considered distinct from both.

Notho'panax m. From Gr. *nŏthŏs*, false; *panax*, ginseng. Evergreen shrubs with palmate leaves, they resemble true *Panax*.

Nothoscor'dum n. Gr. *nŏthŏs*, false; *skŏrdon*, garlic. These bulbous herbs are related to *Allium* (garlic, onion, etc.) but lack its flavour and odour.

no'thus, -a, -um False.

Noto'nia f. In honour of Benjamin Noton, assay master in Bombay, 1812–1835, who collected in southern India.

Notospar'tium n. Gr. *nŏtŏs*, southern; *spartiŏn*, broom. This New Zealand genus is a southern counterpart of the European brooms.

Noty'lia f. Gr. *nŏtŏn*, back; *tylŏs*, hump.

novae-ang'liae Of New England, U.S.A.

novae-caesar'eae Of New Jersey, U.S.A.

novae-zelan'diae Of New Zealand.

noveboracen'sis, -is, -e; novi-bel'gii From New York.

nubi'cola Dweller among the clouds.

nubi'genus, -a, -um Born among the clouds.

nu'cifer, nuci'fera, -um Bearing nuts.

nuda'tus, -a, -um; nu'dus, -a, -um Naked; bare. Also:
  nudicau'lis bare-stemmed
  nudiflo'rus with flowers coming before the leaves

numi'dicus, -a, -um From Algeria.

nummula'rius, -a, -um Resembling a coin, usually applied to plants with small almost circular leaves.

Nu'phar f. Spatter-dock. Yellow water-lily. From the Arabic name.

nu'tans Nodding.

nutka'nus From Nootka Sound, British Columbia.

nuttal'lii In honour of Thomas Nuttall (1786–1839), English botanist, who travelled extensively in U.S.A. in 1811–1834.

nyctagi'neus, -a, -um Night-blooming.

Nyctan'thes f. Gr. *nyx, nyktos*, night; *anthos*, a flower. The flowers of this small Indian tree or shrub open at nightfall and drop at dawn.

Nyctoce'reus m. Gr. *nyx, nyktos*, night; *cereus*, cactus. They are night-blooming.

nymansen'sis Raised at Nymans Gardens, Handcross, Sussex, England.

Nymphae'a f. Water-lily. Named for Nymphe, one of the water nymphs.

Nymphoi'des f. Resembling *Nymphaea*.

Nys'sa f. Tupelo, sour gum. A great favourite of bees. Named for Nysa or Nyssa, one of the water nymphs. The cotton gum, *Nyssa aquatica*, grows in swamps.

O

obcon'icus, -a, -um Shaped like an inverted cone.

obe'sus, -a, -um Fat; succulent.

obfusca'tus, -a, -um Confused; clouded.

obla'tus, -a, -um Flattened at the ends, hence used of rounded shapes broader than long.

obli'quus, -a, -um Lopsided, oblique.

oblitera'tus, -a, -um Erased, suppressed.

oblonga'tus, -a, -um Oblong.

oblongifo'lius, -a, -um With oblong leaves.

oblon'gus, -a, -um Oblong, i.e. with almost parallel sides.

obova'tus, -a, -um Inverted ovate, i.e. egg-shaped with broadest end uppermost.

obscu'rus, -a, -um Dark, dusky, indistinct, uncertain.

obsole'tus, -a, -um Rudimentary, scarcely apparent.

obtusa'tus, -a, -um Blunted.

obtusifo'lius, -a, -um Blunt-leaved.

obtu'sior Blunter.

obtu'sus, -a, -um Blunt.

obvallar'is, -is, -e; obvalla'tus, -a, -um Apparently surrounded by a wall.

occidenta'lis, -is, -e Western.

occita'nus, -a, -um Relating to Occitania, i.e. Languedoc, southern France.

oceanen'sis, -is, -e Growing near the sea; of the sea.

ocella'tus, -a, -um With an eye; having a spot enclosed by another spot of a different colour, or with a circular patch of colour.

Ochaga'via f. In honour of Sylvestri Ochagavia, minister of education in Chile from 1853 to 1854.

Och'na f. Tropical trees or shrubs. Greek name for the wild pear to which the foliage of this genus has some resemblance.

ochra'ceus, -a, -um Ochre-coloured.

ochroleu'cus, -a, -um Yellowish-white.

ocimoi'des Resembling *Ocimum*.

O'cimum n. Basil. From the Greek name *ōkimŏn* for an aromatic herb, possibly this one.

Oco'tea f. Name coined by Aublet, presumably latinized from a French Guiana vernacular name.

octan'drus, -a, -um Having eight stamens.

Octome'ria f. Gr. *okto*, eight; *meris*, a part; in allusion to the eight pollen-bearing masses.

octope'talus, -a, -um Having eight petals.

ocula'tus, -a, -um With an eye, provided with a circular patch of colour.

oculiro'seus, -a, -um With a rose-coloured eye.

ocymoi'des Resembling basil (*Ocimum*).

odessa'nus, -a, -um From Odessa, southern European U.S.S.R.

Odontade'nia f. From Gr. *ŏdŏus, ŏdŏntŏs*, tooth; *adĕn*, gland.

Odon'tia f. Collective name for hybrids between the orchid genera *Miltonia* and *Odontoglossum*.

Odontio'da f. Collective name for hybrids between the orchid genera *Cochlioda* and *Odontoglossum*.

odontochi'lus, -a, -um Having a toothed lip.

Odontoglos'sum n. Gr. *ŏdŏus, ŏdŏntŏs*, a tooth; *glōssa*, a tongue. The lip of these epiphytic orchids is toothed.

Odontone'ma n. Gr. *ŏdŏus, ŏdŏntŏs*, a tooth; *nēma*, a thread. The stamens have toothed filaments.

Odontoso'ria f. Gr. *ŏdŏus, ŏdŏntŏs*, a tooth; *sorus*, a spore case.

odora'tus, -a, -um; odori'fer, -a, -um; odo'rus, -a, -um Fragrant. Also *odoratissimus*, very fragrant.

Oenan'the f. Gr. name for a plant smelling of wine, from *oinos* (*oenos*), wine.

Oenocar'pus m. Wine palm. From Gr. *oinos*, wine; *karpos*, fruit.

Oenothe'ra f. Evening primrose. The Greek name *ŏinōthĕras*, which is sometimes supposed to derive from *oinos*, wine; *thera*, booty, is more likely a corruption of *ŏnŏthĕras*, from *ŏnŏs*, ass; *thĕra*, hunting, chase, pursuit, or *thĕr*, wild beast, comparable to the alternative name *ŏnagra*. Originally, of course, it had nothing to do with the yellow-flowered American plants to which it has been transferred.

officina'lis, -is, -e Sold in shops; applied to plants with real or supposed medicinal properties.

officina'rum Of shops, generally those of apothecaries.

oldha'mii In honour of Richard Oldham (1837-1864), Kew gardener who collected in China.

O'lea f. Olive. The Latin name for this most long-lived and beautiful of fruit trees which from antiquity has been a symbol of peace and good will.

olean'der Specific epithet of a beautiful flowering shrub belonging to the genus *Nerium q.v.*, derived from the Italian *oleandro* and referring to the olive-like leaves.

233

Oleandra

Olean'dra f. In allusion to the long narrow oleander-like fronds of these ferns.

Olea'ria f. Tree aster, daisy bush. Australian and New Zealand trees and shrubs named in honour of Adam Ölschläger (1603–1671), whose name was latinized as Olearius, author of a flora of Halle.

ole'ifer, olei'fera, -um Oil-bearing.

oleifo'lius, -a, -um With leaves like the olive.

oleoi'des Resembling olive.

olera'ceus, -a, -um Of the vegetable garden; applied to vegetables and potherbs.

oligan'thus, -a, -um With few flowers. Also:
  oligocar'pus   with few fruits
  oligophyl'lus   with few leaves
  oligosper'mus   with few seeds

Oli'nia f. In honour of a Swedish medical man Johan Henrik Olin (1769–1824) of Växjö, who was a student of Thunberg's.

olitor'ius, -a, -um Pertaining to culinary herbs.

oliva'ceus, -a, -um Greenish-brown; olive-coloured.

Oliveran'thus m. Named for George Watson Oliver (1857–1923), horticultural expert in United States Department of Agriculture. The genus is separated from *Echeveria*.

olive'ri Of Daniel Oliver (1830–1916), from 1864 to 1890 keeper of the Kew Herbarium.

olivie'ri Of Guillaume Antoine Olivier (1756–1814), French entomologist and plant collector, traveller in the Orient from 1794 to 1798.

Olusa'trum n. L. name *ŏlus atrum*, black herb, from *ŏlus*, garden herb; *ater*, black; in allusion to the black seeds of this umbellifer.

olym'picus, -a, -um Of Mount Olympus, a name used for a number of high mountains in the Balkan Peninsula and Asia Minor and elsewhere.

omeien'sis, -is, -e Of Mount Omi, western Szechwan, China.

omor'ika Local name of the Serbian spruce, of which this is the specific epithet.

Omphalo'des f. Gr. *ŏmphalōdēs*, navel-like, from *ŏmphalŏs*, a navel. The nutlet hollowed out on one side of these herbs resembles the human navel.

Omphalogram'ma n. From Gr. *ŏmphalŏs*, navel; *gramma*, letter, marking.

Onci'dium n. Diminutive of Gr. *ŏnkŏs*, a tumour. The lip of these epiphytic orchids bears swellings.

On'coba f. Latinized form of the Arabic name, *onkob*, for these shrubs or small trees.

Onobry'chis f. Sainfoin. Gr. *ŏnŏbrychis*, apparently from *ŏnŏs*, an ass, and *brychis*, branchlet, or *bryche*, bellowing, braying; asses are said to bray for it.
The English name means holy clover (sometimes saintfoin) because it was said to have been the hay in Christ's manger.

Ono'clea f. Greek name, *ŏnŏklĕia*, for another plant, possibly borage, and connected with *ŏnŏs*, an ass (meaning 'good for asses only'), but applied here because of an alternative meaning: *onos*, a vessel; *kleio*, to close; refers to the closely rolled fertile fronds of these ferns.

Ono'nis f. Rest-harrow. The classical Greek name, *ŏnōnis*, for these plants.

Onopor'dum n. Scotch thistle. The Greek name *ŏnŏpŏrdŏn*, from *ŏnŏs*, an ass; *pŏrdē*, fart; for these thistlelike herbs.

Ono'pteris f. Gr. *ŏnŏs*, an ass; *ptĕris*, fern.

Onos'ma f. Golden drop. Gr. *ŏnŏs*, an ass; *ŏsmē*, a smell; the roots of some of these bristly herbs and sub-shrubs are supposed to smell like an ass.

Onosmo'dium n. False gromwell. Resembles *Onosma* (*q.v.*) to which it is closely related.

Ony'chium n. Gr. *ŏnyx*, a claw. In allusion to the shape of the lobes of the fronds of these ferns.

ophioglossifo'lius, -a, -um With leaves like adder's tongue fern.

Ophioglos'sum n. Adder's tongue fern. Gr. *ŏphis*, a snake; *glōssa*, a tongue; from the shape of the fronds.

Ophiopo'gon m. From Gr. *ŏphis*, snake; *pōgōn*, beard; allusion obscure.

Oph'rys f. The Greek name for a two-leaved orchid.

Ophthalmophyl'lum n. From Gr. *ophthalmŏs*, eye; *phyllŏn*, leaf.

Oplis'menus m. From Gr. *hŏplismos*, equipment for war, weapon. The spikelets of these grasses have awns.

Oplo'panax m. From Gr. *hŏplŏn*, weapon; *panax*, a related genus. In allusion to its ferociously spiny habit, which has earned it the name of 'Devil's Club'.

Opo'panax m. Gr. *ŏpŏs*, a milky juice especially that of the fig-tree; *panax*, a remedy. The sap of these perennial herbs was once used medicinally. Today, gum opopanax is obtained from the root for use in perfumery.

opori'nus, -a, -um Autumnal; from Gr. *ŏpōra*, autumn.

oppositifo'lius, -a, -um With leaves growing opposite to each other on each side of a stem.

opuloi'des Resembling the guelder-rose, *Viburnum opulus.* Also: *opuliflorus* and *opulifolius,* with flowers and leaves, respectively, like the guelder-rose.

Opun'tia f. Prickly pear. Greek name for a different plant which grew around the ancient town of Opus in Greece, the chief town of a tribe called from it the Locri Opuntii; the name *Opus* is supposed to mean the town of figs (see *Opopanax*).

orbicula'ris, -is, -e; orbicula'tus, -a, -um Round and flat; disc-shaped.

orchi'deus, -a, -um; orchioi'des; orchoi'des Orchid-like.

orchidiflo'rus With orchid-like flowers.

Or'chis f. Classical Greek name, from *orchis,* a testicle; in allusion to the paired rounded tubers of some species. For that reason, *Orchis* has been regarded since antiquity as an aphrodisiac.

orega'nus, -a, -um; orego'nus, -a, -um From the State of Oregon, U.S.A., or the old Oregon territory of the Hudson's Bay Company, which included the later States of Washington and Oregon.

Oreodo'xa f. From Gr. *ŏrŏs,* mountain; *doxa,* glory. In allusion to the stately habit of these palms.

Oreo'panax m. Gr. *ŏrŏs,* a mountain; plus *Panax.* These shrubs and trees grow on mountains.

oreo'philus, -a, -um Mountain-loving.

Oreo'pteris From *ŏrŏs,* mountain; *ptĕris,* fern.

ores'bius, -a, -um Gr. *ŏrĕsbiŏs,* living on mountains.

organen'sis, -a, -um From the Organ Mountains of Brazil.

orgya'lis, -is, -e The distance between the tips of the fingers when the arms are extended, i.e. about six feet; from *orgya,* a classical and medieval measure corresponding to the fathom.

orienta'lis, -is, -e From the Orient; Eastern.

origanifo'lius, -a, -um With leaves like marjoram or *Origanum.*

origanoi'des Resembling marjoram.

Ori'ganum n. Marjoram. The classical Greek name *ŏrīganŏn* for these aromatic herbs.

Ori'xa f. From the Japanese name for this shrub.

Or'menis f. Presumably named for Ormenis, granddaughter of Ormenus, a king of Thessaly.

236

Ormo'sia f. Gr. *hŏrmos*, a necklace. The scarlet seeds of *O. coccinea* are strung as beads in Guyana where the tree is indigenous.

ornatiss'imus, -a, -um Very showy.

orna'tus, -a, -um Ornamental, showy.

Ornithi'dium n. Gr. *ornis*, a bird; *eidos*, a shape. The upper lip of the stigma is beaklike in these epiphytic orchids.

ornithoce'phalus, -a, -um Bird-headed.

Ornithochi'lus m. Gr. *ornis*, a bird; *cheilos*, a lip; in allusion to the shape of the lip of these epiphytic orchids.

Ornitho'galum n. Gr. *ornis*, a bird; *gala*, milk. The flowers of these bulbous plants are usually white. The bulbs of star-of-Bethlehem (*O. umbellatum*) are supposed by some to have been the 'dove's dung' of the Bible of which a 'cab' measure was sold for a shekel during the Babylonian siege of Jerusalem.

ornitho'podus, -a, -um; orni'thopus Resembling a bird's foot.

Oroban'che f. Broomrape. From Gr. *ŏrŏbŏs*, a kind of vetch; *anche*, strangle.

Oron'tium n. Golden club. The Greek name, *ŏrŏntiŏn*, now applied to this North American aquatic perennial, is said to have belonged to some plant growing in the Syrian river Orontes.

Oro'ya f. Name of a village in Peru, in the Andes above Lima, where the type-species grows.

Oroxy'lon n. Gr. *ŏrŏs*, a mountain; *xylon*, wood; the tree grows also on plains.

Orphanide'sia f. In honour of Theodoros Georgios Orphanides (1817–1886), professor of botany at Athens, mid-19th century.

ortho- In compound words signifying upright or straight. Thus:

ortho'botrys   with upright clusters
orthocar'pus   with upright fruit
orthocla'dus   with straight branches
orthoglos'sus   with a straight tongue
ortho'pterus   straight-winged
orthose'palus   straight-sepalled

Orthocar'pus m. Gr. *orthos*, upright; *karpos*, fruit; from the small upright pods on these herbs.

Orthrosan'thus m. From Gr. *ŏrthrŏs*, morning; *anthos*, flower. 'From the flowers expanding in the morning, and fading before noon'.

Orychophrag'mus m. From Gr. *ŏrychō*, dig up; *phragma*, fence, partition; in allusion to the minutely pitted septum of the pod.

Ory'za f. Rice. The English name *rice* is ultimately derived from the

Osbeckia

Latin and Greek *ŏrȳza*, which in turn came from an Asiatic name. Rice (*O. sativa*) is one of the staple cereals, world production being of the order of 100 million tons annually. It originated from wild grass in India but is now grown throughout the tropical and warm temperature regions, wherever conditions are suitable. With the exception of the so-called 'hill rice' which is grown on dry land, all varieties need to grow in water for part of their development.

Osbe'ckia f. In honour of Pehr Osbeck (1723–1805), Swedish clergyman, a student of Linnaeus, who travelled to India and China.

Osman'thus m. Sweet olive. Gr. *ŏsmē*, fragrance; *onthŏs*, a flower. All the species of these ornamental shrubs and small trees are very fragrant.

osman'thus, -a, -um With fragrant flowers.

Osmaro'nia f. Osoberry. Gr. *ŏsmē*, fragrance; *Aronia*, choke-cherry. Resembles choke-cherry and is very fragrant.

Osmun'da f. Derivation unknown, often said to be named for Osmundus or Asmund *c.* 1025, a Scandinavian writer of runes who helped prepare the way for the Swedish acceptance of Christianity. *O. regalis*, the Royal fern, is a very beautiful British species.

Osteome'les f. Gr. *ŏstĕŏn*, a bone; *mēlŏn*, an apple; an allusion to the fruit of these evergreen shrubs.

Ostrow'skia f. Giant bellflower. Named for Michael Nicholazewitsch von Ostrowsky, Russian Minister of Imperial Domains *c.* 1884 and a patron of botany.

Os'trya f. Hop-hornbeam. From the Greek name, *ŏstrys*, for this hard-wooded tree.

Otan'thus m. From *ŏus, ōtŏs*, ear; *anthŏs*, flower.

Othon'na f. Greek name for a plant other than this genus, some members of which are succulent.

Ottelia f. Latinized from the Malabar name *ottel-ambel*.

otto'nis In honour of Eduard Otto (1812–1885), botanical collector in Cuba and Venezuela, later curator of Hamburg botanic garden, or his father Friedrich Otto (1782–1856), curator of the Berlin botanic garden.

ouletri'chus, -a, -um With curly hair.

Ouri'sia f. Small but elegant herbs, named for Ouris, governor of the Falkland Islands, where Commerson, Bougainville's naturalist, collected it on their voyage round the world, 1766–1768.

ova'lis, -is, -e Oval, broadly elliptic.

ova'tus, -a, -um Ovate; egg-shaped, with the broad end at the base.

o'vifer, ovi'fera, -um; o'viger, ovi'gera, -um Bearing eggs, i.e. egg-shaped structures.

ovi'nus, -a, -um Relating to sheep; providing fodder for sheep.

Oxa'lis f. Wood sorrel. Gr. *oxys*, acid, sour, sharp. *O. acetosella*, wood-sorrel, is one of the several plants called shamrock (see p. 340). The delicious sorrel soup of France is made from *Rumex acetosa*.

Oxe'ra f. Gr. *ŏxys*, acid. The sap of the shrub is sour.

oxyacan'thus, -a, -um Sharp-spined.

Oxycoc'cos m. From Gr. *ŏxys*, acid; *kokkos*, a round berry.

Oxyden'drum n. Sour-wood. Gr. *ŏxys*, acid; *dendron*, a tree. The foliage is bitter.

oxygo'nus, -a, -um With sharp angles.

Oxylo'bium n. From Gr. *oxys*, sharp; *lobos*, pod.

oxy'lobus, -a, -um With sharp-pointed lobes.

Oxype'talum n. From Gr. *ŏxys*, sharp; *petalum*, in botanical L. a petal.

oxy'philus, -a, -um Loving acid soil.

oxyphyl'lus, -a, -um With sharp-pointed leaves.

oxypterus, -a, -um With sharp-pointed wings.

Oxyria f. Gr. *ŏxys*, referring to the sharp or bitter taste of this northern perennial herb with antiscorbutic properties.

Oxy'tropis f. Gr. *ŏxys*, sharp; *trŏpis*, a keel; an allusion to the form of the flower on these perennial herbs and shrubs.

P

pabula'ris, -is, -e Providing pasture or fodder.

Pachi'ra f. From the native Guyanese name for these trees.

Pachis'tima f. Ornamental evergreen woody plants. From Gr. *pachys*, thick; *stigma*, a stigma; an irregularly formed name like many of Rafinesque's coinages.

pachy- In compound words signifies thick. Thus:

pachyan'thus	thick-flowered
pachycar'pus	with a thick pericarp
pachyphloe'us	with thick bark
pachyphyl'lus	thick-leaved
pachy'podus	with a thick foot, i.e. stalk
pachy'pterus	with thick wings
pachy'trichus	with thick hair

Pachycereus

Pachyce′reus m. Gr. *pachys*, thick; *cereus*, cactus. These cacti have very stout stems.

Pachyphrag′ma n. From Gr. *pachys*, thick; *phragma*, fence, partition; in allusion to the stout ribbed septum of the pod.

Pachy′phytum f. Gr. *pachys*, thick; *phyton*, a plant. Both stems and leaves are thickened on these succulent plants.

Pachyrhi′zus m. Gr. *pachys*, thick; *rhiza*, a root. The roots of these twining herbs are thick and tuberous.

Pachysan′dra f. Mountain spurge. Gr. *pachys*, thick; *aner*, *andros*, in the sense of stamen. These herbs and sub-shrubs have flowers with thickened white filaments.

Pachy′stachys Gr. *pachys*, thick; *stachys*, a spike; in allusion to the dense flower clusters of these tropical perennials.

paci′ficus, -a, -um Of the Pacific Ocean.

padus Greek name of a wild cherry.

Paede′ria f. From L. *paedor*, filth, stench. With reference to the unpleasant smell of the leaves, etc.

Paeo′nia f. Peony. The classical Greek name said to be derived from the name of the physician Paeon, a pupil of Aesculapius to whom Leto (mother of Artemis and Apollo) gave the plant, telling him of its merits. In gratitude for the cures wrought by Paeon, Pluto changed him into the plant. We still speak of a paean of praise.

palaesti′nus, -a, -um From Palestine.

Palafo′xia f. In honour of an unspecified Spaniard, either the prelate Juan de Palafox y Mendoza (1600–1659), or the general José de Palafox y Melzi (1780–1847).

Pala′quium n. Latin version of the Philippine name for these trees.

Pa′laua f. In honour of Antonio Palau (d. 1793), Spanish botanist.

palhin′hae In honour of Ruy Telles Palhinha (1871–1957), Portuguese professor of botany, author of a catalogue of plants of the Azores (1966).

Paliso′ta f. Perennial herbs named for Baron A. M. F. J. Palisot de Beauvois (1752–1820), French botanist and traveller, author of *Flore d'Oware et Benin* (1804–1821) and works on grasses and mosses.

Paliu′rus m. The ancient Greek name for the terribly spiny Christthorn (*Paliurus spina-christi*) from which the crown of thorns is traditionally supposed to have been made.

pal′lens; pal′lidus, -a, -um Pale. Also *pallescens*, rather pale.

pallia′tus, -a, -um Shrouded, cloaked.

papaveraceus

palma'ris, -is, -e A hand's breadth (about 3 in.) high or wide.

palma'tus, -a, -um Palmate, i.e. lobed like a hand with fully out-spread fingers.

palmifo'lius, -a, -um Palm-leaved.

paludo'sus, -a, -um; palus'tris, -is, -e Marsh-loving.

Pamian'the f. In honour of Major Albert Pam (1875–1955) of Wormley-bury, Hertfordshire, author of *Adventures and Recollections* (1945), a financier and specialist in the cultivation of bulbous plants, who re-received this fine member of the *Amaryllidaceae* in 1926 among bulbs collected for him in Peru.

Pa'nax m. Ginseng. From Gr. *panakēs*, all-healing, a panacea; in allusion to the obsessive value placed on it by the Chinese for aphro-disiacal and medicinal purposes. American ginseng is very closely related to the Asiatic ginseng (*P. schinseng* from which the English name derives) and so it became an important item of trade with Canton during the palmy days of the China trade (*c.* 1810–40).

Pancra'tium n. Mediterranean lily, sea daffodil. Greek name for a bulbous plant.

Pan'danus m. Screw-pine. Latinized version of the Malayan name, *pandan*.

Pando'rea f. Ornamental twiners. Named for Pandora. In Greek mythology she was the first woman sent to earth by Zeus in subtle revenge for the theft of fire by Prometheus. The gods endowed her with their choicest gifts, while Zeus himself gave her a box which was to be given to the man she married, whom Zeus intended to be Prometheus. He, however, was too crafty to be caught in this way, but his brother married Pandora and opened the box, releasing all the evils Zeus intended for Prometheus, though he managed to retain Hope. The type-species was associated with an insect plague on Norfolk Island, whence the name. *Podranea* is an anagram of *Pandorea*.

pandura'tus, -a, -um Fiddle-shaped.

panicula'tus, -a, -um With flowers arranged in panicles.

Pani'cum n. Millet. The Latin name.

panno'nicus, -a, -um From Central Europe—Hungary and Yugo-slavia.

panno'sus, -a, -um Tattered.

panormita'nus, -a, -um Of Palermo (ancient Panŏrmŏs), Sicily.

panteumor'phus, -a, -um Altogether beautiful.

Papa'ver n. Poppy. The Latin name.

papavera'ceus, -a, -um Poppy-like.

Paphinia

Paphi'nia f. From Gr. *paphia*, an epithet of Aphrodite (Venus) whose chief seat of worship was at Paphos on the island of Cyprus. A genus of beautiful epiphytic orchids.

Paphiope'dilum From Gr. *paphia* (see above); *pĕdilŏn*, sandal. A group of orchids from tropical and sub-tropical regions.

papi'lio Latin word for butterfly. Specific epithet for the butterfly orchid (*Oncidium papilio*).

papil'liger, papilli'gera, -um; papillo'sus, -a, -um Having papillae, soft protuberances on a surface.

papyra'ceus, -a, -um Papery.

papy'rifer, papyri'fera, -um Paper-bearing.

Papy'rus m. Greek name of the paper made in rolls from the pith of *Cyperus papyrus* in Ancient Egypt.

Paradi'sea f. St. Bruno's lily. Named for Count Giovanni Paradisi (1760–1826) of Modena.

paradi'si; paradisi'acus, -a, -um Belonging to Paradise.

parado'xus -a, -um Paradoxical, contrary to expectation.

Parahe'be f. Gr. *para*, near; *Hebe*, from which genus this species has been cut off.

Paraqui'lega f. Gr. *para*, near, close by, alongside; *Aquilegia*, to which these mountain perennials are akin.

parasi'ticus, -a, -um Parasitic; an epithet sometimes used by early authors for epiphytes as well as parasites.

pardalian'thes Greek name of a plant used for poisoning leopards and other beasts of prey, hence the English name Leopardsbane.

pardali'nus, -a, -um; pardi'nus, -a, -um Relating to leopards, spotted like a leopard.

pardanthi'nus, -a, -um Spotted like Pardanthus (Gr. *pardos*, male panther; *anthos*, flower), a synonym of Belamcanda.

Parieta'ria f. Pellitory. The name derives from the L. *pariētārius*, of walls, from *paries*, a wall, where the plant likes to grow, as Pliny knew when he described it.

Pa'ris f. Herb-Paris. L. *par*, equal; alluding to the regularity of the parts of this small rhizomatous perennial.

pari'shii In honour of the Rev. Charles Samuel Pollock Parish (1822–1897), chaplain at Moulmein, Burma, from 1852 to 1878, if a Burmese plant. Of Samuel Bonsall Parish (1838–1928), American botanist, if a Californian plant.

par'keri In honour of Richard Neville Parker, conservator of forests and forest botanist, author of *Common Indian Trees* (1933).

Parkinso'nia f. Tropical trees or shrubs named for John Parkinson (1567–1650), apothecary of London and author of two important books, *Paradisi in Sole Paradisus Terrestris* (1629) and *Theatrum Botanicum* (1640). Having perpetrated a gross pun in the title of the first (it can be translated as 'Park in sun's earthly paradise'), he proceeds to extol the joys of a garden and to describe in vivid detail the plants available.

Parmentie'ra f. Panama candle-tree. Named for Antoine Augustin Parmentier (1737–1813), French agricultural economist who promoted cultivation of the potato in France. Potage Parmentier (potato soup) is named after him.

parmula'tus, -a, -um Armed with a little round shield (*parmula*).

par'nassi Of Mount Parnassus, Greece.

Parnas'sia f. Grass-of-Parnassus. Shortened version of the 16th-century name *Gramen Parnassi*, referring to Mount Parnassus in Greece. Low-growing moisture-loving herbs.

parnas'sicus, -a, -um Of Mount Parnassus.

parnassifo'lius, -a, -um With leaves like *Parnassia*.

Paro'chetus m. Blue oxalis. Gr. *para*, near; *ŏchĕtŏs*, a brook; from the normal habitat in moist places of this trailing perennial, which occurs on high mountains in both Asia and Africa.

Paro'dia f. In honour of Lorenzo Raimundo Parodi (1895–1966), Argentinian botanist especially noted for his studies of South American grasses.

Parony'chia f. Whitlow-wort. Gr. *parōnychia*, a whitlow (from *para*, near; *onyx*, nail), for which these little herbs were thought to be a cure.

Parro'tia f. Ornamental shrubs and trees named for F. W. Parrot (1792–1841), German naturalist who climbed Mount Ararat in 1834.

Parrotiopsis f. From *Parrotia* and *opsis*, resemblance.

par'ryi In honour of Charles Christopher Parry (1823–1890), American botanical explorer.

Parthe'nium n. American feverfew. Greek name, *parthĕniŏn*, for some plants with white ray flowers.

parthe'nium Specific epithet for *Chrysanthemum* (see *Parthenium*).

Parthenocis'sus n. Boston ivy, Virginia creeper, woodbine, etc. From Gr. *parthenos*, a virgin; *kissos*, ivy; in allusion to its English vernacular name. The state of Virginia was named after England's virgin queen Elizabeth I. *P. quinquefolia*, Virginia creeper, is often listed under *Ampelopsis*.

parti'tus, -a, -um Parted.

parvibractea'tus, -a, -um With small bracts.

par'vus, -a, -um Small. In compound words *parvi-*, as in *parviflorus*, with small flowers; *parvifolius*, with small leaves; also *parvissimus*, very small.

Pas'palum n. Millet. The Greek name.

Passiflo'ra f. Passion-flower. L. *passio*, passion; *flos*, a flower. The name was given by the early missionaries in South America who thought they saw in the parts of the flower various aspects of Christ's crucifixion. The corona became the crown of thorns, the five anthers represented five wounds, the three styles three nails, and so on.

Pastina'ca f. Parsnip. From L. *pastus*, food; in allusion to the edible root. Nowadays often classified under *Peucedanum*.

pastora'lis, -is, -e Relating to shepherds, hence growing in pastures.

patago'nicus, -a, -um From Patagonia, South America.

patavi'nus, -a, -um From the neighbourhood of Padua, (earlier *Patavium*), Italy.

patell'aris, -is, -e; patelliformis, -is, -e Dish-shaped, saucer-shaped.

pa'tens; pat'ulus, -a, -um Spreading.

Patri'nia f. In honour of Eugène L. M. Patrin (1724–1815), French mineralogist who travelled from 1780 to 1787 in Siberia and eastern Asia and collected plants as well as minerals.

pau'cus, -a, -um Few. In compound words *pauci-*. Thus:
pauciflo'rus   with few flowers
paucifo'lius   with sparse leaves
pauciner'vus   with few nerves

Paulli'nia f. Twining shrubs named by Linnaeus for Simon Paulli (1603–1680), professor of anatomy, surgery and botany at Copenhagen, as stated by Linnaeus himself in his *Critica botanica* no. 238 (1737), and not Christian Franz Paullini (1643–1742), a German botanist at Eisenach, as stated by many authors.

Paulow'nia f. Beautiful Chinese and Japanese trees named for Princess Anna Paulowna (1795–1865), daughter of Czar Paul I of Russia.

pauper'culus, -a, -um Poor.

Pavet'ta f. Sinhalese vernacular name for small tropical shrubs and trees from India, Africa and Java.

Pavo'nia f. Tropical herbs and shrubs named for José Antonio Pavón (1754–1840), Spanish botanist and traveller in South America, part-author of the incomplete *Flora Peruviana*, of whose unfortunate history an account will be found in A. R. Steele's *Flowers for the King* (1964).

pavoni'nus, -a, -um Peacock-blue or with a distinct eye recalling that on a peacock's tail feather.

paxto'nii In honour of Sir Joseph Paxton (1801–1865), gardener and architect. Starting as a poor apprentice in the Chiswick garden of the Horticultural Society of London, he became head gardener to the Duke of Devonshire at Chatsworth, when aged only 25; he was manager of the Duke's princely estates and his general adviser and close friend. Having designed and built the huge Chatsworth conservatory of nearly 300 feet in length, he used this experience in glasshouse construction to help him to design the great Crystal Palace, the basic plan of which was prepared in nine days, for the Great Exhibition of 1851. Later, he designed many other important buildings, became a company director and a Member of Parliament. In gardening, he dispatched collectors to many parts of the world and introduced a number of new plants. The plant of the great water lily, *Victoria amazonica* (*V. regia*), which he brought into bloom at Chatsworth in 1849, was obtained from Kew.

Pear With very few exceptions, pears in cultivation are varieties of *Pyrus communis*. Centuries of breeding and tests of seedlings have resulted in much the same conditions that govern apples and for the same reasons (see *Apple*). The range of the pear, however, is much more limited than the apple. Pears cannot be grown, except on a very limited scale, much north of Zone Four in the United States, and only the warmer southern parts of England are really suitable for good growth of fine varieties. Pears can be, and are, grown in less favoured climates but they need special treatment and, generally, sheltered locations.

Few pears ripen well on the tree. Production of the best-quality fruit needs very special attention. Pears should be picked at just the right stage and ripened in cool, dark, dry quarters. For these reasons the finest pears seldom find their way into the markets and then only as speciality items. Indeed, even in private orchards, few people are prepared to take the trouble necessary to put really fine fruit on the table. At least, this is the case in the United States. It may well be different in France and Belgium where most of the very best sorts originated, as is indicated by the many French or Flemish names for the best sorts and the recurrence of the terms *beurré* (e.g. Beurré Bosc) and *doyenné* (e.g. Doyenné de Comice). The first means juicy and melting in the mouth; the second, melting and sugary.

The Romans thought highly of pears and presumably it was they who introduced them into England. At least two fine varieties are thought to antedate the Christian era. Both make small demands on the skill and patience of the grower and will ripen fairly satisfactorily on the tree. Jargonelle is one and is about the earliest of all. The other is

'Bon Chrétien', called 'Williams Bon Chrétien', or just 'Williams' in England and 'Bartlett' in the United States.

The 'Bon Chrétien' pear arrived in England some time in the latter part of the 18th century and acquired the name of Williams who was the great pear breeder of the day. About 1820, Enoch Bartlett of Dorchester, Massachusetts, released it in America under his own name. By the time that it was identified it was too late to change the name.

The finest American contribution to pear growing is the highly flavoured 'Seckel'. This was found on the Seckel farm near Philadelphia by a character called 'Dutch Jacob', a cattle dealer and sportsman, who brought back the pear from one of his expeditions. He kept the source secret but was finally forced to protect his find by buying the piece of land on which the tree stood. It was sent to London in 1819 where the Horticultural Society of London gave it approval. The great period of pear growing in America was in the fifty years after 1820, interest centring in eastern Massachusetts and, particularly, the Massachusetts Horticultural Society. In the middle of the period, Robert Manning of Salem was offering about 1,000 varieties and other growers' lists had many hundreds. Speciality shows with as many as 300 varieties shown continued into the 1900's.

An English pear specialist of the Victorian period was J. D. Blackmore, much better known as the author of *Lorna Doone*.

pectina'tus, -a, -um Comblike.

pecti'nifer, pectini'fera, -um Having a comb.

pectora'lis, -is, -e Relating to the chest, e.g. used in cough mixtures.

peda'lis, -is, -e About one foot (30 cm.) long.

pedati'fidus, -a, -um Cut like a bird's foot, i.e. with a few divisions radiating from the same centre.

peda'tus, -a, -um Like a bird's foot, see above.

pedemonta'nus, -a, -um Of Piedmont, Italy.

Pedicula'ris f. Louse-wort, wood-betony. From L. *pedicularis*, relating to lice, from *pediculus*, a louse. The presence of the plant in fields was supposed to produce lice in sheep.

Pedilan'thus m. Slipper-flower. From Gr. *pĕdilon*, sandal; *anthos*, a flower; from the appearance of the flowers on this succulent shrub.

Pediocac'tus m. Gr. *pĕdion*, a plain; plus *cactus*; from its habitat on the Great Plains.

peduncula'ris, -is, -e; peduncula'tus, -a, -um With a flower stalk or peduncle, usually a well-developed one.

pedunculo'sus, -a, -um With many or especially well-developed flower stalks or peduncles.

Pelargo'nium n. The geranium of the florists. From Gr. *pelargos,* a stork. The fruit has a beak not unlike that of a stork. Sometimes called stork's-bill.

Pelecy'phora f. From *Gr. pĕlĕkys,* a hatchet; *phoreo,* to bear; with reference to the strongly flattened tubercles of this cactus.

pelisseria'nus, -a, -um In honour of Guillaume Pellicier (1490–1568), bishop of Montpellier, the moral force behind the university's academic tolerance when Clusius, Platter, Lobel, J. Bauhin and Rabelais were students there.

Pella'ea f. Cliff-brake. From Gr. *pellaios,* dark; in allusion to the stalks of this fern which are generally dark.

Pellio'nia f. Creeping foliage plants named for Alphonse Odet Pellion (1796–1868), later a French admiral, who when young accompanied the French navigator Freycinet on his second voyage around the world in 1817 to 1820.

pellu'cidus, -a, -um Transparent, can be seen through.

Peltan'dra f. Arrow-arum. Gr. *pĕltē,* a shield; *aner,* a stamen.

Pelta'ria f. Shield-wort. Gr. *pĕltē,* a shield; from the form of the leaf of this garlic-scented herb.

pelta'tus, -a, -um Shield-shaped, i.e. attached on the lower surface away from and not at the margin.

Peltiphyl'lum n. Umbrella plant. Gr. *pĕltē,* a shield; *phyllon,* a leaf; from the form of the leaves of this waterside perennial, formerly known as *Saxifraga peltata.*

Pelto'phorum n. Gr. *pĕltē,* a shield; *phoreo,* to bear; from the shape of the stigma of these tropical trees.

pelvifor'mis, -is, -e Forming a shallow cup.

pemakoen'sis, -is, -e From the province of Pemako, Tibet.

penduli'nus, -a, -um; pen'dulus, -a, -um Hanging.

penicilla'tus, -a, -um Having a tuft of hair like a paint-brush.

peninsula'ris, -is, -e Growing on a peninsula.

Penioce'reus m. Gr. *pēnē,* a thread; *cērĕus,* cactus. The stems are very slender.

penna'tus, -a, -um Feathered, pinnate.

pen'niger, penni'gera, -um Bearing feathers, as featherlike leaves.

Pennise'tum n. From L. *penna,* a feather; *seta,* a bristle. The flower of these grasses has long, feathery bristles.

pennive'nius, -a, -um Pinnately veined.

pen'silis, -is, -e Hanging down.

Penstemon

Penste'mon (also, incorrectly, **Pentstemon**) m. Beard-tongue. Gr. *pĕnte*, five; *stēmōn*, a stamen. The plants have five stamens, one of them sterile.

pensylva'nicus, -a, -um From Pennsylvania.

penta- In compound words signifying five. Thus:

pentade'nius	with five glands
pentago'nus	five-angled
penta'gynus	with five pistils
pentan'drus	with five stamens
pentaphyl'lus	with five leaves
penta'pterus	five-winged

Pentaglot'tis From Gr. *pĕntĕ*, five; *glōtta*, tongue; in allusion to the five tongue-like scales of the corolla.

Pentaptery'gium n. Gr. *pĕntĕ*, five; *pterygion*, a small wing. In this epiphytic shrub the calyx has five small wings.

Pen'tas f. Gr. *pentas*, a series of five; with reference to the flower parts being in fives instead of fours as in related genera.

Pepero'mia f. Gr. *pĕpĕri*, pepper; *hŏmŏiŏs*, resembling. The plants resemble and are closely related to true pepper; see *Piper*.

peramoe'nus, -a, -um Very pleasing.

perbel'lus, -a, -um Very beautiful.

percus'sus, -a, -um Perforated, or apparently so.

peregri'nus, -a, -um Exotic; immigrant.

peren'nans; peren'nis, -is, -e Perennial.

Peres'kia f. Barbados gooseberry. Leafy cacti named for Nicholas Claude Fabre de Peiresc (1580–1637), French naturalist and archaeologist.

Pereskiop'sis f. *Pereskia* (*q.v.*) plus Gr. *opsis*, like. The two genera resemble each other.

Pere'zia f. American herbs named for Lorenzo Perez, 16th-century Spanish apothecary, author of a history of drugs, *Libro de Theriaca* (1573).

perfolia'tus, -a, -um; perfos'sus, -a, -um With the leaf surrounding or embracing the stem; perfoliate.

perfora'tus, -a, -um Having or appearing to have small holes.

pergra'cilis, -is, -e Very slender.

Pericly'menum n. Greek name, *pĕriklymĕnon*, for a honeysuckle.

Perilep'ta f. Gr. *pĕrileptos*, embraced. With reference to the stem-clasping leaf bases.

Peril'la f. Beefsteak plant. Derivation obscure; probably a diminutive of L. *pēra*, bag, wallet, in allusion to the form of the fruiting calyx.

Peri'ploca f. Silk vine. Gr. *pĕriplŏkē*, a twining around, from *pĕri*, around; *plŏkē*, something twisted or woven. Some species intertwine.

Periste'ria f. Dove plant. Gr. *pĕristĕra*, a dove. The column has a dovelike appearance.

Peri'strophe f. Gr. *pĕristrŏphē*, a turning around, from *pĕri*, around; *strŏphē*, turning, twist; in allusion to the twisted corolla tube.

permix'tus, -a, -um Much-mixed.

Pernet'tya (originally **Pernettia**) f. Ornamental shrubs of the heath family named for Antoine Joseph Pernetty (1716–1801), who visited the Falkland Islands and South America with Bougainville, 1763–1764, and wrote about it.

per'nyi In honour of Paul Hubert Perny (1818–1907), French Roman Catholic missionary in China.

Perov'skia f. Semi-woody plants named for V. A. Perovski (1794–c. 1857), Russian general.

perralderia'nus, -a, -um In honour of Henri René le Tourneux de la Perraudière (1831–1861), French naturalist who collected in the Canary Islands and North Africa.

Per'sea f. Avocado or alligator pear. From the Greek name, *pĕrsĕa*, for an Egyptian tree (*Cordia myxa*).

Persica'ria f. Medieval name of a knotweed, from *Persica*, peach, in allusion to the shape of the leaves.

persicifo'lius, -a, -um With leaves like a peach.

per'sicus, -a, -um Persian. The peach, the *pĕrsike* or *persica malus* (Persian apple) of the Ancients, reached Europe from China by way of Persia.

persis'tens Persistent.

Persoo'nia f. Shrubs or small trees named for Christian Hendrick Persoon (1761–1836), South African botanist, especially celebrated as a mycologist, who worked and died in Paris.

perspi'cuus, -a, -um Transparent.

pertu'sus, -a, -um Perforated; thrust through.

perula'tus, -a, -um Provided with perules (bud scales); mostly used when these are conspicuous or persistent.

peruvia'nus, -a, -um Peruvian.

Pescato'ria f. In honour of Monsieur Pescatore, a French orchid-specialist, c. 1852.

petaloi'deus, -a, -um Petal-like.

Petaloste'mum n. Prairie-clover. Gr. *pětalŏn*, a petal; *stēmōn*, a stamen. Petals and stamens are joined.

Petasi'tes m. Butter-bur. The Greek name, from *pětasos*, a hat with a broad brim, with reference to the large leaves of this herb.

petectica'lis, -is, -e Spotted, from Italian *petecchia*, spot produced by a fever.

petiolar'is, -is, -e; petiola'tus, -a, -um Furnished with a leaf-stalk or a particularly long one.

Petive'ria f. Shrubby herbs named for James Petiver (1665–1718), apothecary of London, an ardent collector of botanical specimens now in the British Museum (Natural History).

petr- In Gr. compounds can refer either to rock forming cliffs, ledges (*pětra*) or pieces of rock (*pětrŏs*).

petrae'us, -a, -um Rock-loving.

Pet'rea f. Tropical American woody plants named for Lord Robert James Petre (1713–1743), patron of botany and horticulture, called 'the Phenix of this age' by Collinson; both he and Collinson helped to introduce many plants into European gardens.

pet'riei In honour of Donald Petrie (1846–1925), New Zealand inspector of schools, a tireless amateur investigator of the flora of New Zealand.

Petrocal'lis f. From Gr. *pětrŏs*, rock; *kallis*, beauty; from the habitat and the beauty of the flowers.

Petrocop'tis f. From Gr. *pětrŏs*, a rock; *kopto*, to break. These dwarf plants grow in cracks and broken crevices of rock.

Petroma'rula f. Cretan vernacular name meaning 'rock lettuce'.

Petro'phila f. Gr. *pětrŏs*, rock; *phileo*, to love; from the habitat of these Australian shrubs.

Petrophy'tum n. Gr. *pětrŏs*, rock; *phyton*, a plant. These prostrate shrubs inhabit stony places.

Petrorha'gia f. From Gr. *pětrŏs*, a rock; *rhagas*, rent, chink, from *rhēgnymi*, break asunder. Equivalent to Latin *saxifraga*.

Petroseli'num n. Parsley. Gr. *pětrŏs*, a rock; *sělinŏn*, parsley or celery; i.e. wild parsley.

Pette'ria f. A yellow-flowered shrub named for Franz Petter (1798–1858), Austrian professor at Spalato who wrote on the botany of Dalmatia.

Petu'nia f. Latinized form of the Brazilian name *petun* for tobacco, to which these popular annuals are allied.

Peuce'danum n. Gr. *pĕukĕdanŏn*, parsnip. See *Pastinaca*.

Peu'mus f. Chilean boldo tree. The Chilean name latinized.

Pfei'ffera f. In honour of Louis Carl Georg Pfeiffer (1805–1877), many-sided, extremely industrious German medical man, translator, botanist and conchologist, botanically renowned for his work on cacti and his monumental *Nomenclator botanicus* (1873–1874).

Phace'lia f. California-bluebell. Gr. *phakĕlos*, a bundle; from the bunched flowers of the original species.

Phaedranas'sa f. Gr. *phaidrŏs*, gay; *anassa*, queen.

Phaedran'thus m. Gr. *phaidrŏs*, gay; *anthos*, a flower; with reference to the showy flowers of this Mexican climber.

phaeocar'pus, -a, -um With dark fruit.

pha'eus, -a, -um Dusky.

Phaio'phleps f. From Gr. *phaios*, dark; *phleps*, vein. In allusion to the veining of the flowers of this Iridaceous plant.

Phai'us m. Gr. *phaios*, dusky. The flowers of the type-species of these orchids are dark-coloured.

Phalaeno'psis f. Moth orchid. Gr. *phalaina*, a moth; *ŏpsis*, like.

Phala'ris f. Greek name for a grass of this genus: *P. arundinacea picta* is a ribbon-grass, *P. canarienis* canary grass, sometimes grown for bird feed.

Phase'olus m. Latin diminutive *phăsĕŏlus* or *făsĕŏlus* or *phăsĕlus*, Gr. *phasēlŏs*, the name for a kind of bean (*Vigna*). Cultivated all over the world, this genus includes the French or string bean, limas, and scarlet runners, all of which are of American origin and hence were unknown in antiquity. Other beans are included under *Vicia* (the broad or horse bean) and *Glycine* (soybean).

Pheba'lium n. Said to be from a Greek name for myrtle used by ancient Greek comic poets.

Phego'pteris f. From *phēgŏs*, oak; *ptĕris*, fern.

Phelloden'dron n. Cork-tree. Gr. *phĕllos*, cork; *dĕndrŏn*, a tree; from the corky bark. Commercial cork comes from the cork-oak (*Quercus suber*).

Phellosper'ma n. Gr. *phĕllos*, cork; *sperma*, a seed; because of the corky base of the seed of these cacti.

philadel'phicus, -a, -um Of Philadelphia.

Philadel'phus m. Mock-orange. The Greek name meaning brotherly. In some parts of England the flowers were often used to take the place of orange blossoms in the country bride's wedding bouquet. It is sometimes (wrongly) called syringa as a vernacular name.

Philesia

Phile′sia f. Gr. *phileo*, to love; from the beauty of the flowers of this Chilean shrub.

Philly′rea f. The classical Greek name, apparently derived from a non-Greek name, for these ornamental evergreen shrubs and small trees.

Philoden′dron n. Gr. *phileo*, to love; *dendron*, a tree; from the tree-climbing habits of these ornamental plants.

phleban′thus, -a, -um With veined flowers; from *phleps*, vein; *anthos*, flower.

Phlebo′dium n. From Gr. *phlĕps, phlĕbŏs*, vein; with reference to the numerous veins and veinlets of the fronds.

phleioi′des *Phleum*-like or resembling a reed.

Phle′um n. Timothy. The name for these perennial grasses comes from Gr. *phlĕōs*, a kind of rush or reed.

phlogiflo′rus, -a, -um With flame-coloured flowers; with flowers like *Phlox*.

phlogifo′lius, -a, -um With leaves like *Phlox*.

Phlo′mis f. Jerusalem sage. Greek name, *phlŏmis*, for some plant possibly not of this genus.

Phlox f. Gr. *phlox*, a flame. Also the Greek name for some plant with flame-coloured flowers.

phoeni′ceus, -a, -um Purple-red; from Phoenicia, where the cities of Tyre and Sidon produced the celebrated Tyrian purple dye in antiquity.

phoenicola′sius, -a, -um With purple hairs.

Phoe′nix f. Date palm. The Greek name.

Pholido′ta f. Rattlesnake orchid. Gr. *phŏlidōtos*, scaly, clad in scales. The allusion is probably to the scaly bracts on these epiphytic orchids.

Phor′mium n. New Zealand flax. Gr. *phŏrmiŏn*, mat. The plant provides a very strong fibre which can be woven and is extensively used by the Maoris for their traditional garments and basketry. Sir Joseph Banks brought seeds of it back from Captain Cook's first voyage and hoped to establish it commercially in England. It is an important cordage plant in New Zealand.

Photi′nia f. Apparently from Gr. *phōs, phōtŏs*, light, in allusion to the shining leaves on these shrubs and trees.

Phragmipe′dium n. From Gr. *phragma*, a fence or screen; name formed by analogy with *Cypripedium*.

Phragmi′tes n. Gr. *phragmītēs*, of fences, from *phragma*, a fence or screen; from the way these large grasses form a hedgelike growth along ditches.

phry´gius, -a, -um From Phrygia in the western part of Asia Minor.

phu L. from Gr. *phŏu,* a kind of valerian.

Phyge´lius m. Probably from Gr. *phygē,* flight, avoidance; 'in consequence of its having so long escaped the researches of botanists', according to W. J. Hooker (1855) or *phygē* plus *hēlios,* sun, according to Pfeiffer (1874).

Phy´lica f. Gr. *phyllikos,* leafy; in allusion to the abundant foliage.

Phylla´gathis f. From Gr. *phyllon,* a leaf; *agathis,* ball of thread; in allusion to the large leafy bracts below the compact flower-heads.

Phyllan´thus m. Gr. *phyllon,* a leaf; *anthos,* a flower. In some species the flowers are produced on the edges of the leaf-like branches.

Phylli´tis f. Hart's-tongue fern. Gr. *phyllitis,* from *phyllon,* leaf.

Phyllocac´tus m. Gr. *phyllon,* a leaf; *Cactus.* In allusion to the flattened leaf-like stems.

Phyllo´cladus m. Gr. *phyllon,* a leaf; *klados,* a branch. The branches of these evergreen trees and shrubs are flattened like leaves.

Phyllo´doce f. Heathlike shrubs named for a sea-nymph, Phyllŏdŏce (mentioned by Virgil), in accordance with the custom of naming genera of *Ericaceae* after nymphs and goddesses initiated by Linnaeus when he named *Andromeda.*

phyllomani´acus, -a, -um Producing leafy growth in excessive abundance.

Phyllo´stachys f. Gr. *phyllon,* a leaf; *stachys,* a spike; referring to the leafy inflorescence of these bamboos.

Phy´salis f. Cape gooseberry, ground-cherry. From Gr. *physa,* a bladder; from the inflated calyx of these herbs.

Physocar´pus m. Nine bark. From Gr. *physa,* a bladder; *karpos,* fruit; from the inflated follicles of these deciduous shrubs.

Physochlai´na f. Gr. *physa,* bladder; *chlaina,* cloak, outer garment. With reference to the inflated calyx covering the capsule.

Physosi´phon m. From Gr. *physa,* a bladder; *siphōn,* a tube. The tube of the flower is inflated.

Physoste´gia f. Obedient plant, false dragon-head. Gr. *physa,* a bladder; *stēgē,* roof covering. The fruits of these herbaceous perennials are covered by an inflated calyx.

Phyte´lephas m. Ivory-nut palm. From Gr. *phyton,* a plant; *ĕlĕphas,* elephant; because the nut of this palm is vegetable ivory which in some ways resembles true ivory. It can be carved into various articles as large as billiard balls.

Phyteu´ma n. Rampion. The Greek name for *Reseda phyteuma.*

Phytolac'ca f. Pokeweed, pokeberry, inkberry. From Gr. *phyton*, a plant; modern Latin *lacca*, from Hindi *lakh*, referring to the dye extracted from the lac insect. The allusion is to the staining qualities of the fruit of these herbs, shrubs, or trees.

Pi'cea f. Spruce. L. *pĭcĕa*, pitch-pine, from *pix*, pitch.

pictura'tus, -a, -um Variegated.

pic'tus, -a, -um Painted; brightly coloured.

Pi'eris f. Handsome shrubs named for the Pierides, the generic name of the Muses.

Pi'lea f. L. *pīlĕus*, a cap; from the shape of the female flowers of these herbs.

pilea'tus, -a, -um Furnished with a cap.

pi'lifer, pili'fera, -um With soft short hairs.

Pilosoce'reus m. L. *pilosus*, shaggy; *Cereus*, *q.v.*

pilo'sus, -a, -um Covered with long soft hairs.

pilula'ris, -is, -e; pilu'lifer, piluli'fera, -um With little balls, referring to the globular fruits.

Pimele'a f. Rice-flower. Gr. *pimĕlē*, fat; from its richness in oil.

pimeleoi'des Like *Pimelea*.

Pimen'ta f. Allspice. From Spanish *pimento*. Not to be confused with pimiento (*Capsicum*).

Pimpinel'la f. Anise. Derivation of the name of these herbs is uncertain. The fruit of one species yields oil of anise with a strong, aromatic flavour. Most of the oil of commerce used by liqueur makers is derived from the star-anise of China (*Illicium verum*).

pimpinellifo'lius, -a, -um With leaves like *Pimpinella*.

Pinan'ga f. From the Malayan name *pinina* for these palms.

Pinas'ter Latin for a wild pine, indicating one inferior to the cultivated pine (*Pinus pinea*).

Pinel'lia f. In honour of Giovanni Vincenzo Pinelli (1535–1601), who had a botanic garden in Naples.

pineto'rum Of pine forests.

pi'neus, -a, -um Relating to the pine.

Pingui'cula f. Butterwort. L. *pinguis*, fat; an allusion to the greasy appearance of the leaves of these insectivorous herbs.

pinguifo'lius, -a, -um With fat leaves.

pinifo'lius, -a, -um With leaves like pine.

pinna'tus, -a, -um Featherlike; having leaflets arranged on each side of a common stalk. Also:

pinnati'fidus pinnately cut
pinnatifo'lius with pinnate leaves
pinnati'frons with pinnate fronds
pinnatiner'vis with pinnately arranged nerves

Pi'nus m. Pine. The Latin name, particularly for *Pinus pinea. P. strobus*, the magnificent white pine of North America, now often called Weymouth pine in England, became known early in the 18th-century as 'Lord Weymouth's pine' because of the large and successful plantings on his Longleat estate. In New England the arbitrary marking of all pines over 24 inches diameter as Crown property for mast timber was one of the causes of discontent which led to the American Revolution. As a consequence it is seldom possible to find pine boards more than 23 inches wide in old houses in New England.

Mast timbers were shipped in large quantity from northern New England ports in special ships loaded through bow ports which were closed and battened down for the voyage home.

Pi'per n. Pepper. The Latin name from Gr. *pĕpĕri*, itself derived from an Indian name. Black pepper is made from the whole fruit of *Piper nigrum*, a vinelike climbing plant. For the more expensive white pepper, the external coat is removed and with it much of the flavour.

piperi'tus, -a, -um Pepperlike, tasting like pepper, hot and sharp.

Piptade'nia f. Gr. *pipto*, to fall; *adēn*, a gland.

Piptan'thus m. From Gr. *pipto*, to fall; *anthos*, a flower. The calyx, corolla and stamens fall off together, leaving the young pod without a calyx at base, in these ornamental Asiatic shrubs.

Pique'ria Fragrant perennial herbs and shrubs named for Andrés Piquer, 18th-century Spanish physician.

Pisci'dia f. From L. *piscis*, fish; *caedo*, cut down, slaughter, kill. In allusion to use as a fish-poison; formed by analogy with *hŏmĭcĭda*, man-killer, murderer.

pi'sifer, pisi'fera, -um Pea-bearing.

pisocar'pus, -a, -um With pea-like fruit.

Pista'cia f. Pistachio. From *pistakē*, the Greek word for the nut, derived from a Persian name.

Pis'tia f. Water-lettuce. From Gr. *pistos*, water; in allusion to the watery habitat of this herb.

Pi'sum n. Pea. The Latin name. Peas, known since prehistoric times, have been found in Swiss lakeside dwellings of the Bronze Age.

Pitcair'nia f. Low-growing herbs and shrubs with showy flowers

Pithecellobium

named for William Pitcairn (1711–1791), London physician, who had a botanic garden at Islington.

Pithecello'bium n. From Gr. *pithēkŏs*, ape, monkey; *ĕllŏbiŏn*, ear-ring. The name of this genus was originally published as *Pithecollobium* then amended to *Pithecellobium*. The Brazilian name, Brincos de Sahoy, for these tropical trees and shrubs means monkey's ear-ring.

Pithecocte'nium n. From Gr. *pithēkŏs*, ape, monkey; *ctĕniŏn*, comb. In allusion to the spiny fruits.

Pittos'porum n. Parchment bark, Australian laurel. From Gr. *pitta*, pitch; *spŏra*, seed. The seed of these evergreen woody plants has a sticky resinous coating.

Pityrogram'ma f. Gold fern, silver fern. From Gr. *pityrŏn*, chaff, bran, scurf; *gramma*, writing; from the powder on the fronds.

placa'tus, -a, -um Quiet, calm, gentle.

Plagian'thus m. Ribbon wood. Gr. *plagiŏs*, oblique; *anthos*, a flower. These trees or shrubs have flowers with asymmetrical petals.

Plagioboth'rys f. From Gr. *plagiŏs*, oblique, placed sideways; *bŏthrŏs*, pit. In allusion to the scar on the nutlet.

Pla'nera f. Water-elm. Named for Johann Jakob Planer (1743–1789), German botanist, professor in Erfurt.

plane'tus, -a, -um Wandering.

Planta'go f. Plantain. The Latin name for these herbs and sub-shrubs.

pla'nus, -a, -um Flat. Also:
    planiflo'rus   with flat flowers
    planifo'lius   with flat leaves
    pla'nipes   with a flat stalk

platanifo'lius, -a, -um With leaves like the plane tree (*Platanus*).

platanoi'des Resembling a plane tree (*Platanus*).

Pla'tanus f. Plane tree. The Greek name, *platanŏs*, for the oriental plane. The London plane (*Platanus hybrida*, syn. *P. acerifolia*) is the most widely planted city tree in London, Paris, New York, and other northern cities. It stands almost every kind of abuse, atmospheric and otherwise. It was originally a cross (*P. occidentalis* × *P. orientalis*), first raised, it is believed, in Oxford prior to 1700. Handel's well-known *Largo* is an ode to a plane tree.

platy-
In Greek compound words usually signifies broad, rarely flat. Thus:
    platyacan'thus   with broad spines
    platyan'thus   with broad flowers
    platycar'pus   broad-fruited

platycau'lis	with a broad (flat) stem
platycen'trus	with a broad (flat) spur
platy'cladus	flat-branched
platyglos'sus	broad-tongued
platype'talus	with broad petals
platyphyl'lus	broad-leaved
platy'podus	broad-stalked
platyspa'thus	with a broad spathe
platysper'mus	broad-seeded

Platyca'rya f. From Gr. *platys*, broad; *karyŏn*, a nut; from the form of the fruit of this deciduous Asiatic tree.

Platyce'rium n. Staghorn fern. Gr. *platys*, broad; *kĕras*, a horn; in allusion to the form of the fronds.

Platyco'don m. Balloon flower, Japanese bellflower. Gr. *platys*, broad; *kōdōn*, a bell; from the form of the corolla.

Platycra'ter f. From Gr. *platys*, broad; *kratēr*, bowl. In allusion to the expanded saucer-like calyces of the sterile flowers of this Japanese woody plant.

Platyste'mon m. Gr. *platys*, broad; *stemon*, a stamen. These low annual herbs have broad stamens.

plebe'ius, -a, -um Common.

Plectran'thus m. From Gr. *plēctrŏn*, spur; *anthos*, flower.

Pleiobla'stus m. From Gr *pleios*, more; *blastos*, bud. With reference to the several buds, later shoots at each node of these bamboos.

Plei'one f. Indian crocus. Indian mountain orchids named for *Plēiŏnē*, the mother of the seven Pleiades.

pleioner'vis, -is, -e With many nerves.

Pleiospi'los m. Gr. *pleios*, many; *spilos*, a spot, fleck. The leaves of these succulent plants are conspicuously spotted.

pleniflo'rus, -a, -um With double flowers.

plenis'simus, -a, -um Very double.

ple'nus, -a, -um Double or full.

pleuro'stachys With spikes at the side.

Pleurothal'lis f. Gr. *pleuron*, side, rib; *thallŏs*, branch.

plica'tus, -a, -um Pleated.

plumar'ius, -a, -um; pluma'tus, -a, -um Plumed; feathered.

plumbaginioi'des Like *Plumbago*.

Plumba'go f. Leadwort. The Latin name derived from *plumbum*, lead; *-ago*, a termination of many Latin plant-names used to indicate a resemblance or a property.

plumbeus

plum'beus, -a, -um Relating to lead.

Plume'ria (also, incorrectly, **Plumi'era**) f. Frangipani. Tropical shrubs and trees named for Charles Plumier (1646–1704), French monk of the Franciscan order, botanist and traveller, who in 1689 and 1690 visited Martinique, Guadeloupe and Haiti in the West Indies and made remarkably detailed and accurate drawings and descriptions of their flowering-plants and ferns. He revived the ancient custom of naming genera after persons.

plumo'sus, -a, -um Feathery.

pluriflo'rus, -a, -um Many-flowered.

Po'a f. Gr. *pŏa*, grass. The genus comprises a variety of useful lawn and fodder grasses including Kentucky and Canada blue grass.

poco'phorus, -a, -um Fleece-bearing; from Gr. *pŏcŏs*, wool, a fleece.

poculifor'mis, -is, -e Cup-shaped.

Podachae'nium n. Gr. *pŏus, pŏdŏs*, a foot; modern Latin *achaenium*, an achene or dry one-seeded fruit. The allusion is to the stalked achenes of these tall shrubs.

podag'ricus, -a, -um Gouty; coarsely swollen at base; from Gr. *pŏdagra*, gout in the feet.

Podaly'ria f. South African shrubs named for Podalyrius, son of Aesculapius, celebrated in Greek mythology as a skilful physician.

Podocar'pus m. From Gr. *pŏus, pŏdŏs*, a foot; *karpŏs*, a fruit. The fruits of these evergreen trees related to yew are borne on a fleshy stalk.

Podo'lepis From Gr. *pŏus, pŏdŏs*, a foot; *lĕpis*, a scale. The pedicels of these Australian herbs are scaly.

podo'licus, -a, -um From the Ukraine in south-western Russia (Podolia).

Podophyl'lum n. May-apple. A contraction of *Anapodophyllum*, from L. *anas*, duck; Gr. *pŏus, pŏdŏs*, a foot; *phyllon*, a leaf; with reference to the shape of the leaf in the American species *P. peltatum*. The roots of these herbs have medicinal properties as a cathartic.

podophyl'lus, -a, -um With stout-stalked leaves.

Podra'nea f. Anagram of *Pandorea*, name of the genus of *Bignoniaceae* to which this is related.

poeciloder'mus, -a, -um With spotted skin.

Poellni'tzia f. In honour of Karl von Poellnitz (1896–1945), German specialist on succulent plants.

poeta'rum Of the poets, an epithet applied usually to plants associated with ancient Greek and Roman poets.

258

poe′ticus, -a, -um Pertaining to poets; see above.

Pogonathe′rum n. From Gr. *pōgōn*, beard; *athēr*, awn, barb.

Pogo′nia f. Adder's mouth. Gr. *pōgōnias*, bearded. In most species of these terrestrial orchids the lip is fringed.

pogonosty′lus, -a, -um With a bearded style.

Pogoste′mon m. From Gr. *pōgōn*, beard; *stēmōn*, thread, stamen.

pohuashanen′sis, -is, -e From Po hua shan (Hundred Flower Hill), west of Peiping, China.

Poincia′na f. Flamboyant. Showy-flowered trees named for M. de Poinci, 17th-century French Governor of Martinique; *Poinciana regia*, the flamboyant, is now placed in the genus *Delonix* and *P. pulcherrima*, Barbados pride, in *Caesalpinia*.

Poinset′tia f. Although usually included in *Euphorbia* it will probably always retain the popular name of *Poinsettia*. Joel R. Poinsette (1775–1851), gardener, botanist, and diplomat of South Carolina, was the first American ambassador to Mexico in 1824; he introduced the plant to Charleston friends in 1833. He was bitterly opposed to secession and for that reason his name is often held in small account in the Southern States.

Polemo′nium n. The Greek name for these herbaceous perennials, *pŏlĕmōniŏn*, originally applied to a medicinal plant associated with Polemon of Cappadocia. Jacob's ladder is *P. caeruleum*.

Polian′thes f. Tuberose. Gr. *polios*, grey, whitish; *anthos*, a flower.

polifo′lius, -a, -um Grey-leaved like *Teucrium polium*.

Poliothyr′sis f. Gr. *pŏliŏs*, grey, whitish; *thyrsos*, a panicle; in allusion to the whitish panicles of this tree.

poli′tus, -a, -um Elegant, polished, neat.

poly- In Greek compound words signifies many. Thus:

polyacan′thus	with many thorns
polyan′drus	with many stamens
polyan′themos	with many flowers
polyan′thus	with many flowers
polybot′ryus	with many clusters
polycar′pus	many-fruited
polyce′phalus	many-headed
polychro′mus	of many colours
poly′lepis	with many scales
polymor′phus	of many forms; variable
polype′talus	with many petals
polyphyl′lus	with many leaves
polyrrhi′zus	many-rooted

polyse'palus   with many sepals
polysper'mus   many-seeded
polysta'ch'yus   many-spiked
polystic'tus   many-spotted

Polybot'rya f. Gr. *polys*, many; *botrys*, bunch. With reference to the many clusters of sporangia on the fertile fronds of these ferns.

Poly'gala f. Milkwort. Latin name from the Greek name, *pŏlygalŏn*, from *pŏlys*, much; *gala*, milk. These herbs and shrubs were reputed to aid the secretion of milk.

polygaloi'des Resembling *Polygala*.

Polygo'natum m. Solomon's seal. The Greek name *pŏlygŏnatŏn*, from *polys*, many; *gony*, the knee-joint; in allusion to the many joints of the rhizomes. Once reputed to have value in healing bruises.

Poly'gonum n. Smart or knotweed. From the Greek name, *pŏlygŏnon*, from *polys*, many, much; and either *gŏnŏs*, offspring, seed, in allusion to the numerous seeds, or *gŏny*, knee-joint, in allusion to the swollen joints of the stems.

Polypo'dium n. Polypody fern. Greek name *pŏlypŏdiŏn*, from *polys*, many; *pous*, a foot. The rhizomes are much branched and spread widely.

Polyp'teris f. Gr. *polys*, many; *pteron*, a wing; in allusion to the form of the pappus on the fruits of these North American herbs, now placed under *Palafoxia*.

Polys'cias f. From Gr. *polys*, many; *skias*, canopy like a sunshade, umbel; because of the main umbel being divided into numerous lesser umbels.

Polysta'chya f. Gr. *polys*, many; *stachys*, a spike. The reference is to the many branches or spikes on a single inflorescence in some species of these epiphytic orchids.

Polys'tichum n. Christmas fern. Gr. *polys*, many; *stichos*, a row. The sori of these ferns are in many rows.

poma'ceus, -a, -um Applelike.

Pomader'ris f. Gr. *pōma*, a lid; *dĕrris*, a skin, leathern covering; from the membranous covering of the capsule on these flowering shrubs.

pomeridia'nus, -a, -um Of the afternoon.

po'mifer, pomi'fera, -um Apple-bearing.

pompo'nius, -a, -um Meaning obscure, possibly connected with the Roman name Pomponius.

Ponci'rus m. Hardy orange. Fr. *poncirus*, a kind of citron. This shrub can be used as a root-stock for citrus fruits. It is extremely thorny and hence has been used for hedging.

pondero'sus, -a, -um Heavy.

Ponga'mia f. From the Malabar name, *pongam*, for this tree or shrub.

Pontede'ria Aquatic perennials named for Guilio Pontedera (1688–1757), professor of botany at Padua. *P. cordata* is pickerel-weed.

pon'ticus, -a, -um From the south shore of the Black Sea.

populifo'lius, -a, -um Poplar-leaved.

popul'neus, -a, -um Relating to poplar.

Po'pulus f. Poplar. The Latin name. Includes aspen, cottonwood, popple, and so on.

Pora'na f. Derivation obscure.

porci'nus, -a, -um Relating to pigs; pig-food.

porophyl'lus, -a, -um Having holes in the leaves or seeming to.

porphy'reus, -a, -um Warm reddish colour.

porphyrophyl'lus, -a, -um Purple-leaved.

porrifo'lius, -a, -um With leaves like leek. L. *porrum*, leek.

Por'tea f. In honour of Marius Porte (d. 1866), French explorer and plant-collector in Brazil, Malaya and Philippines.

Porteran'thus m. In honour of Thomas Conrad Porter (1822–1901), American botanist. (See *Gillenia*.)

portenschlagia'nus, -a, -um In honour of Franz von Portenschlag-Ledermayer (1777–1827), Austrian botanist.

Portlan'dia f. Ornamental trees and shrubs named for Margaret Cavendish Bentinck (1715–1785), Duchess of Portland, bluestocking, friend and correspondent of Jean Jacques Rousseau, a keen botanist who had a private botanic garden and museum at Bulstrode, Buckinghamshire.

Portula'ca f. Purslane. The Latin name for *P. oleracea*.

portulaca'ceus, -a, -um *Portulaca*-like.

Portulaca'ria f. Closely related to *Portulaca*.

poscharskya'nus, -a, -um In honour of Gustav Adolf Poscharsky (1832–1914), head gardener in Dresden.

Posoque'ria f. From the native Guianan name for these shrubs or small trees.

Potamoge'ton m. Pondweed. Variant of L. *potamogiton*; Gr. *pŏtamōgeitōn*, from *pŏtamŏs*, river; *geitōn*, neighbour.

potamo'philus, -a, -um Loving rivers or wet places.

potani'nii In honour of Grigori Nikolaevich Potanin (1835–1920),

Russian explorer of central and eastern Asia, who made enormous botanical collections in western China.

potato'rum Of the drinkers or topers; applied to plants used to make beers, etc.

Potentil'la f. Cinquefoil. L. *potens*, powerful; from the reputed medicinal properties.

poteriifo'lius, -a, -um With leaves like *Poterium*.

Pote'rium n. Greek name for another plant (burnet), not these herbs and shrubs.

Po'thos m. From the Sinhalese name *potha* for these climbers, not connected with L. and Gr. *pŏthos*, which was applied to a blue-flowered larkspur and a white-flowered asphodel.

Potina'ra f. Multigeneric hybrid group derived from *Brassavola* × *Cattleya* × *Laelia* × *Sophronitis*, named in honour of M. Potin, president of the French orchid society in 1922.

pot'tsii In honour of John Potts (d. 1822), gardener, who collected in China and Bengal for the Horticultural Society of London, or C. H. Potts (*fl.* 1877) of Lasswade near Edinburgh.

Poute'ria f. Latinized from the Guiana (Galibi) vernacular name *pourama-pouteri*.

praeal'tus, -a, -um Very tall.

prae'cox Very early.

praemor'sus, -a, -um Appearing to be bitten off at the end.

prae'stans Distinguished; excelling.

praete'ritus, -a, -um Past and gone, passed over, omitted.

praeter'missus, -a, -um Neglected, omitted, overlooked.

praetex'tus, -a, -um Bordered.

praever'nus, -a, -um Coming very early, before the spring.

pra'sinus, -a, -um Leek-green, from Gr. *prasŏn*, leek.

praten'sis, -is, -e Of the meadows.

Pra'tia f. Low-growing herbs named for a French naval officer, Ch. L. Prat-Bernon, who sailed with Freycinet on his voyage to the South Pacific in 1817 but who died within a few days of the expedition's setting out.

prattii In honour of Antwerp E. Pratt, English zoologist and traveller in China, 1887–1890.

praviss'imus, -a, -um Very crooked.

precator'ius, -a, -um Relating to prayer, used of seeds made into rosaries.

Prem'na f. Gr. *prĕmnŏn*, the stump of a tree. Most of these trees and shrubs are low growing.

Prenan'thes f. Rattlesnake root. From Gr. *prēnēs*, with face downwards; *anthos*, a flower. The flower heads droop on these perennial herbs.

prep'tus, -a, -um Distinguished, eminent.

Pri'mula f. Primrose. Contraction of medieval name *primula veris* for the daisy, meaning 'firstling of spring', being a diminutive of L. *primus*, first, used for early-flowering herbs.

primuliflo'rus, -a, -um Primrose-flowered.

primulifo'lius, -a, -um Primrose-leaved.

primuli'nus, -a, -um Primrose-coloured, i.e. a pale greenish yellow.

primuloi'des Resembling a primrose.

prin'ceps The most distinguished.

Prinse'pia f. Asian shrubs named for James Prinsep (1799–1840), Secretary of the Asiatic Society of Bengal.

prisma'ticus, -a, -um Shaped like a prism, i.e. with flat surfaces separated by angles.

Pritchar'dia f. Fan palms named for W. T. Pritchard, 19th-century British official in Polynesia; see *Eupritchardia*.

Probosci'dea f. From Gr. *prŏbŏskis*, elephant's trunk, in allusion to the elongated curved ends of the fruit.

probosci'deus, -a, -um Snoutlike; with an elongated proboscis.

proce'rus, -a, -um Tall.

procum'bens Prostrate.

procur'rens Spreading out, extending, projecting.

produc'tus, -a, -um Lengthened, stretched out.

profu'sus, -a, -um Abundant.

pro'lifer, proli'fera, -um Proliferous, i.e. free-flowering or producing side shoots or buds in order to increase.

proli'ficus, -a, -um Very fruitful.

Promenae'a f. Small orchids named after a Greek priestess of Dodona, known to Herodotus.

pro'nus, -a, -um Bending or hanging forwards.

propen'sus, -a, -um Hanging down.

propin'quus, -a, -um Related.

Proserpina'ca f. L. plant-name of obscure application, probably from *proserpo*, creep along.

Proso'pis f. Mesquite. The Greek name for the burdock, but why it was allotted here is obscure. The tree is of great economic value.

Prostanthe'ra f. Mint-bush. Gr. *prŏstithemi*, to append or add to; *prŏsthĕma*, appendage; *anthera*, anther; in allusion to a spur-like appendage on the back of the anther.

prostra'tus, -a, -um Flat on the ground, prostrate.

Pro'tea f. Cape-of-Good-Hope shrubs, trees, and perennials named for Proteus, a potent Greek sea-god.

protru'sus, -a, -um Protruding.

provincia'lis, -is, -e From Provence in France.

pruina'tus, -a, -um; pruino'sus, -a, -um Glistening as though frosted over.

Prunel'la f. Self-heal. In the 15th and 16th centuries German herbalists used the names *Prunella* and *Brunella* interchangeably; they may be derived from the south German word *braun* (from L. *prunum*) meaning purple and referring to the flower-colour, or from the German *Bräune*, meaning quinsy, which these herbs were supposed to cure.

prunelloi'des Resembling *Prunella*.

prunifo'lius, -a, -um With leaves like plum.

Pru'nus f. Plum, cherry. The Latin name.

pruhonicia'nus, -a, -um Of Pruhonice, Czechoslovakia.

pru'riens Causing an itch.

przewal'skii In honour of Nicolai Mikhailovich Przewalski (1839–1888), celebrated Russian explorer of central Asia.

Psammo'phora f. From Gr. *psammŏs*, sand; *phŏrŏs*, bearing. In allusion to their sanded appearance.

Pseuda'corus Gr. *pseudo*, false; *Acorus*, sweet flag; the rhizomes are rather similar in appearance though not in smell.

Pseuderan'themum n. Gr. *pseudo*, false; plus *Eranthemum*. Some of these plants were formerly placed in *Eranthemum*.

Pseudo'larix f. Golden larch. Gr. *pseudo*, false; *Larix*, larch, which it resembles.

Pseudo'panax m. Gr. *pseudo*, false; *Panax*, which this New Zealand genus of shrubs and trees resembles.

Pseudophoe'nix f. Gr. *pseudo*, false; *Phoenix*, the date palm, to which this palm bears a general resemblance.

Pseudosa'sa f. Gr. *pseudo*, false; *Sasa*, a Japanese genus of bamboos to which this is related.

Pseudotsu'ga Douglas fir. Gr. *pseudo*, false; *Tsuga*, hemlock.

Psi'dium n. Guava. Gr. *psidion*, a pomegranate. These evergreen shrubs and trees have edible fruit.

psilan'drus, -a, -um With glabrous stamens.

psilos'tachys With glabrous spike.

psilos'temon With glabrous stamens.

psittaci'nus, -a, -um Parrotlike, i.e. with green or contrasting colours.

psittacor'um Of parrots.

Psophocar'pus m. Gr. *psŏphŏs*, a noise; *karpos*, fruit. The seed capsules of these twiners explode noisily when ripe.

Psora'lea f. Scurfy pea. Gr. *psōralĕos*, scabby. These herbs are covered with glandular dots.

Psycho'tria f. Wild coffee. Variant coined by Linnaeus of Gr. *psychŏtrŏphŏn*, applied by Patrick Browne to Jamaican plants but by the Greeks to betony, because found in cold places.

psyco'des Butterfly-like.

ptar'mica Greek name *ptarmikē* for a plant which caused sneezing.

Pte'lea f. Greek name *ptĕlĕa* for an elm tree, presumably applied to this genus on account of its flattened, almost circular, fruit having a wing all around like that of an elm.

pteran'thus, -a, -um With winged flowers.

Pteri'dium n. Bracken. Brake. Diminutive of Gr. *ptĕris*, fern, from *ptĕron*, wing, in allusion to the form of the fronds of these ferns. Commonly known as *Pteris*.

pteridoi'des Resembling *Pteris*.

Pteridophyl'lum n. From Gr. *ptĕris*, fern; *phyllŏn*, leaf.

Pte'ris f. Gr. *ptĕris*, fern; from the feathery fronds. (See *Pteridium*.)

Pteroca'rya f. Gr. *ptĕron*, a wing; *karyon*, a nut. These ornamental trees are called wing-nuts.

Pteroce'phalus m. Gr. *ptĕron*, a wing; *kephale*, a head. The fruiting head of these herbs and shrubs appears to be covered with feathers.

pteroneu'rus, -a, -um With winged nerves.

Pterosper'mum n. Gr. *ptĕron*, a wing; *sperma*, a seed. The seeds of these trees or shrubs are winged.

Pterosty'rax f. Gr. *ptĕron*, a wing; plus *Styrax*. The seeds of one species of these *Halesia*-like trees and shrubs have winged seeds like *Styrax*.

Ptychosper'ma n. Gr. *ptyche*, a fold; *sperma*, a seed; in allusion to a technical characteristic of the seed of these palms.

pubens; pubes'cens; pu'biger, pubi'gera, -um Downy.

pud'dum Variant of the Hindi vernacular name *padam* for a beautiful Himalayan cherry.

pudero'sus Very bashful.

pu'dicus, -a, -um Bashful.

pue'lii After Timothée Puel (1812–1890), French botanist and medical man.

Puera'ria f. Herbs and shrubs. Climbers and twiners, named for Marc Nicolas Puerari (1766–1845), Swiss botanist, born at Geneva, who spent most of his life as a professor at Copenhagen, but gave his important herbarium to De Candolle of Geneva.

pugionifor'mis, -is, -e Dagger-shaped.

pulchel'lus, -a, -um; pul'cher, pul'chra, pulchrum Pretty.

Pule'gium n. Latin name, from L. *pulex*, flea.

Pulica'ria f. Latin name, from L. *pulex*, flea.

pul'lus, -a, -um Dark-coloured.

Pulmona'ria f. Lungwort. L. *pulmo*, the lung. In accordance with the old Doctrine of Signatures the plant was considered to be an effective remedy for diseases of the lung because the spotted leaves were supposed to resemble diseased lungs.

Pulsatil'la f. Pasqueflower. From L. *pulso*, strike, set in violent motion, its relevance uncertain.

Pultena'ea f. Shrubs named for Richard Pulteney (1730–1801), English botanist, author of *A General View of the Writings of Linnaeus* (1781) and *Historical and Biographical Sketches of the Progress of Botany in England* (1790).

pulverulen'tus, -a, -um Powdered as with dust.

pulvina'tus, -a, -um Cushionlike.

pu'milus, -a, -um Dwarf.

puncta'tus, -a, -um Spotted.

pun'gens Sharp-pointed.

Pu'nica f. Pomegranate. The Latin name contracted from *punicum malum*, Carthaginian apple.

puni'ceus, -a, -um Reddish-purple.

pur'gans Purgative, purging.

purpura'tus, -a, -um Made purple.

purpu'reus, -a, -um Purple. Also *purpurascens*, tending to purple.

purpu'sii In honour of Carl Albert Purpus (1853–1941), or his brother

Joseph Anton Purpus (1860–1932), German plant-collectors in central America.

Pur'shia f. A spreading shrub named for Frederick Traugott Pursh (1774–1820), German explorer, collector, horticulturist, author, who lived in the United States from 1799 to 1811 and wrote an important *Flora Americae Septentrionalis* (1814).

Puschki'nia f. Striped squill. Bulbous plants named for the Russian Count Mussin-Puschkin (d. 1805), who collected in the Caucasus; *Nepeta mussinii* also commemorates him.

pusil'lus, -a, -um Very small.

pustula'tus, -a, -um Appearing as though blistered.

Putofria f. From L. *putor*, foul smell.

Puy'a f. The Chilean vernacular name.

Pycnan'themum n. Mountain mint. From Gr. *pyknŏs*, dense; *anthŏs*, a flower. The flowers of these North American perennial herbs are densely crowded.

Pycno'stachys f. From Gr. *pyknŏs*, dense; *stachys*, a spike. These perennial herbs have dense flower spikes.

Pyg'mea f. A genus of New Zealand dwarf plants so-named from their habit of growth.

pygma'eus, -a, -um Pygmy.

Pyracan'tha f. Fire-thorn. Greek name *pyrakantha*, from *pyr*, fire, *akantha*, a thorn; in allusion to the thorny branches and the showy crimson fruit of these shrubs.

pyramida'lis, -is, -e; pyramida'tus, -a, -um Pyramid-shaped.

pyrenae'us, -a, -um; pyrena'icus, -a, -um From the Pyrenees.

Py'rethrum n. Feverfew. Gr. *pyrĕthrŏn*, from *pyr*, fire; in allusion to the bitter roots of *Anacyclus pyrethrum* which were regarded as a remedy for fevers and by their hotness promoted a flow of saliva, whence the alternative Latin name *salivaris*. Now classified under *Chrysanthemum*, but the older name remains in familiar use. The dried heads of one species (C. *cinerariifolium*) provide an effective insecticide, and the plant is grown as a crop for this.

pyrifo'lius, -a, -um Pear-leaved.

pyrifor'mis, -is, -e Pear-shaped.

Py'rola f. Shinleaf. The Latin diminutive of *Pyrus*, pear, from the pear-shaped leaves of these low perennial herbs.

Pyroste'gia f. From Gr. *pyr*, fire; *stĕgē*, a roof; from the colour and form of the upper lip of the flowers on these ornamental vines.

pyrrhoan'thus, -a, -um With flaming red flowers.

Pyrro'sia f. Gr. *pyrrŏs*, flame-coloured, reddish yellow, tawny.

Py'rus f. Pear. The Latin name.

Pyxidanthe'ra f. Flowering moss. Gr. *pyxis*, a small box; *anthera*, anther. The anthers of this evergreen creeping plant open like the lid of a box.

pyxida'tus, -a, -um Provided with a pyxis or a lid.

Q

quadra'tus, -a, -um In fours or in four. Thus:

quadrangula'ris	with four angles
quadrangula'tus	with four angles
quadriauri'tus	with four ears
quadri'color	four-coloured
quadridenta'tus	four-toothed
quadri'fidus	cut into four
quadrifo'lius	four-leaved
quadriparti'tus	parted four ways
quadrival'vis	with four valves

Qua'moclit f. Derivation of the name of these vines is obscure. Possibly Gr. *kuamos*, bean.

Quas'sia f. Name given by Linnaeus to honour Graman Quasi, a Negro slave of C. G. Dahlberg, who used the bark of this tree of Surinam as a remedy for fever. It is used today in medicine as a bitter tonic and a vermifuge. The name like the West Indian Quashie is from the (Tui)*Kwasi*, a personal name given to a boy born on the first day of the week, *Kwasida* or Sunday.

quercifo'lius, -a, -um With leaves like *Quercus* or oak.

querci'nus, -a, -um Relating to the oak.

Quer'cus f. Oak. The Latin name. The oak has long been considered a symbol of strength and is celebrated in legend and mythology in many lands. It was sacred to Jupiter and Thor, while the Druids thought that it inspired prophecy and built their altars under its branches. The timber of most species is valuable, the bark of some is used in tanning, while the acorn is used both for animal and human food. In southern France pigs are fattened on acorns.

Quilla'ja f. Soap-bark tree. The Chilean name for these evergreen trees and shrubs.

quina'tus, -a, -um In fives. Thus:

quinqueflo'rus	with five flowers
quinquefo'lius	five-leaved
quinquener'vis	five-nerved
quinquepuncta'tus	five-spotted
quinquevul'nerus	with five marks

Quisqua'lis f. Rangoon Creeper. L. *quis*, who? *qualis*, what? This curious name is the result of a play on words by Rumphius (1627–1702). The Malay name for this plant is *Udani*, rendered as *hoedanig*, meaning in Dutch 'how, what', which Rumphius translated into Latin as *quis qualis*, expressive of his surprise at the variability of the plant's growth. The flowers open white, become pink and end blood-red.

quiten'sis, -is, -e Referring to Quito, Ecuador.

R

racemiflo'rus, -a, -um; racemo'sus, -a, -um With flowers in racemes, stalks, as in lily-of-the-valley.

rad'ians Radiating outward.

radia'tus, -a, -um With rays or of radiating form.

ra'dicans Having rooting stems. Also:

radica'tus	having conspicuous roots
radico'sus	with many roots
rad'icum	of roots

radio'sus, -a, -um Having many rays.

rad'ula Rasp, file.

Raffenal'dia f. In honour of Alire Raffeneau Delile (1778–1850), professor of botany at Montpellier, one of the scientists who accompanied Napoleon to Egypt, also commemorated by the genus *Lilaea*.

Raffle'sia f. Appropriately named in honour of Sir Thomas Stamford Raffles (1781–1826), founder of Singapore, a great patron of science as well as statesman, *Rafflesia* has the world's largest flower but, owing to its parasitic habit, unfortunately no horticultural potentialities.

ragusi'nus, -a, -um Relating to Dubrovnik (formerly Ragusa), Jugoslavia.

ramiflo'rus, -a, -um With a branching inflorescence.

Ramon'da f. Stemless herbs named for Louis Francis Ramond, Baron de Carbonnière (1753–1827), French botanist and traveller, in the Pyrenees.

ramondioi'des Resembling *Ramonda*.

ramo'sus, -a, -um Branched. Also, *ramulosus*, twiggy; *ramosissimus*, much branched.

Ran'dia f. After Isaac Rand (d. 1743), London apothecary, director of the Chelsea Physic Garden.

Rane'vea f. A dwarf palm named for Louis Renevé, horticulturist at Berlin, *c.* 1870.

ranunculi'nus, -a, -um; ranunculoi'des Resembling *Ranunculus*.

Ranun'culus m. Buttercup, crowfoot. The Latin name from the diminutive of *rana*, a frog, because many species grow in damp places.

Ranza'nia f. In honour of Ono Ranzan (1729–1810), celebrated Japanese naturalist.

Raou'lia f. Tufted or creeping perennial herbs named for Edouard Raoul (1815–1852), French naval surgeon who collected in New Zealand from 1840 to 1846 and published on New Zealand plants in 1846.

rapa'ceus, -a, -um Relating to turnips.

Ra'phanus m. Radish. The Latin name, from Gr. *rhaphanis*, for this vegetable which has been known from antiquity.

Ra'phia f. Raffia—the source of the almost indispensable garden tying fibre. From Gr. *haphis*, a needle; in allusion to the beaked fruit of these palms.

Raphio'lepis f. See *Rhaphiolepis*.

rapunculoi'des Specific epithet, resembling *Rapunculus*, the now obsolete name for a group of bellflowers, meaning a little turnip and referring to the roots.

rariflo'rus, -a, -um With scattered flowers.

ra'rus, -a, -um Rare, uncommon.

Rathbu'nia f. Cacti named for Dr. Richard Rathbun (1852–1918), Assistant Secretary of the Smithsonian Institute, Washington, D.C.

Rati'bida f. Meaning obscure. See *Lepachys*.

Rauvol'fia f. Trees and shrubs named for Leonhart Rauwolf (d. 1596), German physician and traveller in the Orient. Long familiar to Oriental medicine, certain species have now achieved respectability by their acceptance in Western medicine, especially in the treatment of nervous disorders.

Ravena'la f. Traveller's tree. From the native name in Madagascar.

ra'vus, -a, -um Greyish or tawny.

Rebu'tia f. Cacti named for P. Rebut, French dealer in cacti, *c.* 1900.

Rechsteine'ria In honour of a Swiss-German clergyman and botanist, Rechsteiner (1797–1858).

reclina'tus, -a, -um Bent backward.

rec'tus, -a, -um Upright.

recurva'tus, -a, -um; recur'vus, -a, -um Curved backward.

recuti'tus, -a, -um Circumcised.

redivi'vus, -a, -um Brought back to life, used of plants reviving after a period of desiccation.

reduc'tus, -a, -um Drawn back, reduced, hence made small.

reevesia'nus, -a, -um In honour of John Reeves (1774–1856), from 1812 to 1831 tea inspector at Macao and Canton, who was instrumental in the introduction of Chinese plants into British gardens.

refle'xus, -a, -um; refrac'tus, -a, -um Bent sharply backward.

reful'gens Shining brightly.

rega'lis, -is, -e Of outstanding merit; regal.

regi'nae Of the queen. As an epithet in *Tradescantia* honouring Queen Marie of the Belgians, wife of Leopold II.

re'gius, -a, -um Royal.

reh'deri In honour of Alfred Rehder (1863–1949), American dendrologist of German origin, from 1898 until his retirement in 1940 at the age of 76 on the staff of the Arnold Arboretum, Massachusetts. Author of the celebrated *Manual of Cultivated Trees and Shrubs* (1940) and after retirement of the *Bibliography of Cultivated Trees and Shrubs* (1949).

Rehman'nia f. Perennial herbs named for Joseph Rehmann (1753–1831), German physician, who settled in Leningrad.

Reine'ckea f. A perennial herb named for Johann Heinrich Julius Reinecke (1799–1871) of Berlin.

Reinwar'dtia f. A North Indian flowering shrub named for Caspar Georg Carl Reinwardt (1773–1854), professor at Harderwijk, then Leyden, founder of Bogor (Buitenzorg) botanic garden, Java, Indonesia.

religio'sus, -a, -um Used for religious ceremonies; sacred.

remo'tus, -a, -um Scattered.

Remusa'tia f. In honour of Abel Remusat (1788–1832), celebrated French scholar, professor of Chinese at Paris.

Renanthe'ra f. From L. *rēnes*, kidneys; *anthera*, anther; in reference to the kidney-shaped pollinia of these epiphytic orchids.

renar'dii In honour of Charles Claude (Karl Ivanovich) Renard (1809–1886) of Moscow.

renifor'mis, -is, -e Kidney-shaped.

repan'dus With slightly wavy margins.

re'pens; rep'tans Creeping.

replica'tus, -a, -um Doubled back; folded.

Rese'da f. Mignonette. The Latin name derived from *resedo*, to heal, the plant having once been used as a remedy for bruises.

resedifo'lius, -a, -um With leaves like Mignonette.

resi'nifer, resini'fera, -um; resino'sus, -a, -um Resinous.

resupina'tus, -a, -um Bent back, put on its back, applied to organs turned upside down by a twist in their support.

Reta'ma f. Latinized form of the Arabic name *retem* for these broomlike shrubs.

reticula'tus, -a, -um Netted; with a net-like pattern.

retrofle'xus, -a, -um; retor'tus, -a, -um; retrofrac'tus, -a, -um Twisted back.

retu'sus, -a, -um With a rounded, slightly notched tip.

rever'sus, -a, -um Reversed; turned around.

revolu'tus, -a, -um Rolled backward, as some leaves are at the edges.

rex King.

Rhabdotham'nus m. Gr. *rhabdŏs*, a rod; *thamnŏs*, a bush. A New Zealand shrub with a twiggy habit and elegant flowers.

rhabdo'tus, -a, -um Striped.

rhae'ticus, -a, -um From the Rhaetian Alps of Switzerland and Austria.

Rhago'dia f. Salt-bush. Gr. *rhagos*, a berry; in reference to the fruit. Sheep delight in this drought-resistant Australian shrub and grow very fat on it. Few plants, however, could appear less inviting.

rhamnifo'lius, -a, -um With leaves like *Rhamnus*.

rhamnoi'des Resembling *Rhamnus*.

Rham'nus m. Buckthorn. The Greek name of various spiny shrubs. Many of the species have medicinal value, the dried bark of *R. purshianus* being the source of *Cascara sagrada*.

Rhaphido'phora f. Gr. *rhaphis*, needle; *phŏrŏs*, carrying. In allusion to the needle-like points on the fruit.

Rhaphio'lepis f. Gr. *rhaphis*, a needle; *lĕpis*, a scale. From the very narrow persistent bracteoles on the inflorescence.

Rhaphitham'nus m. Gr. *rhaphis*, a needle; *thamnos*, a shrub; with reference to the spiny habit of these small evergreen trees.

Rhapidophyl'lum n. Blue palmetto, needle palmetto. A genus of dwarf fan palms, with a resemblance to *Rhapis* in producing suckers freely.

Rha'pis f. Ground rattan. Gr. *rhapis*, a needle; with reference to the needlelike leaf segments of these sucker-producing palms.

Rhazy'a f. In honour of Abu Bekr-el-Rasi (d. *c.* 932), Arabian physician, author of a medical encyclopaedia, who was known in the West as Rhazes.

Rhektophyl'lum n. From Gr. *rhēktŏs*, rent, penetrable; *phyllon*, a leaf. The leaves of these African climbers are cut and perforated.

Rhe'um n. Rhubarb. The Greek name was *rhēŏn* or *rha*, whence the names *Rhaponticum*, meaning the *rha* of Pontus, and *Rhabarbarum*, the *rha* of the foreigners or barbarians. Like it or not, for most of us rhubarb means the first truly fresh dessert of spring. Many species, however, have considerable medicinal value. In the palmy days of the China trade (1800–1850), the best medicinal rhubarb was exported from Canton. In fact, the Chinese Imperial government of the day staked much on the theory that Westerners could not do without it.

Rhe'xia f. Meadow-beauty, deer-grass. Name used by Pliny, of uncertain origin, for some other plant.

Rhipo'gonum See *Ripogonum*.

Rhipsalidop'sis f. From *Rhipsalis*; *opsis*, likeness.

Rhip'salis Gr. *rhips*, wicker-work; from the pliant interlacing twigs of these epiphytic cacti.

rhizophyl'lus, -a, -um With rooting leaves.

rhodan'thus, -a, -um With rose-coloured flowers.

rhoden'sis, -is, -e; rhodius, -a, -um From the island of Rhodes, Aegean Sea.

Rhodochi'ton n. From Gr. *rhŏdŏ-*, red; *chiton*, a cloak; the calyx is large and red.

Rhododen'dron Gr. *rhŏdŏdĕndrŏn*, a name for the rose-flowered oleander, from *rhŏdŏn*, a rose; *dĕndrŏn*, a tree; transferred to these handsome woody plants.

Rhodohypo'xis f. Gr. *rhŏdŏ-*, red; *Hypoxis*, the genus to which this is related.

Rhodomyr'tus f. Gr. *rhŏdŏ-*, red; *myrtos*, myrtle; from the rose-coloured, myrtle-like flowers of these shrubs.

Rhodospa'tha f. Gr. *rhŏdŏ-*, red; *spatha*, spathe.

Rhodotham'nus m. Gr. *rhŏdŏ-*, red; *thamnos*, a shrub; from the shrub's rosy flowers.

Rhodotypos

Rhodo'typos m. Gr. *rhŏdŏn*, a rose; *typŏs*, a type. From the resemblance of this shrub's flowers to a rose.

Rho'eo f. Derivation of the name of this tropical American herb is unknown.

Rhoicis'sus f. Presumably from L. *rhŏicus*, belonging to Sumach (Rhus); *cissus*, ivy.

rhom'bicus, -a, -um; rhomboi'deus, -a, -um Diamond-shaped; rhomboidal.

Rhombophyl'lum n. A genus once included in *Mesembryanthemum*, named for its rhomboid leaves.

Rhopalosty'lis f. Gr. *rhŏpalŏn*, a club; *stylis*, a small pillar; from the club-shaped spadix of these palms.

Rhus f. Sumach. The Greek name for one species, *Rhus coriaria*. *R. typhina*, stag-horn sumach, was also known to the early North American colonists as dyer's sumach from the yellow dye which could be made from it. This genus of shrubs also includes poison-oak (*R. toxicodendron*) and poison-ivy (*R. radicans*).

Rhynchely'tron n. From Gr. *rhynchos* beak; *ělytron*, cover, husk. In allusion to the beaked flower-parts (upper glume and lower lemma) of these grasses.

Rhyncho'sia f. Gr. *rhynchos*, a beak; from the beaked keel of the flowers of these leguminous plants.

Rhynchosty'lis f. Gr. *rhynchos*, a beak; *stylis*, a small pillar; in allusion to the beaked column of these epiphytic orchids.

Rhytico'cus f. Gr. *rhytis*, a wrinkle; *Cocos*, coconut. Closely related to *Cocos*, the name derives from the seed of these palms.

rhytidophyl'lus, -a, -um With wrinkled leaves.

Ri'bes n. Currant, gooseberry, etc. A large and useful genus, the name of which apparently derives from the Arabic or Persian name *ribas*, acid-tasting.

ricasolia'nus, -a, -um In honour of Vincenzo Ricasoli, 19th-century Italian general and patron of horticulture.

Richar'dia f. In honour of Richard Richardson (1663–1741).

ricinifo'lius, -a, -um With leaves like *Ricinus*.

ricinoi'des Resembling *Ricinus*.

Ric'inus m. Castor bean. L. *ricinus*, a tick; from the appearance of the seeds. Possibly best known for its medicinal properties, castor oil is of great commercial value in the manufacture of soap, margarine and lubricants, and is widely grown on this account. Not many years back it had considerable value as an ingredient in high-altitude lubricants for aircraft. The seeds themselves are very poisonous, and must be harvested with care, since as soon as the capsules are ripe they dehisce audibly, shooting the seeds many feet.

Rico'tia f. These herbs are named for Ricot, a little-known French botanist.

ri'gens; rig'idus, -a, -um Rigid; stiff.

rigidifo'lius, -a, -um Rigid-leaved.

rin'gens Gaping, e.g. the mouth of an open two-lipped corolla.

ripa'rius, -a, -um Of the banks of rivers.

Ripo'gonum n. Gr. *rhips*, wicker-work; *gonu*, a knee; from the many-jointed stalks on these climbing liliaceous shrubs that make formidable tangles in the New Zealand bush.

riva'lis, -is, -e Growing by streams.

Rivi'na f. Rouge plant. Herbs with attractive berries named for August Quirinus Rivinus (1652–1722), professor of botany, Leipzig.

rivula'ris, -is, -e Brook-loving.

rob'biae In honour of Mrs Robb of Liphook, Hampshire, England, who introduced *Euphorbia robbiae* from Turkey by bringing back rooted plants in her hat-box some time in the 1890's; whence the name 'Mrs Robb's Bonnet'.

Robi'nia f. Yellow locust, false acacia. Trees and shrubs named for Jean Robin (1550–1629), of Paris, gardener to Henri IV and Louis XIII of France.

ro'bur Latin word for oak-wood, also strength. Specific epithet of English oak (*Quercus robur*).

robustispi'nus, -a, -um With strong spines.

robus'tus, -a, -um Stout; strong in growth.

Ro'chea f. Succulents named for Daniel de la Roche (1743–1813), Swiss physician, part-author of *Pharmacopaea Genevensis* (1780), who settled in Paris in 1782 and became medical officer to the Swiss Guard. His son François de la Roche (1782–1814), likewise a physician in Paris, was also a distinguished botanist and wrote a monograph on *Eryngium*.

ro'ckii In honour of Joseph F. C. Rock (1884–1962), originally Josef Franz Karl Rock. American botanist, plant-collector, explorer and orientalist, noted for his work on plants of Hawaii, western China and Eastern Tibet.

Rodger'sia f. Herbaceous perennials named for Rear-Admiral John Rodgers (1812–1882), distinguished American naval officer who

275

commanded in 1852–1856 a Pacific expedition during which the first species of this genus was discovered.

Rodrigue'zia f. Epiphytic orchids named for Manuel Rodriguez, 18th-century Spanish physician and botanist.

Roel'la f. In honour of G. Roelle, 18th-century professor of anatomy in Amsterdam.

Roeme'ria f. Annuals named for Johann Jakob Roemer (1763–1819), professor of botany, Zürich.

Roh'dea f. A foliage plant named for Michael Rohde (1782–1812), physician of Bremen.

roma'nus, -a, -um Roman.

Romanzof'fia f. Perennial herbs named for Prince Nicholas Romanzoff, who financed a round-the-world expedition, 1816–1817.

Romne'ya f. In honour of Thomas Romney Robinson (1792–1882), celebrated Irish astronomer.

Romu'lea f. In honour of Romulus, legendary king of the Romans and founder of Rome.

Rondele'tia f. Evergreen shrubs and trees named for Guillaume Rondelet (1507–1566), chancellor of the University of Montpellier, France, and a very influential professor of natural history and medicine.

Rorip'pa f. Water-cress. Latinized from a Saxon (East German) vernacular name, Rorippen, mentioned by Euricius Cordus.

Ro'sa f. Rose. The Latin name. See *Rose*.

rosa'ceus, -a, -um Roselike.

Rosche'ria Derivation of the name of this palm is obscure.

Rosco'ea f. In honour of William Roscoe (1753–1831) of Liverpool, lawyer, opponent of the slave-trade, historian, promoter of the arts, a founder of the Liverpool botanic garden and author of a magnificent book on the ginger family (*Zingiberaceae*).

Rose No flower of the Western world has had such great significance for so long as the rose. As a symbol it recurs constantly in religion, art, literature, and heraldry. The cytological and genetical researches of Dr. C. C. Hurst, at the Cambridge University Botanic Garden, between the two world wars, brought a new understanding of their classification as to species and varieties, and elucidated the parentage of modern roses with accuracy.

Prior to the latter part of the 16th century, only four species of rose were known to the Western world. These were *Rosa rubra*, the highly scented Apothecary's rose, or the rose of Provins; *Rosa phoenicia*, the Damask or Crusaders' rose; *Rosa moschata*, the musk rose; and *Rosa canina,* the dog rose. All these roses bloomed but once a year, except *Rosa bifera,* the so-called Virgil's rose of Paestum, which bloomed twice. All these roses were red or white or pink.

Some time in the 16th century a Persian yellow rose arrived on the scene which became known as the Austrian brier rose and is a parent of all the 19th-century Pernet hybrids. In the 16th century also, the Dutch triumphed in producing *Rosa centifolia,* the cabbage rose, which is a hybrid of all the four then-known species. The popular little moss rose is a sport of this rose.

The revolution in rose breeding came about the year 1800 with the introduction of fine China rose species which had been in cultivation in Chinese and Indian gardens for many centuries. There were four recognized China stud roses; Slater's Crimson; Parson's Pink; Hume's Blush; and Park's Yellow, all ever-blooming and tea-scented. Here began today's great groups of garden roses—the teas, the hybrid teas; the hybrid perpetuals, etc. Roses of high scent and fine colour with excellent growth habits could now be produced to bloom all summer long.

The advent of the China roses marked the beginning of a vogue for rose breeding which soon reached very large proportions. The nursery-man and breeder Vibert achieved fame by saving no fewer than ten thousand seedlings from damage by the victorious armies advancing from Waterloo. Rose growing was greatly stimulated by the ex-Empress Josephine, to whom Napoleon had given La Malmaison as a kind of toy. It was an expensive one with millions poured into the development of house and garden.

Among other things, Josephine was determined to possess the greatest collection of roses in the world. To this end, she employed the best gardeners to be found, including a Scotsman, Thomas Blaikie, who continued in her service, quietly breeding roses, war or no war. Also, in the first year or two of the century, John Champney, a wealthy rice planter of Charleston, South Carolina, achieved his Champney's Pink Cluster with a skilful cross of the musk rose and Parson's Pink China. Noisette, a French nurseryman in Charleston, sent seedlings to his brother in Paris, who gave them his own name, by which they are still known. The Noisette rose appears in the parentage of such famous roses as Maréchal Niel, Gloire de Dijon, and Rêve d'Or. It is also a progenitor of the rose named for the great American rose fancier, William Allen Richardson. The introduction of Parson's Pink China into the Ile de Bourbon (Réunion) resulted in a natural cross with the only other rose then growing on the island, Pink Autumn Damask. This marked the beginning of the group of Bourbon roses from which stem all of today's hybrid perpetuals.

The Chinese and Japanese multiflora roses which became available about the middle of the 19th century gave rise to the polyantha roses, in the later development of which the name of the Danish breeder, Poulsen, is probably the best known.

ro'sea Rose colour. As a feminine noun used as a specific epithet it does not have to agree with the gender of the genus, e.g. *Sedum rosea*. This use comes to us from the old herbalists.

ro'seus, -a, -um Rose-coloured.

rosmarinifo'lius, -a, -um With leaves like rosemary.

Rosmari'nus m. Rosemary. The Latin name for this aromatic shrub, derived from *ros*, dew; *marinus*, maritime. It is found wild on sea cliffs in southern Europe.

rostra'tus, -a, -um Beaked.

rosula'ris, -is, -e Having rosettes.

Rota'la f. From *rŏtalis*, wheeled, wheel-like. With reference to the whorled leaves.

rota'tus, -a, -um Wheel-shaped; used of circular, almost flat, or saucer-like flowers.

rotunda'tus, -a, -um; rotun'dus, -a, -um Rounded.

rotundifo'lius, -a, -um With round leaves.

Rou'pala f. From the native Guyanese name for these ornamental trees and shrubs.

roxburghii In honour of William Roxburgh (1751–1815), Scottish doctor, and superintendent of Calcutta botanic garden 1793–1814.

Roye'na f. Evergreen trees and shrubs named for Adrian van Royen (1705–1779), professor of botany, Leyden.

Roysto'nea f. Palms named for General Roy Stone (1836–1905) American army engineer in Puerto Rico.

rubel'lus, -a, -um; rubes'cens Reddish-coloured.

ru'bens; ru'ber, rub'ra, rub'rum Red. In compound words, *rubri-*.

Ru'bia f. Madder. L. *ruber*, red; in allusion to the reddish dye obtained from the roots of these herbs.

rubigino'sus, -a, -um Rusty.

rubioi'des Resembling *Rubia*.

rubri'caulis, -is, -e Red-stemmed.

rubropilo'sus, -a, -um Red-haired.

Ru'bus m. Blackberry, bramble, raspberry. The Latin name.

rudbaren'sis, -is, -e; rudba'ricus, -a, -um Of Rudbar, north-west Iran, near Resht.

Rudbe'ckia f. Cone-flower, black-eyed Susan. Named for Olof Rudbeck (1630–1702), 'one of the most remarkable Swedes of all time', anatomist, botanist and antiquarian, founder of the Uppsala botanic garden. The woodblocks illustrating some 11,000 plants and intended for a great work on which he had spent much of his life were nearly all destroyed in the terrible Uppsala fire of 1702, and he did not long survive this disaster. His son Olof Rudbeck the younger (1660–1740), also a professor in Uppsala, befriended Linnaeus when he was a poverty-stricken student. *Rudbeckia* commemorates both.

ru'dis, -is, -e Rough, untilled, coarse.

Ruel'lia f. Herbs and shrubs named for Jean Ruel (1474–1537), herbalist to François I of France.

rufes'cens Becoming reddish.

ru'fus, -a, -um Red.

rugo'sus, -a, -um Wrinkled, rugose.

Ru'mex f. Sorrel, dock. The Latin name. *R. acetosa* is a plant more used in France than in America; it forms the basis of one of the more delectable soups.

runcina'tus, -a, -um Saw-toothed, with the teeth pointing towards the base.

rupes'tris, -is, -e Rock-loving.

rupi'cola A dweller among rocks.

rupi'fragus, -a, -um Rock-breaking, i.e. growing in clefts of the rock.

Rus'chia f. Named after a South African farmer, E. Rusch, of Windhoek.

ruscifo'lius, -a, -um Having leaves like butcher's broom or *Ruscus*.

Rus'cus m. Butcher's broom. The Latin name.

Ruspo'lia f. In honour of Prince Eugenio Ruspoli (1866–1893), Italian explorer of Somaliland killed by an elephant.

russa'tus, -a, -um Russet.

Russe'lia f. In honour of Dr. Alexander Russell (*c.* 1715–1768), physician to the English Factory at Aleppo (*c.* 1740 to 1753) and the author of a *Natural History of Aleppo* (1756).

russellia'nus, -a, -um In honour of John Russell (1766–1839), 6th Duke of Bedford, a zealous promoter of botany and gardening.

russotin'ctus, -a, -um Red-tinged.

rustica'nus, -a, -um; rus'ticus, -a, -um Pertaining to the country.

Ru'ta f. Rue. The Latin word. As in English, the same word means bitterness or unpleasantness. The leaves of these perennial herbs have a bitter taste.

ruthen'icus, -a, -um From Ruthenia, a region of European Russia inhabited by the Ruthenians; the name was, however, used in a general sense for much of European Russia.

ru'tilans Reddish.

Rut'tya f. In honour of John Rutty (1697–1775), physician in Dublin. The name *Ruttyruspolia* covers the hybrids of *Ruspolia* and *Ruttya*.

S

Sa'bal f. Palmetto. Possibly from the South American name for these spineless palms.

Saba'tia f. Rose pink, American centaury. Showy herbs named for an Italian botanist, Liberato Sabbati (b. *c.* 1714), curator of the Rome botanic garden.

saba'tius, -a, -um Of Savona, Riviera, north-west Italy.

sabau'dus, -a, -um Of Savoy, south-eastern France.

sacca'tus, -a, -um Resembling a bag; provided with pronounced sacs or nectar-producing pits; see *saccifer.*

sacchara'tus, -a, -um Sugared or sugary, i.e. sweet-tasting or looking as if sprinkled with sugar.

saccha'rifer, sacchari'fera, -um Sugar-producing.

sacchari'nus, -a, -um Sugary.

saccharoi'des Resembling sugarcane.

Sac'charum n. Sugarcane. Gr. *sakcharon,* the sweet juice of sugarcane, from an Asiatic word, seemingly the Malay *singkara.* Sugarcane has been known since antiquity in Asia whence it passed to Africa. It was introduced into Sicily around A.D. 800 and by the Moors into Spain before A.D. 1000. The cane was taken by Columbus to the New World on his second voyage where it quickly became widely cultivated and led to the establishment of negro slavery. Manufactured sugar was imported into Europe from very early times. In the Middle Ages it was handled by apothecaries as a medicine. It became cheap and readily available to all with the advent of steam navigation and manufacturing machinery.

It is cultivated throughout the tropical and semitropical regions of the world.

sac'cifer, saccif'era, -um Bag-bearing, i.e. provided with hollowed-out organ or sac.

Saccola'bium n. From L. *saccus*, bag; *labium*, lip; in allusion to the lip of these orchids.

sachalinen'sis, -is, -e From the island of Sakhalin, U.S.S.R., north of Japan.

sacro'rum Of sacred places.

Sadle'ria f. Tree ferns named for Joseph Sadler (1791–1841), professor of botany, Budapest.

Sagi'na f. Pearlwort. L. *sagina*, fodder; from the fattening qualities of spurrey (*Spergula sativa*) on which sheep quickly gain condition and which was once placed in this genus of tufted herbs.

sagitta'lis, -is, -e; sagitta'tus, -a, -um Arrow-shaped.

Sagitta'ria f. Arrowhead. L. *sagitta*, an arrow; from the form of the leaves of these perennial aquatic herbs.

sagittifo'lius, -a, -um Arrow-leaved.

Saintpau'lia f. African violet. Named for a German, Baron Walter von Saint Paul-Illaire (1860–1910), who discovered the African violet (*S. ionantha*) in East Africa.

salicifo'lius, -a, -um Willow-leaved.

salici'nus, -a, -um Willow-like.

Salicor'nia f. Glasswort, marsh samphire. L. *sal*, salt; *cornu*, a horn; from the hornlike branches of this genus of salt-marsh herbs.

salicornioi'des Resembling *Salicornia*.

salig'nus, -a, -um Resembling willow.

sali'nus, -a, -um Growing in salty places.

Sa'lix f. Willow. The Latin name.

Willow bark was, until the early part of this century, the sole source of salicylic acid which has great medicinal value, especially in the treatment of rheumatism. German chemists discovered before 1914 how to synthesize the acid which became the principal ingredient of aspirin, closely covered by patents held by the Bayer interests. Their monopoly was broken as a result of the first World War and aspirin came into wide general use at modest prices.

The discovery of the value of salicylic acid for the treatment of rheumatism resulted from application of the once widely held belief that any endemic ill had its remedy close by. Rheumatism being associated with living in damp places, it was natural to seek the remedy close at hand.

In many country places in England, pussy willows are eagerly sought by children for the silky catkin-bearing twigs called 'palms' to be used for church decoration on Palm Sunday. Similarly, in northern Russia the week before Palm Sunday was formerly called 'willow week'. The 'palms', tied with bright ribbons, were sold for Palm Sunday.

salmia′nus, -a, -um; sal′mii In honour of Prince Joseph Salm-Reifferscheid-Dyck (1773–1861), German prince who became a leading authority on succulent plants; see also *Dyckia*.

Salpi′chroa f. From Gr. *salpinx*, a trumpet; *chroa*, a skin; in allusion to the pretty trumpet-shaped flower of these shrubs and sub-shrubs.

Salpiglos′sis f. From Gr. *salpinx*, a trumpet; *glōssa*, a tongue; with reference to the elongated clarinet-shaped style.

Sal′sola f. Russian thistle. From L. *sal*, salt. They are generally salt-soil plants, often prickly but not true thistles.

salsugino′sus, -a, -um Growing in salt marshes.

saluenen′sis, -is, -e of the Salween (Salwin) River of Burma and China.

Sal′via f. Sage. The Latin name, itself derived from *salvus*, safe, well, sound; from the supposed medicinal values of the plants. *S. sclarea*, clary or clear-eye, was particularly valued for affections of the eye, hence the English name.

salviifo′lius, -a, -um *Salvia*-leaved.

Salvin′ia f. A floating plant named for Antonio Maria Salvini (1633–1729), professor of Greek at Florence, who helped Micheli in his botanical studies.

Sama′nea f. Rain-tree or monkey-pod. From the native South American name, *saman*. It is called rain-tree because the leaflets fold up in cloudy weather and at night.

sambucifo′lius, -a, -um With leaves like elder or *Sambucus*.

sambu′cinus, -a, -um Resembling *Sambucus* or elder.

Sambu′cus f. Elder. The Latin name, perhaps connected with *sambuca*, a kind of harp.

Sa′molus m. The Latin name, probably of Celtic origin, for a plant of damp places gathered by Druids.

Samue′la f. Date-yucca. Named for Samuel Farlow Trelease (1892–1958), American botanist, son of William Trelease (1857–1945). Both flowers and fruit of these large plants are eaten locally in Mexico.

Sanche′zia f. Handsome shrubs and herbs named for José Sanchez, a 19th-century professor of botany at Cadiz.

sanc′tus, -a, -um Holy.

Sanderso′nia f. In honour of John Sanderson (1820 or 1821–1881), who discovered this remarkable plant in Natal in 1851.

Sanguina'ria f. Bloodroot. From L. *sanguis*, blood. All parts of this low-growing perennial have copious yellowish-red sap.

sanguin'eus, -a, -um Blood-red.

Sanguisor'ba f. Burnet. From L. *sanguis*, blood; *sorbeo*, to soak up. This herb was supposed to have styptic qualities and was used as an infusion to stop bleeding from dysentery, etc.

Sansevie'ria f. Bow-string hemp and, in the florists' shops, snake plant or leopard lily. Named for an 18th-century Italian patron of horticulture.

San'talum n. Sandalwood. Gr. *santalon*, sandalwood tree, derived via the form *sandanon* from Sanskrit *chandana*, fragrant. For most of us sandalwood means the fragrant, carved boxes and fans from India. The wood, however, is employed in funeral and religious rites wherever Buddhism prevails. Also, ground to a paste, it is one of the pigments used by Brahmans in making their caste marks.

Santoli'na f. Lavender cotton. L. *sanctum linum*, holy flax, an old name for one species of these shrubby herbs, *S. virens*.

Sanvita'lia f. Yellow-flowered herbs apparently named for Federico Sanvitali (1704–1761), professor at Brescia, Italy.

sa'pidus, -a, -um Pleasant to taste.

sapien'tum Pertaining to wise men; used of the better forms of banana in contrast to those called *troglodytarum* considered fit only for apes or cave-dwellers.

Sapin'dus m. Soap-berry. From L. *sapo*, soap; *indicus*, Indian. The pulp of these trees and shrubs lathers like soap and was used as such by Indians.

Sa'pium m. Latin name for a resinous pine. The stems of these trees, which are not pines, exude a sticky sap. *S. sebiferum*, the Chinese tallow tree, has seeds with a waxy covering which can be used in making soap and candles.

sapona'ceus, -a, -um Soapy.

Sapona'ria f. Soapwort, bouncing Bet. L. *sapo*, soap. The roots of *S. officinalis* lather and were once used as a substitute for soap.

Sapo'ta f. From native South American name. The sapodilla or naseberry (*Manilkara zapota*, synonym *Sapota achras*) has a delicious fruit but commercially it is most important as the source of chicle which is the basis of chewing gum. It should not be confused with the sapote or mammee sapote (*Pouteria sapota*, synonym *Calocarpum sapota*) which has a much larger seed.

Sa'raca f. Apparently a corruption of an East Indian name *Asoka*, commonly applied to this tree which is remarkable among *Leguminosae* in having brightly coloured sepals but no petals.

Sarcan'thus m. From Gr. *sarx, sarkos,* flesh; *anthos,* a flower. The flowers of these orchids are fleshy.

Sarcochi'lus m. Gr. *sarx, sarkos,* flesh; *cheilos,* a lip; in allusion to the fleshy lip of the flower of these epiphytic orchids.

Sarcococ'ca f. Sweet box. Gr. *sarkos,* flesh; *kokkos,* a berry; in allusion to the fleshy fruits of these ornamental shrubs.

sarco'des Fleshlike.

sargentia'nus, -a, -um; sargen'tii In honour of Charles Sprague Sargent (1841–1927), leading American dendrologist, first director of the Arnold Arboretum of Harvard University.

sarma'ticus, -a, -um From Sarmatia, i.e. eastern Poland and south-western European Russia.

sarmento'sus, -a, -um Having or bearing runners.

sarnien'sis, -is, e Of Guernsey (ancient Sarnia), Channel Islands.

Sarotham'nus From Gr. *sarŏn,* broom; *thamnŏs,* shrub.

Sarrace'nia f. Pitcherplant. Named for Michel Sarrasin (1659–1734), French physician and botanist of Quebec, who sent the first of these handsome carnivorous plants to Europe.

Sa'sa f. The Japanese name for these bamboos.

sasan'qua Japanese name *sasankwa* for a popular native species of *Camellia.*

Sas'safras n. Derivation of the name for these handsome trees is dubious, but it probably comes from an American Indian name used in Florida. The roots and bark are aromatic and are the source of oil of sassafras which is used in perfumery. During the 17th and 18th centuries sassafras was an important export from the New World, ranking second only to gold among the commodities which ships' captains sailing for America were called on to seek. Oil of sassafras is used in medicine and at one time was regarded as the most reliable cure for syphilis. The tree grows to considerable size under the proper conditions. The wood is white, strong, close-grained, and elastic.

sati'vus, -a, -um Cultivated.

satura'tus, -a, -um Saturated.

Sature'ja (also at one time, **Satureia**) f. Savory. The Latin name for this herb which was well known to the ancients. It is an excellent bee plant and was recommended for planting around hives by Virgil in the *Georgics.*

sauroce'phalus, -a, -um Lizard-headed.

Sauro'matum n. An old name used by Dodoens and derived from Gr. *sauros*, a lizard.

Sauru'rus m. Gr. *sauros*, a lizard; *oura*, a tail; from the form of the inflorescence of these perennial marsh herbs.

Saussu'rea f. Perennial mountain herbs named for Horace Bénédict de Saussure (1740–1799), Swiss philosopher and scientist, for many years professor of philosophy at Geneva, a pioneer investigator of the geology, meteorology and botany of the high Alps, celebrated for his ascent in 1787 of Mont Blanc, Europe's highest mountain, first conquered in 1786 by Paccard and Balmat through his encouragement.

saxa'tilis, -is, -e Found among rocks.

Saxegothae'a f. Prince Albert's yew. A Chilean evergreen conifer named for Prince Albert (1819–1861) of Saxe-Coburg-Gotha, Prince Consort of Queen Victoria, and a zealous promoter of the arts and sciences.

saxi'cola Growing in rocks.

Saxif'raga f. Saxifrage. L. *saxum*, a rock; *frango*, to break. Growing in rock crevices this herb was supposed to be capable of breaking rocks. Hence, by deduction, it was accorded a medicinal quality of breaking up stone in the bladder.

saxo'sus, -a, -um Full of rocks, hence growing among rocks.

sca'ber, scab'ra, scab'rum Rough.

Scabio'sa f. Mourning bride, pincushion flower, scabious L. *scabies*, the itch, which, by the analogy of the roughness of the leaves, the plant was supposed to cure.

scabiosifo'lius, -a, -um With leaves like *Scabiosa*.

scabrifo'lius, -a, -um Rough-leaved.

scan'dens Climbing.

sca'piger, scapi'gera, -um Scape-bearing.

scapo'sus, -a, -um With scapes, i.e. with leafless flower-stems.

scario'sus, -a, -um Shrivelled, used of thin, dry organs.

scelera'tus, -a, -um Wicked, hurtful.

scep'trum A sceptre.

Schaue'ria f. Shrubby herbs with ornamental foliage named for Johann Conrad Schauer (1813–1848), professor at Greifswald, a specialist in the study of *Myrtaceae* and *Verbenaceae*.

Schee'lea f. Palms named for Karl Wilhelm Scheele (1742–1786), celebrated Swedish chemist of German origin, the discoverer of many substances, including Scheele's Green.

Scheffle'ra f. Trees or shrubs named for J. C. Scheffler, 19th-century botanist of Danzig.

Schi'ma f. Derivation obscure; according to Backer, from Gr. *skiasma,* shadow, shelter, with reference to the thick leafy crown of the tree.

schin'seng Chinese word for ginseng for which this is the specific epithet (*Panax schinseng*).

Schi'nus m. Pepper-tree. Gr. *schinos,* the mastic tree, which this genus resembles in that the trees are resinous and yield a mastic-like juice.

schipkaen'sis, -is, -e Referring to the Shipka Pass, Bulgaria.

Schisan'dra f. Gr. *schizo,* to divide; *aner,* male; in allusion to the well separated anther cells.

Schismatoglot'tis f. Gr. *schismas,* cleft, division; *glōtta,* a tongue. The tongue-like blade of the spathe becomes cut off from the more persistent tubular basal part.

schisto'calyx With split calyx.

Schivere'ckia f. Dwarf perennials named for S. B. Schivereck (1782–1815) of Krakow, Poland.

Schizae'a f. Comb or rush fern. Gr. *schizo,* to divide. The fronds are divided into a fan shape.

Schizan'dra See *Schisandra.*

Schizan'thus m. Butterfly or fringe flower. Gr. *schizo,* to divide; *anthos,* a flower. In these annual herbs the corolla is deeply cut.

Schizocen'tron n. Gr. *schizo,* to divide; *kentron,* a spur; from the bifid spur of the larger stamens.

Schizoco'don m. Gr. *schizo,* to divide; *kōdōn,* a bell; the segments of the bell-shaped corolla are elegantly cut into a deep fringe.

Schizolo'bium n. Gr. *schizo,* to divide; *lōbŏs,* pod; the thin inner coat of the pod splits away from the firm outer coat.

Schizope'talon n. Gr. *schizo,* to divide; *petalon,* a petal. These annual herbs have flowers with deeply cut petals.

schizope'talus, -a, -um With cut petals.

Schizophrag'ma n. Gr. *schizo,* to divide; *phragma,* a fence or screen. Parts of the vase-shaped fruits on these climbing shrubs fall away leaving them skeletonized.

schizophyl'lus, -a, -um With cut leaves.

Schizosty'lis f. Kaffir lily. Gr. *schizo,* to divide; *stylis,* a column, style. The style of these South African herbs is divided into three parts.

schlippenba'chii In honour of Baron Alexander von Schlippenbach,

Russian naval officer of Baltic German origin, who collected in Korea in 1854.

Schlumber'gera f. Cacti named for Frederick Schlumberger, Belgian horticulturist, *c.* 1900.

schoenopra'sum Specific epithet for chives (*Allium schoenoprasum*) compounded from Gr. *schoinos*, rush; *prasŏn*, leek; with reference to its rush-like leaves.

schola'ris, -is, -e Relating to school. *Alstonia scholaris* was so named because children wrote their lessons on boards made from its wood.

Schomburg'kia f. Epiphytic orchids named for Sir Robert Hermann Schomburgk (1804–1865), German-born, naturalized British explorer who found the *Victoria amazonica* (*V. regia*) water-lily in Guyana in 1837. It had, however, been discovered in the Rio Mamore, a tributary of the Amazon, in 1801 by Thaddäus Haenke, also German-born, then employed by the Spanish government. Moritz Richard Schomburgk (1811–1890), his brother, accompanied him to Guyana in 1840 and was later the director of the Adelaide botanic garden.

Scho'tia f. Small trees and shrubs named for Richard van der Schot (d. 1819), head gardener at the palace of Schönbrunn, near Vienna.

Schran'kia f. Perennial herbs and shrubs named for Franz von Paula von Schrank (1747–1835), German botanist, professor at Munich.

Schuber'tia f. In honour of Gotthilf Heinrich von Schubert (1780–1860), German doctor and naturalist, 1827 onwards professor of natural history in Munich, who collected in 1837 in Egypt, Palestine and Greece.

Sciado'pitys f. Umbrella pine. Gr. *skias, skiados*, umbel; *pitys*, a fir tree. The spreading whorls of needles resemble the ribs of an umbrella or parasol, for which the Greek name was *skiadĕiŏn*.

Scil'la f. Squill. The Greek name *skilla* for sea-squill (*Urginea scilla*). The English bluebell used to be named S. *nutans*, but is now placed in the genus *Endymion* as *E. non-scriptus*.

scilloi'des Resembling *Scilla*.

Scindap'sus m. Ivy arum. From the Greek name for a plant resembling ivy applied to these climbing perennials.

scintil'lans Gleaming.

Scir'pus m. Bulrush. The Latin name. The biblical bulrush was the related *Cyperus papyrus*.

Scleran'thus m. From Gr. *sklēros*, hard, dry, harsh; *anthos*, flower.

Sclerocac'tus m. Gr. *sklērŏs*, harsh, cruel, plus *Cactus*; in allusion to the sharp spines.

sclerocar'pus, -a, -um With hard fruits.

Scolopen'drium n. Gr. *skŏlŏpĕndrion*, from *skŏlŏpĕndra*, millipede; in allusion to the rows of sori.

Sco'lymus m. Latin name for Spanish oyster-plant. This biennial plant is edible and closely related to salsify.

scopar'ius, -a, -um Broomlike.

Scopo'lia f. In honour of Giovanni Antonio Scopoli (1723–1788), Italian botanist, zoologist and geologist, professor in Pavia.

scopulo'rum Of cliffs, crags, projecting rocks.

scorodopra'sum Gr. *skŏrŏdŏprasŏn*, a kind of *Allium* between the garlic, *skŏrŏdŏn* or *skŏrdŏn*, and the leek, *prasŏn*.

scorpioi'des Resembling a scorpion.

Scorpiu'rus m. Gr. *skorpios*, a scorpion; *oura*, a tail; from the form of the pods on these herbs.

Scorzone'ra f. Black salsify. From Old French *scorzon*, Italian *scorzone*, viper. The root of the herb was once regarded as a cure for snakebite.

sco'ticus, -a, -um Scottish.

scotophyl'lus, -a, -um From Gr. *skotos*, dark; *phyllon*, leaf.

scottia'nus, -a, -um In honour of Munro B. Scott (1889–1917), Kew botanist killed in action in the first World War.

scrobicula'tus, -a, -um Pitted with minute depressions.

Scrophula'ria f. Fig-wort. L. *scrofule*, which it was supposed to cure. The roots of some species of these herbs look like scrofulous tumours.

scul'ptus, -a, -um Carved.

scuta'tus, -a, -um Armed with a shield, shield-shaped.

Scutella'ria f. Skullcap. L. *scutella*, a small dish or saucer, from the pouch on the fruiting calyx of these herbs.

Scutica'ria f. L. *scutica*, a whip, lash; from the shape of the long leaves of these orchids.

scu'tum A shield.

Scyphan'thus m. From Gr. *skyphŏs*, cup, beaker; *anthos*, flower.

Scyphoste'gia f. From Gr. *skyphŏs*, beaker; *stĕgē*, covering, shelter.

se'bifer, sebi'fera, -um Tallow-bearing.

sebo'sus, -a, -um Full of grease or tallow.

Seca'le n. Rye. Latin name for some cereal grain, perhaps rye.

288

secali'nus, -a, -um Resembling rye.

sechella'rum Of the Seychelle Islands in the Indian Ocean.

Se'chium n. From the West Indian name for this tuberous-rooted climbing herb.

seclu'sus, -a, -um Hidden.

secundiflo'rus, -a, -um; secun'dus, -a, -um One-sided, secund. Applied to leaves or flowers arranged on one side of a stalk only.

Securi'daca f. L. *securis*, an axe; from the shape of the wing at the end of the pods of these tropical shrubs, often climbers.

secu'riger, securi'gera, -um Axe-bearing.

Securi'gera f. Hatchet vetch. L. *securis*, an axe; *gero*, to bear; from the shape of the pods.

Se'dum n. Stonecrop, wall-pepper. Name used by the ancients for various succulents, mostly species of *Sedum* or *Sempervivum*, from *sedo*, to sit; from the manner in which some species grow on rocks and walls.

segeta'lis, -is, -e; se'getum Of cornfields.

seinghkuen'sis, -is, -e From the Seinghku Valley, Upper Burma.

Selaginel'la f. Club-moss. Diminutive of *Selago*, the name of another moss-like plant (*Lycopodium selago*).

selaginoi'des Resembling club-moss.

Selenice'reus m. Night-blooming cereus. Gr. *sĕlēnē*, the moon, plus *Cereus*. The flowers of these cacti are mostly nocturnal.

Selenipe'dium n. From Gr. *sĕlēnē*, moon; *pĕdilŏn*, sandal; referring to the lip of these orchids.

selen'sis, -is, -e From the Sie-la pass, Yunnan.

Sellie'ra f. In honour of François Noël Sellier (1737–c. 1800) of Paris, who engraved botanical illustrations for Cavanilles and Desfontaines between 1780 and 1800.

sel'loi, sellovia'nus, -a, -um; sello'wii In honour of Friedrich Sellow (1789–1831), German traveller and naturalist, who made extensive collections in Brazil and Uruguay. The family name was 'Sello' but when in Brazil he altered it to 'Sellow', hence the varied spelling of epithets of South American plants commemorating him.

Se'mele f. A climbing shrub named for Sĕmĕle, the daughter of Kadmos and mother of Dionysos.

semi- In Latin compound words signifies half.

semperflo'rens Ever-blooming.

semper'virens Evergreen.

Sempervivel'la f. Diminutive of *Sempervivum*.

sempervivoi'des Resembling *Sempervivum*.

Sempervi'vum n. House-leek. L. *semper*, always; *vivus*, alive, living. Hen-and-chickens, *S. tectorum* (of roofs), was regarded as effective against lightning and for that reason this fleshy herb was planted on roofs.

Sene'cio m. Groundsel, ragwort. L. *senex*, an old man; from the hoary pappus of these herbs.

senecioi'des Resembling *Senecio*.

senes'cens Growing old.

seni'lis, -is, -e Aged, hence white-haired.

sensib'ilis, -is, -e; sensiti'vus, -a, -um Sensitive; responding quickly to touch, changes in light, etc.

sentico'sus Thorny, full of thorns.

sepiar'ius, -a, -um; se'pium Growing in hedges or used for hedging.

sept- In compound words signifying seven. Thus:

septangula'ris	seven-angled
septem'fidus	with seven cuts
septem'lobus	with seven lobes
septempuncta'tus	seven-spotted

septentriona'lis, -is, -e Northern.

sepulcra'lis, -is, -e Growing in burial places.

sepul'tus, -a, -um Buried.

Sequoi'a f. Redwood. The name commemorates the inventor of the Cherokee alphabet Sequoia or Sequoiah (1770–1843), otherwise George Gist, the son of a German-American merchant and an American Indian girl. The name *sequoia*, in Cherokee the name of the opossum, was used as a nickname for a half-breed. He was born at a village by the Cherokee (now Tennessee) River and brought up as a full-blooded Indian without knowledge of English. The adoption of his syllabary by the Cherokees in 1821 quickly made the tribe literate.

Sequoiaden'dron n. Big tree. *Sequoia* plus Gr. *dendron*, tree. In England *Wellingtonia* is the commonly accepted synonym of *S. gigantea*. This is sometimes claimed as the biggest and oldest living thing known to man. The first seeds seem to have been collected in the early 1850's by G. H. Woodruff, a New Yorker and down-and-out gold miner in California, who found himself in a shower of seeds dropped by squirrels feeding on the cones. He filled his snuff-box and sent the seeds by pony express at a cost of $25 to the nursery of Ellwanger and Barry of Rochester, New York. They obtained about 4,000 seedlings, of which

400 went to England where they were eagerly bought. Europeans were soon taking all they could lay their hands on, for every great house and botanic garden wanted an avenue. Woodruff's share of the profits came to $1,030.60. The tree was originally listed as '*Washingtonia gigantea*, . . . *Wellingtonia* of the English and *Sequoia* of the French'.

Sera'pias f. From the Egyptian god Serapis.

Sere'noa f. Palms named for Sereno Watson (1826–1892), of Harvard University, distinguished American botanist who named and described many of the new species found during the pioneer botanical exploration of western and middle North Africa.

serican'thus, -a, -um With silky flowers.

seri'ceus, -a, -um Silky.

seri'cifer, seri'cifera, -um; seri'cofer, seri'cofera, -um Silk-bearing.

Sericocar'pus f. Gr. *sērikon*, silk; *karpŏs*, fruit. The dry fruits of these aster-like herbs are covered with silky hairs.

Seris'sa f. From the East Indian name for this shrub.

Serja'nia f. Named in 1703 by Plomier in honour of the Rev. Father Philippe Sergeant of Caux, France, a monk skilled in botany and medicine, for many years domiciled in Rome.

sero'tinus, -a, -um Late in flowering or ripening.

ser'pens Creeping.

serpenti'nus, -a, -um Serpentine; relating to snakes, or to serpentine rocks.

serpyllifo'lius, -a, -um With leaves like *Thymus serpyllum*.

serratifo'lius, -a, -um With serrated or saw-toothed leaves.

Serra'tula f. L. *serrula*, a small saw; from the serrated leaves of these thistlelike perennials.

serra'tus, -a, -um Saw-toothed.

serrula'tus, -a, -um With small saw-like teeth.

Se'samum n. Sesame. The Greek name *sēsamŏn*, from a Semitic name for this important oil plant sometimes commercially called gingelly.

Sesba'nia f. From the Arabic name for these herbs and shrubs.

Se'seli n. Gr. *Sěsěli*, ancient name of an umbelliferous plant.

Sesle'ria f. In honour of Leonardo Sesler (d. 1785), doctor at Venice, who had a private botanic garden there.

sesquipeda'lis, -is, -e One and one-half feet in length or height, applied to a Madagascar species of *Angraecum* because of the length of the spur of this orchid, which led Darwin to predict that a moth with a tongue (proboscis) long enough to reach the nectar in it would later be discovered; such a moth was indeed later discovered and named *Xanthopan morgani* subsp. *praedicta.*

ses'silis, -is, -e Stalkless; sessile. In compound words *sessili-*.

seta'ceus, -a, -um Bristled.

Seta'ria f. Foxtail millet. L. *seta*, a bristle; from the bristly awns of these grasses.

seti- In compound words signifies bristled. Thus:
setifo'lius	with bristly or bristle-like leaves
se'tifer	bearing bristles
se'tiger	bearing bristles
seto'sus	full of bristles, bristly
setulo'sus	full of small bristles

Severi'nia f. A shrub or small tree related to the orange, named for Marco Aurelio Severino (1580–1656), professor of anatomy at Naples.

sexangula'ris, -is, -e With six angles.

shal'lon Rendering of an American Indian name for *Gaultheria shallon.*

Shepher'dia f. Ornamental shrubs or small trees named for John Shepherd (1764–1836), curator of the Liverpool botanic garden.

sherri'ffii Specific epithet of many Himalayan plants named after Major George Sheriff (1898–1967), who explored south-eastern Tibet in company with F. Ludlow.

Shor'tia Evergreen stemless herbs named for Dr. Charles W. Short (1794–1863), Kentucky botanist. There are two species, the American *S. galacifolia*, sometimes called Oconee bells, and the Japanese *S. uniflora*, Nippon bells. The former came to the notice of Asa Gray in 1839 when he went to Paris to study André Michaux's North American herbarium made some fifty years before but which had not been put completely in order. Michaux's specimen lacked flowers but Gray, who had never before seen the plant, was much struck by it. Michaux had collected it somewhere in Carolina, in 1788, but it was not rediscovered until 1887. Meanwhile the Japanese species, at first named *Schizocodon uniflorus*, had been discovered and Gray, examining Japanese specimens of this in 1858, recognized its affinity to his own then obscure *Shortia galacifolia*. Sargent in 1888 found the place in the mountains of Carolina where Michaux had camped in December 1788 and there grew the *Shortia.*

shwelien'sis, -is, -e From Shweli, western China.

sia'meus, -a, -um Of Siam (Thailand).

Sibbal'dia f. In honour of Sir Robert Sibbald (1641–1722), first professor of medicine in the University of Edinburgh and author of *Scotia Illustrata*.

Sibirae'a f. From the Siberian habitat of one species of these *Spiraea*-like shrubs.

Sibthor'pia f. Perennial herbs named for Humphrey Sibthorp (1713–1797), English botanist and professor of botany at Oxford from 1747 to 1783, during which period he gave only one lecture; he was the father of John Sibthorp (1758–1796), professor of botany at Oxford from 1783 to 1795, celebrated for his travels in Greece, Cyprus and Asia Minor, commemorated in plants named *sibthorpii*.

Sica'na f. From the Peruvian name for these vines of the cucumber family.

siculifor'mis, -is, -e Dagger-shaped.

si'culus, -a, -um Of Sicily.

Si'cyos m. Bur-cucumber. Greek for cucumber, to which this is closely related.

Si'da f. Gr. name for a water plant.

Sidal'cea f. False mallow. From *Sida* and *Alcea*, both related genera.

Side'rasis f. Said by the author of the name, Rafinesque, to mean 'rusty fur', in allusion to the dense reddish hair-covering, presumably from Gr. *sidĕrŏs*, iron; *sisura*, goat-skin.

Sideri'tis f. The Greek name *sidērītis* for plants used as wound dressings, etc., from *sidĕrŏs*, iron. These herbs and shrublets were supposed to heal wounds caused by swords and other iron weapons.

siderophloi'us, -a, -um Iron-barked.

siderophyl'lus, -a, -um With rusty leaves.

Siderox'ylon n. Miraculous berry of West Africa. From Gr. *sidĕrŏs*, iron; *xylŏn*, wood; from the hardness of the heartwood. The fruit is acid-sweet but, after eating it, acid and sour foods like limes taste sweet, and the effect can last for two or three hours.

sie'beri In honour of Franz Wilhelm Sieber (1789–1844) of Prague, natural history collector and traveller, who himself visited Italy, Crete, Egypt, Palestine, Mauritius, Australia and South Africa and sent collectors elsewhere.

siebol'dii In honour of Philipp Franz van Siebold (1796–1866), German doctor, in Japan from 1823 to 1830, during which period he introduced many Japanese plants into European gardens and collected material for his important publications on Japan; he was again in Japan from 1859 to 1863.

Sigesbeckia

Sigesbe'ckia f. A weedy small-flowered composite named by Linnaeus after his bitter opponent Johann Georg Siegesbeck (1686–1755).

signa'tus, -a, -um Well-marked.

Sile'ne f. Catchfly, campion. The Greek name for another plant (*Viscaria*), now applied to these common herbs.

sili'ceus, -a, -um Growing in sand.

Sil'phium n. Rosinweed. Perennial American herbs to which has been transferred the Greek name *silphiŏn* of an umbelliferous plant, a species of *Ferula* probably now extinct but in ancient times of great economic importance in Cyrenaica as the source of an aromatic gum and hence portrayed on coins.

silva'ticus, -a, -um; silves'tris, -is, -e Growing in woods or growing wild.

silvi'cola An inhabitant of woods.

Si'lybum n. Lady's thistle, milk thistle, blessed thistle. The Greek name, *silybŏn*, for some thistlelike plant. The English names apply to *S. marianum*, also called Our Lady's thistle from the white spots on the leaves supposed to have been the result of milk dropped on them by the Virgin.

sim'ilis, -is, -e Similar; like.

sim'plex Simple, unbranched. Also:
simplicicau'lis with simple stems
simplicifo'lius with simple leaves

sim'sii In honour of John Sims (1749–1831), medical man and botanist, editor of *Curtis's Botanical Magazine* from 1800 to 1826.

sim'ulans Resembling.

Sina'pis L. name, also spelled *sinapi* and *sinape*, for mustard plant.

Sinarundina'ria f. From *sino-*, Chinese, plus *Arundinaria*; with reference to the Chinese habitat of these bamboos.

sinen'sis, -is, -e; si'nicus, -a, -um Chinese.

Sinnin'gia f. Tropical herbs named for Wilhelm Sinning (1794–1874), head gardener, University of Bonn. This is the gloxinia of the florists (*S. speciosa*).

Sinome'nium n. Chinese moonseed. A shrubby vine. From Gr. *sinai*, the Chinese; *mēnē*, the moon.

Sinowilso'nia f. A shrub or tree like *Hamamelis* named for Ernest Henry Wilson (1876–1930), of the Arnold Arboretum, Boston. Known as Chinese Wilson (Gr. *sinai*, L. *sinae*, the Chinese), he travelled widely in China and introduced more than a thousand plants, including *Lilium regale*, to the gardens of England and America. English-born and Kew-trained, he brought unusual imagination and perception to all he did, including a remarkable ability to get along with Chinese at all levels. He was sent out to China in 1899 by the firm James Veitch and Sons of Chelsea primarily to introduce into cultivation *Davidia involucrata,* but his expedition proved so successful in revealing the wealth of plants available only in China that he made further Chinese expeditions in 1903 to 1905, 1907 to 1908 and 1910 to 1911. In 1914 and 1918 he visited Japan. He escaped innumerable dangers in the Far East only to be killed in a motor accident in the United States. Among his books, *China, Mother of Gardens,* is well worth reading both for its botanical descriptions and as a story of adventurous travel in the distant parts of China.

sinua′tus, -a, -um; sinuo′sus, -a, -um With a wavy margin.

Siphocam′pylus m. From Gr. *siphōn,* tube; *kampylŏs,* curve. With reference to the curved corolla of these typical American shrubs.

Siphonosman′thus m. From Gr. *siphōn,* tube plus *Osmanthus,* with reference to the long tube of the corolla which distinguishes these shrubs from typical *Osmanthus.*

sisala′nus, -a, -um Pertaining to Sisal, on the coast of Yucatan, Mexico.

Sisymb′rium n. Old Greek name, *sisymbriŏn,* for various plants.

Sisyrin′chium n. The ancient Greek name for another plant. These iridaceous plants are native to North and South America, and naturalized elsewhere.

sitchen′sis Of Sitka, Alaska.

Si′um n. Ancient Greek name, *siŏn.*

Skim′mia f. From the Japanese name for these ornamental shrubs.

smarag′dinus, -a, -um Emerald-green.

Smilaci′na f. False Solomon's seal, false spikenard. Diminutive of *Smilax.*

smilaci′nus, -a, -um Relating to *Smilax.*

Smi′lax f. Greenbrier. The Greek name for these climbers although the smilax of florists is generally *Asparagus asparagoides.*

Smithian′tha f. In honour of Matilda Smith (1854–1926), botanical artist at Kew, who provided drawings for the *Botanical Magazine* from 1878 to 1923.

smi′thii Of the many botanists with the surname Smith, the first to be commemorated was Sir James Edward Smith (1759–1828) of Norwich, England, purchaser of Linnaeus's collections and founder and first president of the Linnean Society of London. Among others are

Christen Smith (1785–1816), Norwegian botanist who died on a Congo expedition; John Smith (1798–1888), Scottish gardener and pteridologist at Kew; and Joannes Jacobus Smith (1867–1947), Belgian botanist many years at Buitenzorg.

Smyr'nium n. Greek name, *smyrniŏn*, from *smyrna*, myrrh, with reference to its smell.

sobo'lifer, soboli'fera, -um Having creeping rooting stems.

Sobra'lia f. Terrestrial orchids named for Francisco Martinez Sobral, Spanish physician, *c.* 1790.

socia'lis, -is, -e Forming colonies.

socotra'nus, -a, -um Of the island of Socotra.

sodo'meus, -a, -um Of the Dead Sea area; from *Sŏdoma*, Sodom.

Solan'dra f. Showy-flowered vines named for Daniel Carl Solander (1736–1782), a pupil of Linnaeus, who emigrated to England in 1760, sailed with Joseph Banks as botanist on Captain Cook's first voyage of exploration in the *Endeavour* in 1768 to 1771 and became keeper of the Natural History Department of the British Museum, London, in 1773. He was a very able and industrious biologist as is shown by his numerous descriptions of animals as well as plants made on Cook's voyage.

Sola'num n. Latin name for some plant, probably *Solanum nigrum*, assigned to this huge and varied genus which includes such plants as the potato, Jerusalem cherry and woody nightshade.

sola'ris, -is, -e Of the sun; growing in sunny situations.

Soldanel'la f. Apparently a diminutive of Italian *soldo*, a small coin; in allusion to the rounded leaves of these alpine herbs.

soldanelloi'des Resembling a *Soldanella*.

Soleiro'lia f. In honour of Joseph François Soleirol (d. 1863) who made vast collections of specimens of Corsican plants in the first half of the 19th century. *Soleirolia soleirolii* is the latest name for the little creeping herb usually known as *Helxine soleirolii* or 'Mind-your-own-business'.

Solida'go f. Goldenrod. From *solido*, to make whole; in allusion to the reputed healing qualities of these perennial herbs.

Solidas'ter m. Hybrid between *Aster* and *Solidago*.

so'lidus, -a, -um Solid; dense.

Soli'sia f. In honour of Octavio Solis of Mexico City, a cactus-specialist *c.* 1923.

Sol'lya f. Evergreen climbers named for Richard Horsman Solly (1778–1858), English plant physiologist and anatomist.

solstitia'lis, -is, -e Relating to midsummer.

som'nifer, somni'fera, -um Sleep-producing.

Son'chus m. Sowthistle. The Greek name *sŏnchŏs*.

Sone'rila f. From the native Malabar name, *soneri-ila*, for these herbs and small shrubs.

songa'ricus, -a, -um From Dzungaria, central Asia.

Sopho'ra f. From the Arabic name for these ornamental plants.

Sophro'nitis f. From Gr. *sōphrōn*, modest; in allusion to the small but pretty flowers; *Sophronia* is a genus of orchids to which *Sophronitis* is related.

Sorba'ria f. False *Spiraea*. Deciduous shrubs. From the Latin, meaning resembling *Sorbus*.

sorbifo'lius, -a, -um With leaves like *Sorbus*.

Sor'bus f. Mountain ash, rowan. L. *sorbum*, the fruit of the service tree (*Sorbus domestica*). Hybrids with *Pyrus* are called *Sorbopyrus*.

sor'didus, -a, -um Dirty.

Sorg'hum n. Latinized from Italian *sorgo* but origin obscure.

soro'rius, -a, -um Sisterly, i.e. very closely related.

soulangia'nus, -a, -um In honour of the chevalier Etienne Soulange-Bodin (1774–1846), horticulturist at Fromont near Paris, raiser of a celebrated hybrid Magnolia.

sou'liei In honour of Jean André Soulie (1858–1903), French missionary in China.

spachia'nus, -a, -um; spa'chii In honour of Edouard Spach (1801–1879) Alsatian botanist, from 1829 onwards at the Musée d'Histoire Naturelle, Paris, noted for careful and precise work on the flora of the Near East and other publications which somewhat paradoxically split genera and lump species.

spadi'ceus, -a, -um Date-brown, dark chestnut-brown.

Spara'xis f. From Gr. *sparasso*, to tear. The bracts appear torn.

Sparga'nium n. Bur-seed. Greek name *sparganion* used by Dioscorides.

Sparman'nia f. Shrubs or trees named for Dr. Andreas Sparrman (1748–1820), Swedish botanist who sailed with Captain Cook in the *Resolution* on his second voyage of exploration (1772–1775), and travelled with Thunberg in South Africa.

spar'sus, -a, -um Few; far-between.

spar'teus, -a, -um Pertaining to *Spartium*, i.e. like *Spartium* in growth.

Spar'tium n. Weavers' broom. Gr. *spartion*, from *spartŏn*, i.e. esparto grass (*Stipa tenacissima*), a kind of grass used for weaving and cordage.

Spathiphyl'lum n. Gr. *spathe; phyllon*, a leaf; from the leaflike spathe of these low herbs.

Spatho'dea f. From Gr. *spathe*; *-ōdēs*, of the nature of; in allusion to the spathe- or boat-like calyx of these evergreen trees. The Scarlet Bell or Fountain Tree (*Spathodea campanulata*) is among the showiest of tropical trees.

Spathoglot'tis f. Gr. *spathe*; *glotta*, a tongue; in allusion to the mid-lobe of the lip of these terrestrial orchids.

spathul'atus, -a, -um Like a spatula.

specio'sus, -a, -um Showy.

specta'bilis, -is, -e Spectacular; showy.

Specula'ria f. Venus' looking-glass. L. *speculum*, a mirror; from the old name for these annuals, *speculum Veneris*, literally, the looking glass of Venus.

specula'tus, -a, -um Shining (as if with mirrors).

Spence'ria f. In honour of Spencer Le Marchant Moore (1850–1931), botanist first at Kew, then at the British Museum (Natural History).

Sper'gula f. Spurrey. Origin obscure.

sphacela'tus, -a, -um Withered as if dead.

Spha'cele f. Gr. *sphakos*, sage; which these shrubs resemble in foliage.

Sphaeral'cea f. Globe-mallow. Gr. *sphaira*, a globe; *Alcea*, mallow; in allusion to the globose fruit of these malvaceous plants.

sphae'ricus, -a, -um Spherical. Also:
sphaerocar'pus   with round fruits
sphaeroce'phalus   with a round head
sphaeroi'des   resembling a sphere

spheco'des See *sphegodes*.

sphego'des Wasp-like. In medieval Latin *c* and *g* were sometimes interchanged; hence the old spelling *sphegodes* should not be altered to *sphecodes*, from *sphēx, sphēkos*, wasp.

sphenanthe'rus, -a, -um With wedge-shaped anthers.

sphenochi'lus, -a, -um With wedge-shaped lip.

Sphenogy'ne f. From Gr. *sphēn*, wedge; *gynē*, woman; hence ovary or pistil.

Spheno'meris f. From Gr. *sphēn*, wedge; *mĕris*, part. In allusion to the wedge-shaped segments of the fronds of these ferns.

spi'cifer, spici'fera, -um Spike-bearing.

spicifor'mis, -is, -e Spike-shaped. Also:
spi'ciger   spike-bearing
spiculiflo'rus   with flowers in small spikes

Spige'lia f. Indian pink, pink-root. North American herbs named for Adrian van der Spiegel (1578–1625), professor of anatomy at Padua and author of *Isagoges* (1606) on botany.

Spilan'thes f. From Gr. *spilŏs*, spot, fleck; *anthē*, flower; in allusion to the white flowers of the type-species being flecked with black pollen.

Spina'cia f. Spinach. Origin obscure, latinized from Italian *spinace* or Spanish *espinaca* and possibly from L. *spina*, a spine; in allusion to the spiny husk of the fruit.

spines'cens; spi'nifer, spini'fera, -um; spin'ifex; spino'sus, -a, -um Spiny.

spinu'lifer, spinuli'fera, -um Bearing little spines.

Spirae'a f. Bridal wreath, spiraea. Deciduous flowering shrubs. Gr. *speiraira*, a plant used for garlands.

spira'lis, -is, -e Spiral.

Spiran'thes f. Ladies' tresses. Gr. *speira*, spiral; *anthos*, a flower; from the spiral inflorescence of these terrestrial orchids.

Spirone'ma n. From Gr. *speira*, spiral; *nēma*, a thread; in allusion to the spiral cells in the stamen filaments of this Mexican herb.

splen'dens; splen'didus, -a, -um Splendid.

Spon'dias f. Greek for the plum. Assigned to this genus of tropical trees because of the fruit. It includes the Jamaica plum, mombin, etc.

spoo'neri In honour of Herman Spooner, botanist on staff of Messrs. Veitch at Chelsea, later of Imperial Institute, Kensington, London.

Spreke'lia f. Bulbous plants named for Johann Heinrich von Sprekelsen (1691–1764), a lawyer of Hamburg, who possessed a fine garden and botanical library visited by Linnaeus in 1735.

spuma'rius, -a, -um Frothing.

spu'rius, -a, -um False.

Spyri'dium n. Gr. *spyris*, a basket; *eidos*, like; from the form of the calyx of these Australian shrubs.

squa'lens, -a, -um; squa'lidus, -a, -um Dirty.

squama'tus, -a, -um Squamate; with small scalelike leaves or bracts.

squamo'sus, -a, -um Full of scales.

squarro'sus, -a, -um With parts spreading or recurved at the ends.

stachyoi'des Resembling *Stachys*.

Sta'chys f. Woundwort. The Greek name for various members of the *Labiatae*. *S. lanata* is variously known as lamb's tails, lamb's tongues, bunnies' ears, on account of its densely haired grey leaves.

Stachytar'pheta f. Gr. *stachys*, a spike; *tarphys*, thick; from the thick flower spikes of these herbs and shrubs.

Stachyu'rus m. Gr. *stachys*, a spike; *oura*, a tail; from the form of the racemes of these shrubs.

stah'lii In honour of Christian Ernst Stahl (1848–1919), German botanist of Alsatian origin, professor at Jena.

stami'neus, -a, -um With prominent stamens.

standi'shii In honour of John Standish (*c*. 1809–1875), English nurseryman, partner in Standish and Noble of Bagshot, Surrey, who raised the Chinese and Japanese plants introduced by Robert Fortune 1848 to 1860.

Stange'ria f. In honour of William Stanger (1812–1854), English naturalist, surveyor and medical man who discovered the remarkable cycad *S. paradoxa* in Natal and sent it to the Chelsea Physic Garden in 1851.

Stanho'pea f. Epiphytic orchids named for Philip Henry Stanhope (1781–1855), 4th Earl Stanhope, President of the Medico-Botanical Society of London, 1829 to 1837.

stans Erect, upright.

Stape'lia f. Carrion-flower. *Cactus*-like plants named by Linnaeus for Johannes Bodaeus van Stapel (d. about 1636), of Amsterdam, editor of Theophrastus's major work on plants.

stap'fii In honour of Otto Stapf (1857–1933), botanist of Austrian origin, from 1890 to 1908 Assistant in the Kew Herbarium, 1909 to 1921 Keeper of the Herbarium, editor of *Index Londinensis* and *Botanical Magazine* (1922–1933).

Staphy'lea f. Bladdernut. Gr. *staphyle*, a cluster; from the arrangement of the flowers of these shrubs or small trees.

Sta'tice f. Marsh rosemary, sea lavender, sea pink, thrift. Greek name for thrift or sea pink, now abandoned owing to its application by some authors to the sea pinks (*Armeria*), by others to the sea lavenders (*Limonium*), thus creating confusion and ambiguity. Thrift or sea pink has been assigned to *Armeria* and sea-lavender or marsh rosemary to *Limonium*. Gardeners, however, will probably long continue to call the latter *Statice*.

Staunto'nia f. Evergreen climbing shrubs named for Sir George Leonard Staunton (1737–1801), Irish medical man and naturalist, who accompanied Lord Macartney on his futile and costly mission to the Emperor of China in 1792.

stauracan'thus, -a, -um With crosswise spines.

Steirone'ma n. Loosestrife. Gr. *steiros*, sterile; *nema*, a thread; referring to the staminodes (sterile stamens) alternating with the fertile stamens.

Ste'lis f. Gr. *stĕlis*, a name for the mistletoe (*Viscum*) transferred to these epiphytic orchids because they also grow on trees though not parasitic.

Stella'ria f. Chickweed, stitchwort, starwort. From L. *stella*, a star; from the starry flowers of these herbs.

stella'ris, -is, -e; stella'tus, -a, -um Starry, star-like, with spreading leaves or petals star-wise.

Stel'lera f. In honour of Georg Wilhelm Steller (1709–1746), German naturalist and traveller in Siberia, after whom the extinct Steller's sea-cow is also named.

stel'liger, stelli'gera, -um Star-bearing.

Stenan'drium n. From Gr. *stĕnŏs*, narrow; *aner*, man. These herbs with ornamental foliage have narrow stamens.

Stenan'thium n. From Gr. *stĕnŏs*, narrow; *anthŏs*, a flower; in allusion to the narrow sepals and petals of these bulbous plants.

ste'naulus With narrow furrows.

steno- In compound words signifies narrow. Thus:
stenocar'pus	with narrow fruits
stenope'talus	with narrow petals
stenophyl'lus	narrow-leaved
steno'pterus	narrow-winged
stenosta'chyus	narrow-spiked

Stenocar'pus m. From Gr. *stĕnŏs*, narrow; *karpos*, fruit; from the flat, narrow fruits of these trees.

Stenochlae'na f. Gr. *stĕnŏs*, narrow; *chlaina*, a cloak. In these ferns there is no indusium—that part which normally covers the sori.

Stenoglot'tis f. From Gr. *stĕnŏs*, narrow; *glōtta*, a tongue; in allusion to the narrow division of the lip of these terrestrial orchids.

Stenosperma'tion n. Gr. *stĕnŏs*, narrow; *spermation*, a little seed, diminutive of *sperma*, seed.

Stenota'phrum n. Gr. *stĕnŏs*, narrow; *taphros*, a trench; from the cavities on the stem of the inflorescence on which the flower spikelets of these grasses are situated. A variegated form of *S. secundatum* is cultivated as a glasshouse basket-plant.

Stephanan'dra f. Gr. *stephanŏs*, a crown; *aner*, a man. The numerous

Stephanotis

short stamens of these ornamental shrubs persist as a rather obvious crown.

Stephano'tis f. Wax-flower, Madagascar jasmine. Gr. *stĕphanōtis*, a name of the myrtle which was used for making crowns (Gr. *stephanŏs*, a crown) but applied to these evergreen twining shrubs also, in reference to the auricles in the staminal crown.

Stercu'lia f. Trees named for Sterculius, Roman god of privies (L. *stercus*, dung). The flowers and leaves of some species stink.

ste'rilis, -is, -e Infertile; sterile.

Sternber'gia f. Winter daffodil. Bulbous herbs named for Count Kaspar von Sternberg (1761–1838), Austrian botanist, a founder of the Bohemian National Museum in Prague.

Stetso'nia f. Named in honour of Francis Lynde Stetson of New York in 1920.

steve'nii In honour of Christian von Steven (1781–1863), Finnish botanist for many years in the Crimea.

Stevenso'nia f. A palm named for Sir William Stevenson, Governor of Mauritius, 1857–1863.

Ste'via f. Herbs or sub-shrubs named for Pedro Jaime Esteve (d. 1566) of Valencia, Spanish botanist and physician.

Stewar'tia f. Deciduous flowering shrubs and trees named for John Stuart (1713–1792), 3rd Earl of Bute, British Prime Minister 1762 to 1763, a zealous patron of botany and horticulture, for whom *Butea* was also named.

stewartia'nus, -a, -um In honour of Laurence Baxter Stewart (1876–1934), curator of Edinburgh Botanic Garden.

stictocar'pus, -a, -um From Gr. *stiktos*, spotted; *karpos*, fruit.

Stigmaphyl'lon n. From Gr. *stigma*, stigma; *phyllon*, a leaf. The stigma of these woody vines is broad and somewhat leaf-like.

Sti'pa f. Feather-grass. Gr. *stuppē*, tow; from the feathery inflorescence on these perennial grasses.

stipula'ceus, -a, -um; stipula'ris, -is, -e; stipula'tus, -a, -um Having stipules or well-developed stipules, the outgrowths found at the base of the leaf-stalks of many flowering plants.

Stizolo'bium n. Velvet-bean. Gr. *stizo*, to prick; *lŏbŏs*, a lobe, pod. The pods of some species of these ornamental vines have stinging hairs.

Stoke'sia f. Stokes' aster. A perennial herb named for Dr. Jonathan Stokes (1755–1831) of Chesterfield, English botanical author.

stolo'nifer, stoloni'fera, -um Having stolons or rooting runners.

Stoma'tium n. From Gr. *stŏma*, mouth, opening.

stra'gulus, -a, -um Covering like a mat or carpet, hence mat-forming.

strami'neus Straw-coloured.

strangula'tus, -a, -um Constricted.

Stranvae'sia f. Handsome evergreen trees and shrubs named for William T. H. Fox-Strangways (1795–1865), 4th Earl of Ilchester, English diplomatist and botanist, also commemorated in the genus *Foxia*.

Stratio'tes m. Water soldier. Gr. *stratiōtēs*, name for the water-lettuce (*Pistia stratiotes*) from *stratiōtēs*, soldier, and hence transferred to this aquatic herb on account of its sword-shaped leaves.

Strelit'zia f. Bird-of-paradise flower. Perennial herbs named for Charlotte of Mecklenburg-Strelitz (1744–1818), who in 1761 became Queen to George III.

stre'pens Creaking, rattling, clattering; usually referring to seeds in a capsule.

strepto- In compound words signifies twisted. Thus:

streptocar'pus with twisted fruits
streptope'talus with twisted petals
streptophyl'lus with twisted leaves
streptose'palus with twisted sepals

Streptocar'pus m. Cape primrose. From Gr. *strĕptŏs*, twisted; *karpos*, fruit. The seed capsules of these herbs are twisted in a spiral.

Strep'topus m. Twisted stalk. Gr. *strĕptŏs*, twisted; *pous*, a foot; with reference to the twisted flower-stalks.

Streptoso'len m. From Gr. *strĕptŏs*, twisted; *sōlēn*, pipe, tube; referring to the twisted corolla-tube.

stria'tus, -a, -um Striped.

stric'tus, -a, -um Erect, upright.

strigillo'sus, -a, -um With short oppressed bristles.

strigo'sus, -a, -um With stiff bristles.

striola'tus, -a -um Faintly striped or with fine lines.

Strobilan'thes f. Gr. *strŏbilŏs*, a cone; *anthos*, a flower; from the form of the flower-head on these herbs or shrubs.

strobi'lifer, strobili'fera, -um Cone-bearing.

Stroman'the f. From Gr. *strōma*, a bed; *anthos*, a flower; from the form of the inflorescence on these perennial herbs.

Strombocactus

Strombocac'tus m. Gr. *strŏmbŏs*, a spinning-top; plus *Cactus*; from the shape of the plants.

Strongy'lodon m. From Gr. *strŏngylŏs* (*strŏggulŏs*) round; *ŏdŏus*, tooth; in allusion to the rounded teeth of the calyx.

strongylophyl'lus, -a, -um Round-leaved.

Strophan'thus m. From Gr. *strŏphŏs*, twisted band or cord; *anthos*, flower. In allusion to the long appendages of the corolla.

struma'rius, -a, -um; strumo'sus, -a, -um Having cushionlike swellings.

Strych'nos f. Greek name for various poisonous plants and applied here by Linnaeus because so many of these shrubs and trees are poisonous. The genus includes *S. nux-vomica* or strychnine. One or two species are the source of some of the most lethal arrow poisons known.

Stuartia See *Stewartia*.

Styli'dium n. Trigger-plant. Gr. *stylos*, a column. The stamens are united with the style to form a column in these plants, which are mostly Australian.

Stylo'phorum n. Celandine poppy. Gr. *stylos*, style; *phoros*, bearing; with reference to the long columnar style.

stylo'sus, -a, -um With a prominent or well-developed style.

styracif'luus, -a, -um Flowing with gum.

Sty'rax f. Snowbell, storax-tree. The classical Greek name derived from a Semitic name for these ornamental woody plants.

suave'olens Sweet-scented.

sua'vis, -is, -e Sweet.

sub- In compound words possibly the most used prefix, signifying somewhat; almost; rather; slightly; partially; under; etc.

subacau'lis, -is, -e Without much of a stem.

subalpi'nus, -a, -um Growing in the lower mountain ranges.

subauricula'tus, -a, -um Somewhat eared.

subcaeru'leus, -a, -um Slightly blue.

subca'nus, -a, -um Greying.

subcarno'sus, -a, -um Rather fleshy.

subcorda'tus, -a, -um Rather heart-shaped.

subdenta'tus, -a, -um Nearly toothless.

subdivarica'tus, -a, -um Somewhat spreading.

subelonga'tus, -a, -um Somewhat elongated.

suberec'tus, -a, -um Almost upright.

subero'sus, -a, -um Cork-barked.

subfalca'tus, -a, -um Somewhat curved or hooked.

subglau'cus, -a, -um Somewhat glaucous, i.e. with a somewhat blue-green or grey-green appearance.

subhirtel'lus, -a, -um Somewhat hairy.

subluna'tus, -a, -um Rather crescent-shaped.

submer'sus, -a, -um Submerged.

subpetiola'tus, -a, -um With a very short leaf-stalk.

subscan'dens Tending to climb.

subterra'neus, -a, -um Underground.

subula'tus, -a, -um Awl-shaped.

subvillo'sus, -a, -um With rather soft hairs, or somewhat villous.

Succi'sa f. Devil's bit. From L. *succido*, cut off below. With reference to the end of the rhizome, looking as if bitten off by the Devil.

succi'sus, -a, -um Cut off below.

succotri'nus, -a, -um From the island of Socotra, Indian Ocean.

succulen'tus, -a, -um Fleshy; juicy.

sude'ticus, -a, -um Referring to the Sudetenland of Czechoslovakia and Poland.

sue'cicus, -a, -um Swedish.

suffrutes'cens; suffrutico'sus, -a, -um Somewhat shrubby.

suio'num Of the Swedes.

sulca'tus, -a, -um Furrowed.

sulfu'reus, -a, -um; sulphu'reus, -a, -um Sulphur-yellow.

sulta'ni Of the Sultan, referring in *Impatiens sultani* to the Sultan of Zanzibar.

sumatra'nus, -a, -um From Sumatra, Indonesia.

super'bus, -a, -um Superb.

supercilia'ris, -is, -e Eyebrowlike.

supi'nus, -a, -um Prostrate.

supranu'bius, -a, -um Above the clouds, i.e. growing high in mountains.

surculo'sus, -a, -um Producing suckers or shoots.

susia'nus, -a, -um From Susa in western Iran (Persia).

suspen'sus, -a, -um Hanging.

sutchuenen'sis, -a, -um From Szechwan province, China.

Sutherlan'dia f. Shrubs named for James Sutherland (*c.* 1639–1719), superintendent, botanic garden, Edinburgh, and professor of botany, Edinburgh, 1676 to 1715.

Sutto'nia f. Evergreen trees and shrubs, named for the Reverend Charles Sutton (1756–1846), clergyman at Norwich, England, a keen-eyed amateur botanist. This genus is sometimes included in *Myrsine*.

Swain'sona f. Darling River pea. Herbs and sub-shrubs named for Isaac Swainson (1746–1812), London physician who had a private botanic garden at Twickenham.

Swer'tia f. Herbaceous plants of the gentian family commemorating Emanuel Sweert, 16th-century Dutch florist.

Swi'da f. Czech name for dogwood; see *Cornus*.

Swiete'nia f. Mahogany tree. Named for Gerard van Swieten (1700–1772), Dutch botanist who settled in Vienna in 1745 and became physician to the Empress Maria Theresa, reforming medical education in Vienna. This is the original mahogany of Central America and the West Indies. African and Philippine mahogany, commercially so called, are members of several other genera although the wood, in most cases, bears some resemblance to that of *Swietenia*.

Syco'psis f. From Gr. *sykōn*, fig; *ŏpsis*, appearance.

sylva'ticus, -a, -um; sylves'ter; sylves'tris, -is, -e Growing in woods, forest-loving, wild.

sylvi'cola An inhabitant of woods.

Sym'pagis f. From Gr. *sympagēs*, joined together. In allusion to the basal membrane joining the stamens.

Symphoricar'pos m. Snowberry. From Gr. *symphorein*, bear together; *karpŏs*, fruit. The berries on these shrubs are borne in clusters.

Symphyan'dra f. Pendulous bellflower. Gr. *symphuio*, to grow together; *aner*, anther. The anthers of these perennial herbs are joined.

Sym'phytum n. Comfrey. The ancient Greek name of this herb which was reputed to heal wounds.

Symplocar'pus m. Skunk-cabbage. Gr. *symplŏkē*, combination, connection, embrace; *karpŏs*, fruit. The ovaries of this swamp-loving perennial herb grow together to make one fruit.

Sym'plocos f. Gr. *symplŏkē*, combination. In these ornamental woody plants the stamens are united at the base.

Synade'nium n. Gr. *syn*, with, together; *adēn*, a gland; from the united glands of the inflorescence.

Synechan'thus m. Gr. *synĕchēs*, holding together, continuous; *anthos*, a flower; from the form of the flower heads on these palms.

Syngo'nium n. From Gr. *syn*, with, together; *gŏne*, reproductive organs, womb. In allusion to the united ovaries.

Synne'ma n. From *syn*, together; *nēma*, thread. In allusion to the united stamens.

Syn'thyris f. Gr. *syn*, together; *thyris*, a small door; in allusion to a characteristic of the fruits of these perennial herbs.

syria'cus, -a, -um Syrian.

Syrin'ga f. Lilac. Originally also applied to *Philadelphus*. The name is still very commonly applied to mock-orange as a vernacular name. The name derives from Gr. *syrinx*, a pipe, and refers to the hollow stems.

syringan'thus, -a, -um With flowers like lilac.

syringifo'lius, -a, -um With leaves like lilac.

Syzy'gium n. From Gr. *suzugos*, joined. With reference to the paired leaves and branchlets of a Jamaican species (*Calyptranthes suzygium*) for which the name was used originally.

szovitsia'nus, -a, -um In honour of Johann Nepomuk Szovits (d. 1830), Hungarian apothecary who collected in the Caucasus.

T

Tabebu'ia f. From the native Brazilian name *tabebuia* or *taiaveruia* for these ornamental trees.

Tabernaemonta'na f. Evergreen flowering trees or shrubs named for Jakob Theodor von Bergzabern (d. 1590), personal physician to the Count of the Palatine at Heidelberg, West Germany. He latinized his name as Tabernaemontanus and is also commemorated by species named for him in *Amsonia, Potentilla* and *Scirpus*. He was the author of a celebrated herbal *Neuw Kreuterbuch* (1588–1591), of which the illustrations were issued separately at Frankfurt-am-Main in 1590 under the title *Eicones Plantarum*. The woodcuts were mostly copied from those in other herbals but make an attractive book. The London printer, John Norton, acquired them from the Frankfurt printer Nicolaus Bassaeus and used them in 1597 to illustrate Garard's *Herball*.

tabula'ris, -is, -e Flat, like a table or board; growing on Table Mountain, South Africa.

tabulifor'mis, -is, -e Flat, like a table or board.

tacamahac′ca American vernacular name said to be of Aztec origin converted to Hackmatack and used for both *Populus balsamifera* and *Larix laricia.*

Tac′ca f. From the Indonesian name, *taka*, for these perennial herbs.

tae′diger, taedi′gera, -um Torch-bearing.

Tage′tes f. Marigold. African and French. Named for an Etruscan deity, Tages, said to have sprung from the earth as it was being ploughed and to have taught the Etruscans the art of divination. An extract from the roots is inimical to soil nematodes.

taggia′nus, -a, -um In honour of Harry Frank Tagg (1874–1933), botanist at the Edinburgh botanic garden.

tagliabua′nus, -a, -um In honour of the brothers Alberto Linneo Tagliabue and Carlo Tagliabue, Italian nurserymen at Lainate near Milan in 1858. The hybrid between *Campsis grandiflora* and *C. radicans* is named after them.

taiwanen′sis, -is, -e From Taiwan (Formosa).

Taiwa′nia f. From Taiwan (Formosa) where the one species of this genus, *T. cryptomerioides*, was originally found.

talien′sis, -is, -e From the Tali Range, Yunnan, west China.

Tali′num n. Derivation of the name for these fleshy herbs is obscure.

Tamarin′dus m. Tamarind. From the Arabic name *tamar* (date); *hindi* (Indian). The acid pulp of the seed-pods is used in medicine.

tamariscifo′lius, -a, -um With leaves like tamarisk.

Ta′marix f. Tamarisk. The Latin name, of which *tamariscus* is another form, for these showy deciduous shrubs and trees.

Ta′mus f. Black bryony. From the Latin name, *tamnus*, for another climbing plant.

tanacetifo′lius, -a, -um Tansy-leaved.

Tanace′tum n. Tansy. From the medieval Latin name *tanazita*, still used in some European places, ultimately derived from Gr. *athanasia*, immortality. Once regarded as a specific for intestinal worms, it was formerly used in Europe as well as in some rural areas of New England in funeral winding sheets, as a discouragement to worms; see p. 342.

Tanakae′a f. In honour of a pioneer Japanese botanist Yoshio Tanaka (1838–1916).

tanasty′lus, -a, -um Long-styled.

tancarvil′leae, tankervil′leae In honour of Lady Emma Tankerville (d. 1836), wife of Charles, Earl of Tankerville. The title is derived from the earldom of Tancarville in Normandy, created in 1419 for John Grey.

tangu'ticus, -a, -um From Kansu, north-west China; named for the Tangut people of this region.

Tapeinochi'los m. From Gr. *tapeinos*, low; *cheilos*, lip; in allusion to the short labellum.

tapetifor'mis, -is, -e Carpet-like.

taraxacifo'lius, -a, -um With leaves like a dandelion.

Tara'xacum n. Dandelion. From a very ancient name traceable through Latin and Arabic to Persian *talkh chakok*, meaning bitter herb.

tardiflor'us, -a, -um Late-flowering.

tardi'vus, -a, -um; tar'dus, -a, -um Late.

taronen'sis, -is, -e From the Taron Gorge, Yunnan, China.

tarta'reus, -a, -um With a rough surface.

tarta'ricus, -a, -um From Central Asia.

tasma'nicus, -a, -um From Tasmania, named in honour of the Dutch navigator Abel Janszoon Tasman (1603–1659), who discovered it in 1642 or 1643.

tatsienen'sis, -is, -e From Tatsienlu (K'angting), western China.

tau'ricus, -a, -um From the Crimea.

tauri'nus, -a, -um From the neighbourhood of Turin, Italy.

taxifo'lius, -a, -um With leaves like yew.

Taxo'dium n. Swamp cypress. L. *Taxus*, yew; Gr. *eidos*, resemblance; from a similarity of leaf shape in these deciduous and semi-evergreen trees and shrubs.

Ta'xus f. Yew. The Latin name for these evergreen trees and shrubs.

tazet'ta Italian vernacular name used for a *Narcissus*, meaning a small cup (*tazza*), from the form of the corona.

tchonos'kii In honour of a Japanese botanical collector, Tchonoski, who between 1862 and 1867 made extensive collections of Japanese plants for Carl Maximowicz, which included many new species.

tech'nicus, -a, -um Used for a special purpose.

Teco'ma f. Abbreviated from the Mexican name *tecomaxochitl* for these ornamental shrubs.

Tecoma'ria f. Cape honeysuckle. From *Tecoma* which it closely resembles.

Tecophilae'a f. In honour of an Italian botanical artist Tecophila Billotti, apparently the daughter of the Italian botanist Luigi Colla (1766–1848), of Turin, whose works she illustrated. *T. cyanocrocus* has a crocus-like flower of an intense and wonderful blue; it is native to Chile.

Tectona

Tec'tona f. From the Tamil name for the teak tree (*T. grandis*). This magnificent forest tree, growing up to 150 feet, supplies the teak of commerce. The wood is extremely durable and proof against the ravages of white ants. It is not particularly hard but is not easy on tools because the silicate content of the grain quickly blunts the best edges. Before the war teak was exported in very large quantity from Rangoon. Nowadays the teak which has been popularized for fine furniture by the Danes and others seems mostly to be of the Siamese variety, distinguished by the handsome dark striations in the grain. The almost black hardwood stools and other furniture made in China are certainly not teak wood.

tecto'rum Of the roofs of houses. Specific epithet of *Sempervivum* (house-leek) which grows freely on stone and slate roofs and which, in some places, is thought to avert lightning by acting as a conductor of electricity.

tec'tus, -a, -um Concealed; covered.

Tele'kia f. In honour of a Hungarian nobleman, Samuel Teleki de Szék, patron of the botanist J. C. Baumgarten (1765–1843).

tele'phium Specific epithet of *Sedum*, of uncertain meaning, possibly connected with Telephus, a king of Mysia in Asia Minor.

Tel'lima f. Anagram of *Mitella* to which these herbaceous perennials are closely related.

telmate'ius, -a, -um Of marshes.

telonen'sis, -is, -e From Toulon (ancient Telenium), France.

Telo'pea f. Gr. *tēlōpŏs*, seen from afar. The crimson flowers of these tall shrubs are visible at a distance.

Telos'ma f. From Gr. *tēlĕ*, far; *osmē*, fragrance, smell. In allusion to the great distance over which the scent carries.

teme'nius, -a, -um Gr. *tĕmĕniŏs*, belonging to a sacred place.

Templeto'nia f. In honour of John Templeton (1766–1825) of Belfast, Irish botanist who gave special attention to mosses, fungi, lichens and algae.

temulen'tus, -a, -um; tem'ulus, -a, -um Drunken.

te'nax Strong; tough; matted.

tenebro'sus, -a, -um Of shaded places.

tenel'lulus, -a, -um; tenel'lus, -a, -um Tender; delicate.

te'nens Holding, persistent, enduring.

ten'uis, -is, -e Slender; thin. Also:
  tenuicau'lis   slender-stemmed
  tenuiflo'rus   with slender flowers

tenuifo'lius   slender-leaved
tenuipe'talus   with slender petals

Tephro'sia f. Gr. *tĕphrŏs*, ash-coloured; from the appearance of the leaves of these perennial herbs.

terebinthina'ceus, -a, -um; terebinthi'nus, -a, -um Pertaining to turpentine, from Gr. *tĕrĕbinthŏs*, itself probably of pre-Greek origin, the name of *Pistacia terebinthus* and its resin.

te'res Cylindrical; circular in section.

Termina'lia f. Indian almond. L. *terminus*, end. The leaves of these tropical trees are borne at the ends of the shoots.

termina'lis, -is, -e Terminal; relating to boundaries; coloured forms of *Cordyline terminalis* are sometimes planted as boundary markers.

ter'minans Ending.

terna'teus, -a, -um From the island of Ternate in the Moluccas.

terna'tus, -a, -um In clusters of three.

Ternstroe'mia f. Evergreen trees or shrubs named for Christopher Tärnström (1703–1746), Swedish student of Linnaeus, who went out in 1745 to investigate the natural history of China but died at Pulo Condor off the coast of Indochina.

terres'tris, -is, -e Of the ground; growing in the ground as opposed to growing on trees or in water.

tessella'tus, -a, -um Checkered in squares.

testa'ceus, -a, -um Brick-coloured.

testicula'ris, -is, -e; testicula'tus, -a, -um Like testicles.

testudina'rius, -a, -um Like, for example, tortoise shell with the curved surface divided into areas.

testu'do A tortoise.

tetra- In compound words signifies four. Thus:

tetracan'thus   four-spined
tetrago'nus   with four angles
tetran'drus   with four anthers
tetran'thus   four-flowered
tetraphyl'lus   with four leaves
tetra'pterus   four-winged

Tetracen'tron n. Gr. *tĕtra*, four; *kĕntrŏn*, a spur. The fruit of this deciduous ornamental tree is four-spurred.

Tetracli'nis f. Gr. *tĕtra*, four; *kline*, a bed. The leaves of this evergreen tree are grouped in fours.

Tetrago'nia f. New Zealand spinach. Gr. *tĕtra*, four; *gōnia*, an angle; from the form of the fruits of this succulent, edible annual.

Tetragono'lobus m. From Gr. *tetra,* four; *gōnia,* an angle; *lŏbŏs,* pod. In allusion to the four-angled or -winged pods.

tetra'lix Old Greek name, *tĕtralix,* transferred to this heath having leaves in fours.

Tetrane'ma n. From Gr. *tetra,* four; *nēma,* thread. In allusion to the four stamens.

Tetra'panax m. Rice paper tree. Gr. *tetra,* four, plus *Panax.* The flowers are in fours and have some resemblance to *Panax.*

Tetrathe'ca f. Gr. *tetra,* four; *thēkē,* case. The anthers of these heath-like shrubs are often four-lobed.

teucrioi'des Resembling *Teucrium.*

Teu'crium n. Germander. The Greek name, probably named for Teucer, first king of Troy, who first used the plant in medicine.

texa'nus, -a, -um Texan.

tex'tilis, -is, -e Used in weaving.

Tha'lia f. Aquatic or marsh herbs named for Johannes Thal (1542–1583), German physician at Nordhausen whose Flora of the Harz region, *Sylva Hercynia,* was posthumously published in 1588.

thalictroi'des Like *Thalictrum.*

Thalic'trum Meadow rue. The Greek name for a plant which may have been one of these elegant perennial herbs.

The'a Tea. From the Dutch rendering *thee* of the Chinese (Amoy) word for tea, *t'e,* which in Mandarin is *ch'a. Thea sinensis* is now placed in the genus *Camellia.* The cultivation of tea was virtually a Chinese monopoly until in the 1840's Robert Fortune exported both tea plants and skilled labour to India to establish the industry there. This, however, was not the first attempt to grow tea outside China. In the Western world, as early as 1772, tea-growing was tried in Georgia. For the next hundred years several attempts were made to establish the industry in the mountains from North Carolina southward, some of them large scale and government sponsored. They seem to have failed, not because the plants could not be grown, but because the skills of manufacture were never mastered.

theba'icus, -a, -um Of Thebes.

the'ifer, thei'fera, -um Tea-bearing.

Thelesper'ma n. From Gr. *thēlē,* a nipple; *sperma,* a seed; in allusion to the form of the seeds in some species of these glabrous herbs or sub-shrubs.

Thelocac'tus m. From Gr. *thēlē,* a nipple, plus *Cactus.*

Theobro'ma n. Cacao. A name coined by Linnaeus, from *thĕŏs*, a god; *brōma*, food; to replace the name *Cacao*, from the Mexican *caca-hoaquahuitl* and its variants, which Linnaeus rejected as a barbarous generic name although he retained it as a specific epithet. The source of both cocoa and chocolate, and a highly profitable crop both on a large and on a small scale, cacao (*Theobroma cacao*) is grown throughout the tropics wherever the right climatic and other conditions prevail. The cacao tree requires protection of shade from other trees, hence a plantation may often take on a rather unkempt look. The fruit is borne on the bark of the larger branches. It is shaped like a rugby football, with pointed ends, and may be as long as ten inches. The seeds are about the size of lima beans and are embedded in a whitish pulp which becomes ill-smelling as the beans are fermented.

The genus *Theobroma* has about twenty-two species.

The cocoa-butter of commerce is derived from coconut and has no connection with cacao.

Theophras'ta f. In honour of the Ancient Greek philosopher and botanist Theophrastos (371–*c*.287 B.C.), disciple and successor of Aristotle at Athens.

therma'lis, -is, -e Of warm springs.

Thermop'sis f. Gr. *thermos*, lupin; *opsis*, like. The flower heads of these herbaceous perennials resemble yellow lupins.

Thespe'sia f. Bhendi tree. Gr. *thĕspĕsiŏs*, divine; with reference to *Thespesia populnea* 'being regarded as a sacred plant in Tahiti at the time of Captain James Cook's visit in 1769 and being grown around places used for worship'.

Theve'tia f. Small trees and shrubs named for André Thevet (1502–1592), French monk, who travelled in Brazil and Guiana.

Thladian'tha f. From Gr. *thladias*, eunuch (from *thlao*, crush, bruise); *anthos*, flower; the suppression of stamens and their conversion into staminodes in the female flowers suggesting that they had been castrated.

Thlas'pi f. Penny-cress. The Greek name for a cress.

Thoma'sia f. Mostly Australian shrubs and small trees named in 1821 for a Swiss family of plant-collectors, Peter Thomas and his brother Abraham Thomas and the latter's sons Philip (d. 1831), Louis (d. 1823) and Emanuel.

thomso'nii Of Thomas Thomson (1817–1878), a Scots doctor who became a surgeon in the Bengal Army and was superintendent of the Calcutta Botanic Garden, 1854–1861. Many Himalayan plants commemorate him.

Thri'nax f. Thatch palm. Fan palm. Gr. *thrinax*, a trident.

Thryal'lis f. Greek name for another plant.

Thu'ja (also **Thuya**) f. Arbor-vitae, Northern white cedar. The Greek name for *thuia*, a kind of Juniper.

Thujo'psis f. *Thuja*; *opsis*, like. There is a similarity in these evergreen trees to *Thuja*.

Thunber'gia f. Flowering shrubs and climbers named for Carl Peter Thunberg (1743–1828), a student of Linnaeus who was persuaded by Dutch lovers of new plants to enter the service of the Dutch East India Company as a doctor and send back plants from Japan to Europe. He travelled in South Africa and Japan and became professor of botany at Uppsala. Black-eyed Susan is a familiar name for *T. alata*.

thu'rifer, thuri'fera, -um Incense-bearing.

thuyoi'des Resembling *Thuja*.

thymifo'lius, -a, -um Thyme-leaved.

thymoi'des Like Thyme.

Thy'mus m. The ancient Greek name for these aromatic herbs and sub-shrubs.

thyrsiflo'rus, -a, -um With flowers in a thyrse, a many-flowered kind of inflorescence.

Thysano'tus m. Gr. *thysanōtŏs*, fringed, with reference to the three inner perianth segments which are beautifully fringed with long hairs.

Tiarel'la f. False mitrewort, foamflower. Greek diminutive of *tiara*, a small crown, in reference to the form of the fruit.

tibe'ticus, -a, -um Of Tibet.

tibici'nus, -a, -um Of a flute-player; used in *Schomburgkia* to refer to the long narrow pseudobulbs.

Tibouchi'na f. Spider-flower. From the native name in Guiana for these tropical South American flowering shrubs and herbs.

Tigrid'ia f. Tiger-flower. L. *tigris*, a tiger; because of the spots on the flowers (see *tigrinus*) on these Central American plants.

tigri'nus, -a, -um Either striped like the Asiatic tiger or spotted like the jaguar, known as a tiger in South America. Also, tiger-toothed.

Ti'lia f. The Latin name for the linden or lime tree.

tilia'ceus, -a, -um Linden-like.

tiliifo'lius, -a, -um Linden-leaved.

Tillan'dsia f. Spanish moss. American herbs, mostly epiphytic, named for Elias Til-Landz (d. 1693), a Swedish botanist and professor of medicine at Abo (Turku), Finland.

Tinan'tia f. Erect herbs named for François A. Tinant (1803–1858), forester in Luxemburg, author of *Flore Luxembourgeoise* (1836 and 1855).

tinctor'ius, -a, -um Used in dyeing.

tinctor'um Of dyers.

tinc'tus, -a, -um; tin'gens Coloured.

tingita'nus, -a, -um Of Tangiers.

Tin'nea f. A genus of herbs or sub-shrubs, hairy or woolly, whose name commemorates a scientific expedition up the Nile undertaken in 1861 by three courageous Dutch ladies, Madame Henrietta L. M. Tinne, her daughter Alexandria Tinne, and her sister Adrienne van Capellen.

ti'nus Latin name for the laurustinus, *Viburnum tinus*.

Tipua'na f. From the South American name for these showy trees.

tipulifor'mis, -is, -e Shaped like a daddy-long-legs, which belongs to the genus *Tipula*.

tirolen'sis, -is, -e Of the Tyrol.

titanius, -a, -um Very large.

Titano'psis f. L. *Titān*, the sun god; Gr. *ŏpsis*, like; from the resemblance of the flower of these stemless succulents to the sun.

Titho'nia f. Annual herbs named for Tithonus, a young man much loved by Aurora, the dawn-goddess.

Toco'ca f. From the native name for these shrubs in Guyana.

To'dea f. In honour of Heinrich Julius Tode (1733–1797), German botanist specially interested in the fungi of Mecklenburg, north Germany.

Tolmie'a f. A perennial herb named for Dr. William Fraser Tolmie (1830–1886), Scottish physician, fur-trader, ethnologist and botanist in N.W. America.

Tol'pis f. Yellow hawkweed. The name of these showy daisylike herbs was coined by Adanson, who gave no derivation.

tomento'sus, -a, -um Densely woolly; with matted hairs.

tommasinia'nus, -a, -um In honour of Muzio Giuseppe Spirito de Tommasini (1794–1879), magistrate and botanist at Trieste, celebrated for his work on the Dalmatian flora.

tonduzi In honour of J. F. Adolphe Tonduz (1862–1921), Swiss botanist from 1888 to 1919 in Costa Rica.

ton'sus, -a, -um Sheared; smooth-shaved.

Torenia

Tore'nia f. Pretty annuals named for the Reverend Olof Torén (1718–1753), chaplain to the Swedish East India Company at Surat, India, and in China.

tormina'lis, -is, -e Relating to or effective against colic.

toro'sus, -a, -um Cylindrical, with contractions at intervals.

Torrey'a f. Ornamental evergreens named for Dr. John Torrey (1796–1873), one of the giants of American botany and co-author with Asa Gray of *The Flora of North America.* He was regarded as an ultimate authority and in his lifetime described many thousands of plants brought back by such explorers as Fremont and Pickering.

tor'ridus, -a, -um Growing in hot, dry places.

tor'tilis, -is, -e; tor'tus, -a, -um Twisted.

Tournefor'tia f. In honour of Joseph Pitton de Tournefort (1656–1708), celebrated French botanist who travelled with Aubriet and Gundelsheimer in the Levant and was the first botanist properly to define genera. He was an important fore-runner of Linnaeus.

Townsen'dia f. Rocky Mountain herbs named for David Townsend (1787–1858) of Pennsylvania.

toxica'rius, -a, -um; to'xicus, -a, -um; to'xifer, toxi'fera, -um Poisonous.

Trache'lium n. Throatwort. Gr. *trachēlos,* a neck. These perennial herbs were supposed to be good for afflictions of the throat (trachea).

Trachelosper'mum n. From Gr. *trachēlos,* a neck; *sperma,* a seed.

Trachycar'pus m. From Gr. *trachys,* rough; *karpŏs,* a fruit. The allusion is to the fruit of some species of these palms.

Trachyme'ne f. From Gr. *trachys,* rough; *mēninx,* a membrane; in reference to the fruits of some species of these annual and perennial herbs.

Trachyste'mon m. Gr. *trachys,* rough; *stēmōn,* a stamen.

Tradescan'tia f. Widow's-tears, wandering-Jew, spiderwort. Perennial herbs named for John Tradescant (1608–1662), gardener to King Charles I; he visited Virginia in 1654.

Tragopo'gon m. Goatsbeard. Gr. *tragŏpōgōn* from *tragŏs,* goat; *pōgōn,* beard. The pappus of these biennial or perennial herbs is silky.

translu'cens Translucent.

transpa'rens Transparent.

transylva'nicus, -a, -um From Transylvania, Rumania.

Tra'pa f. Water-chestnut. Aquatic floating herbs with edible fruit. A contraction of L. *calcitrappa,* called a caltrop or crow's foot—a weapon of defensive war consisting of four sharp iron points which, thrown on the ground, has one point always pointing upward particularly to pierce the hooves of cavalry horses. The reference here is to the similarly four-pointed fruit.

trapezifor'mis, -is, -e With four unequal sides.

Trautvette'ria f. Perennial herbs named for Ernst Rudolf von Trautvetter (1809–1889), Russian botanist.

tremuloi'des Resembling the quivering poplar.

tre'mulus, -a, -um Quivering; trembling.

Treve'sia f. Showy tropical Asian shrubs and small trees named for the family of Treves de Bonfigli of Padua, 18th-century supporters of botanical research.

tri- In compound words signifies three. Thus:

triacan'thus	three-spined
trian'drus	with three stamens
triangula'ris ⎱ **triangula'tus** ⎰	with three angles
tricauda'tus	three-tailed
trice'phalus	three-headed
tricho'tomus	three-branched
tricoc'cus	with deeply three-lobed fruit, each division usually one-seeded
tri'color	three-coloured
tricor'nis	with three horns
tricuspida'tus	three-pointed
tri'dens ⎱ **tridenta'tus** ⎰	three-toothed
tri'fidus	cut in three
triflo'rus	three-flowered
trifolia'tus ⎱ **trifo'lius** ⎰	with three leaves or leaflets
trifurca'tus ⎱ **trifur'cus** ⎰	three-forked
trigonophyl'lus	with three-cornered leaves
triloba'tus	three-lobed
trimes'tris	of three months
triner'vis	three-nerved
trinota'tus	three-spotted or -marked
tripe'talus	three-petalled
triphyl'lus	three-leaved
tri'pterus	three-winged
trisper'mus	with three seeds

tri'stachyus three-spiked
triterna'tus thrice in threes

Trichan'tha f. From Gr. *thrix, trichos,* hair; *anthos,* flower. In allusion to the profuse long hairs on the corolla.

Trichi'lia f. Gr. *tricha,* three parts. In these tropical trees and shrubs the ovary and capsule are usually three-celled.

tricho- In compound words signifies hairy or hair-like, from Gr. *thrix, trichos,* hair; *trichion,* small hair. Thus:

tricho'calyx	with a hairy calyx
tricho'gynus	with a hairy ovary
tricho'phorus	bearing hairs
trichophyl'lus	with hairy leaves
trichosper'mus	hairy-seeded
trichosto'mus	with a hairy mouth

Trichoce'reus m. From Gr. *thrix,* a hair, plus *Cereus.* The flowering parts of these cacti are hairy.

Trichodiade'ma n. From Gr. *thrix,* hair; *diadēma,* band around the head, diadem. In allusion to the crown of bristles at the tips of leaves and sepals of these succulent plants formerly placed in *Mesembryanthemum.*

Tricholae'na f. From Gr. *thrix,* a hair; *chlaina,* a cloak. The spikelets of these ornamental grasses have a hairy coating.

Trichoma'nes f. Gr. *trichŏmanĕs,* name of a fern mentioned by Theophrastus.

Trichopi'lia f. From Gr. *thrix,* a hair; *pilion,* a cap. The anthers of these epiphytic orchids are concealed under a cap.

Trichosan'thes f. Snakegourd. Gr. *thrix,* hair; *anthos,* a flower. The corolla of these climbers is conspicuously fringed.

Trichoste'ma n. Bluecurls, bastard pennyroyal. From Gr. *thrix,* hair; *stēma,* penis, stamen. The stamens of these annual herbs have very long and slender filaments.

Tricyr'tis f. Japanese toad-lily. From Gr. *tri-,* three; *kyrtos,* humped; because the bases of the three outer petals are swollen and sacklike.

Tri'dax f. Greek name for another plant used as a vegetable, not these perennial herbs.

Trienta'lis f. Star flower. Latin word meaning one-third of a foot in height, approximately that of these small perennials.

Trifo'lium n. Clover. The Latin name, from *tri-,* three; *folium,* a leaf; because of the trifoliolate leaves of these annual and perennial herbs.

Trigonel'la f. Fenugreek. Diminutive of L. *trigōnus,* three-cornered, triangular, from Gr. *tri-*, three; *gōnia*, angle. The flowers of these herbs appear triangular.

Tri'lisa f. Carolina vanilla. Anagram of *Liatris*, indicating a close relationship to that genus.

Tril'lium n. Wake-robin, birthroot. From L. *tri-*, three, triple. Leaves and other parts of these rhizomatous perennial herbs are in threes.

Trime'za f. From Gr. *tri-*, three; *mĕizōn*, greater (comparative of *mĕgas*, great). The three outer flower segments of these bulbous plants are bigger than the inner.

Trios'teum n. Horse-gentian. From Gr. *tri-*, three; *ŏstĕŏn*, a bone; from the three very hard seeds of these perennial herbs.

Tripha'sia f. Limeberry. Gr. *triphasios*, triple. In these small ornamental shrubs or trees, the parts of the flowers are in threes.

Trip'laris f. Gr. *triplex*, triple. The parts of the flowers of these trees are in threes.

Tripleurosper'mum n. From Gr. *tri-*, three; *pleuron*, rib; *sperma*, seed. In allusion to the three-ribbed achenes (cypselae).

Tripogan'dra f. From Gr. *tri-*, three; *pōgōn*, beard; *anēr*, *andrŏs*, *andra*, man, hence stamen. In allusion to the three longer stamens being hairy, but the three shorter ones glabrous.

Trip'sacum n. Origin obscure.

Triptery'gium n. Gr. *tri-*, three; *pteryx*, a wing. The fruits of these shrubs are equipped with three membranous wings.

Trise'tum n. From L. *tri-*, three; *seta* (*saeta*) bristle. In allusion to the three awns of the lemnas of these grasses.

Trista'nia f. Evergreen flowering shrubs and trees named for Jules Tristan (1776–1861), French botanist.

tri'stis, -is, -e Dull; sad.

Trithri'nax f. Gr. *tri-*, three; *thrinax*, trident; from the form of the leaves of these palms.

Tri'ticum n. Wheat. The classical Latin name. Next to rice, wheat is the most important cereal crop.

tritifo'lius, -a, -um With polished leaves.

Trito'nia f. According to Gawler the name is derived from *Triton*, in the signification of a weather-cock and alluding to the variable direction of the stamens of the different species.

trium'phans Splendid; triumphant.

trivia'lis, -is, -e Common; ordinary.

Trochodendron

Trochoden′dron n. Gr. *trŏchōs*, a wheel; *dĕndrŏn*, a tree; in allusion to the spreading stamens on this aromatic evergreen tree.

trolliifo′lius With leaves like *Trollius*.

Trol′lius m. Globe flower. From the Swiss German name, *Trollblume*, with the same meaning. Sometimes called double buttercup.

Tropa′eolum n. The nasturtium of gardeners, though not that of botanists. Named by Linnaeus, from Gr. *tropaion*, trophy; L. *tropaeum*. The plant growing on a post reminded Linnaeus of a classical trophy, with round shields and golden helmets hung on a pillar, that is a sign of victory set upon a battlefield.
The word *nasturtium* is Latin for a pungent-tasting plant.

tro′picus, -a, -um Of the tropics.

trullipe′talus, -a, -um With petals shaped somewhat like a bricklayer's trowel.

trunca′tus, -a, -um Cut off square.

tsarongen′sis, -is, -e From Tsarong province, Mekong-Salwin divide, south-eastern Tibet.

Tsu′ga f. Hemlock. From the Japanese name for these coniferous evergreen trees.

tubaefor′mis, -is, -e Trumpet-shaped.

tuba′tus, -a, -um Trumpet-shaped.

tubercula′tus, -a, -um Tuberculate, i.e. covered with wart-like excrescences.

tuberculo′sus, -a, -um Covered with wartlike excrescences; tubercled.

tuberge′nii In honour of Messrs. C. G. van Tubergen, bulb-growers of Haarlem, Holland.

tubero′sus, -a, -um Tuberous.

tu′bifer, tubi′fera, -um; tubulosus, -a, -um Tubular; pipelike.

tubiflo′rus, -a, -um Tubular-flowered.

tugurio′rum Of native or peasant huts, that is growing over or around them.

Tulba′ghia f. Perennial herbs in the lily family named for Rijk Tulbagh (1699-1771), Dutch Governor of the Cape of Good Hope.

Tu′lipa f. Tulip. The Latin version of the Arabic for a turban. With many species native to the Caucasus, Anatolia, and the Middle East generally, tulips have been grown in Turkish gardens for centuries and were widely used as a decorative motif in their ceramics. They seem to have been introduced to the West by Ogier Ghiselin de Busbecq, Ambassador of the Holy Roman Empire to Suleiman the

Magnificent, who saw them growing on his way to Constantinople in 1554. Conrad Gessner described tulips he saw growing in Augsburg in 1559. It is likely that Clusius sent tulips to England around 1578. He took bulbs to Holland in 1593 when he became professor of botany at Leyden where tulips must already have been known and appreciated, for he soon lost his bulbs by theft. A few years later there developed in Holland and, afterwards, in Turkey the extraordinary hysteria known as the 'Tulipomania'. It reached its most extravagant heights and sudden collapse in Holland in 1634–1637. Fabulous prices were asked and paid—up to 100,000 florins for a single bulb. Regular markets were established, and the bulbs were sold by weight in 'perits' of less than a gram. One bulb of 'Admiral Liefken' of less than 440 perits fetched 4,400 florins; 'Semper Augustus' at 200 perits fetched 5,500 florins. Normally shrewd merchants who were making fortunes in a boom period in the spice and East India trades plunged deeply until finally the government stepped in with restrictive legislation to bring these speculations to a sudden and, for many, a shattering halt.

tuli'pifer, tulipi'fera, -um Tulip-bearing.

tu'midus, -a, -um Swollen.

Tu'nica f. A synonym of *Petrorhagia*.

tuolumnen'sis From Tuolumne county, California, U.S.A., so named from the Talmalamne Indians who once lived there.

Tupidan'thus m. Gr. *tupis*, a mallet; *anthos*, a flower; from the shape of the flower buds on this tall climber.

turbina'tus, -a, -um Shaped like a spinning top.

tur'gidus, -a, -um Inflated; full.

Turnera f. In honour of William Turner (*c.* 1508–1568), the 'Father of English Botany', clergyman, physician and herbalist, who spent several years in exile on account of his ardent Protestant views.

Turra'ea f. Trees or shrubs named for Giorgia della Turre (1607–1688), professor of botany at Padua.

Tussa'cia f. Tropical American plants named for Richard de Tussac (1751–1837), French botanist and author of a book on the flora of the Antilles.

Tussila'go f. Coltsfoot. L. *tussis*, a cough. The leaves of this herb with flowers like dandelions were supposed to cure coughs.

Ty'pha f. Cat-tail, reed-mace. The Greek name. In England *Typha* is commonly called bulrush which is properly a *Scirpus*. The biblical bulrushes were almost certainly *Cyperus papyrus*, the source of papyrus and the material used for making small rafts. The infant Moses is reported to have been found by Pharaoh's daughter in an ark of bulrushes laid in the flags by the river's brink.

typhi'nus, -a, -um Resembling *Typha.*

ty'picus, -a, -um Typical; agreeing with the type of a group.

U

uitewaalia'nus, -a, -um In honour of Adriaan Joseph Antoon Uitewaal (1889–1963), Dutch photographer, editor of *Succulenta* from 1947 to 1956 and specialist on the genus *Haworthia.*

U'lex m. Gorse. The ancient Latin name for these ornamental spiny evergreen shrubs.

ulici'nus, -a, -um Resembling *Ulex.*

uligino'sus, -a, -um Of swamps and wet places.

Ul'lucus m. From the native Peruvian name for these fleshy herbs.

Ulma'ria f. From L. *ulmus*, elm; referring to the leaflets of the meadow-sweet (*Filipendula ulmaria*).

ulmifo'lius, -a, -um With leaves like elm.

ulmo'ides Resembling elm.

Ul'mus f. Elm. The Latin name for these deciduous trees. The Romans were accustomed to use pollarded elms in vineyards over which to grow their vines. Trees are still so used in parts of Italy and Portugal.

ulva'ceus, -a, -um Resembling the seaweed *Ulva.*

umbella'tus, -a, -um Furnished with umbels.

Umbellula'ria f. California laurel. L. *umbella*, an umbel; from the shape of the inflorescence on this evergreen tree.

Umbili'cus m. L. for navel; in allusion to the sunken centre of the leaves of these succulents.

umbona'tus, -a, -um Having a stout projection at the centre.

umbracu'lifer, umbraculi'fera, -um Umbrella-bearing.

umbrel'la Of umbrella-like shape.

umbro'sus, -a, -um Shade-loving.

uncina'tus, -a, -um Hooked at the end.

unda'tus, -a, -um; undula'tus, -a, -um Waved, wavy, as in *undulifolius,* wavy-leaved.

U'nedo Latin name for the strawberry tree and its attractive fruit, rarely ripe enough in England to be edible.

unger'nii In honour of Baron Franz Ungern-Sternberg (1808–1885), professor at Dorpat (now Tartu).

Ungna'dia f. Mexican buckeye. A small tree or shrub named for Baron David von Ungnad, Austrian Ambassador at Constantinople (1576–1582), who sent horse-chestnut and other seeds to Clusius.

unguicular'is, -is, -e; unguicula'tus, -a, -um Furnished with claws, i.e. contracted into a long narrow base.

unguipe'talus, -a, -um With claw-shaped petals.

unguispi'nus, -a, -um With hooked spines.

uni- In compound words signifies one. Thus:

uni'color	of one colour
unicor'nis	with one horn
unidenta'tus	with a single tooth
uniflo'rus	one-flowered
unifo'lius	with one leaf
unilatera'lis	one-sided
uniner'vis	one-nerved
uniseria'tus	in one row
univitta'tus	with one stripe

Uni'ola f. Latin name for an unidentifiable plant, not these perennial grasses.

uplan'dicus From Uppland, Sweden.

urba'nus, -a, -um; ur'bicus, -a, -um Belonging to towns.

urceola'tus, -a, -um Urn-shaped.

Urceo'lina f. From L. *urceolus*, a little pitcher, in allusion to the shape of the flowers.

u'rens Stinging; burning.

U'rera f. Cow-itch. L. *ure*, burn or sting. One member (*U. baccifera*) of this genus of trees and shrubs is also called chichaste and described as 'one of the most dangerous plants of Central America. . . . When one is struck by the coarse hairs, the effect is almost like that of an electric shock and there often follows the most intense pain. . . . The plant is often used for hedges, which few larger animals care to penetrate.' (Standley, *Flora of Costa Rica*.)

Urgi'nea f. From the name of an Arab tribe in Algeria, the Beni Urgin. These bulbous herbs, closely related to *Scilla*, are the source of the commercial 'squill'.

ur'niger, urni'gera, -um Urn-bearing, for example with urn-shaped fruits.

urophyl'lus, -a, -um With leaves having a long tail-like tip.

Ursi′nia f. Annuals, perennials, and sub-shrubs named for Johannes Heinrich Ursinus (1608–1667) of Regensburg, a botanical author.

ursi′nus, -a, -um Like a bear, in shagginess or other respects. Also, sometimes, Northern (from the Great Bear).

Urti′ca f. Stinging nettle. The Latin name. All species sting but some are occasionally so virulent (e.g. *U. urentissima*) as to cause death. The power to sting lasts even in the dried plant. For instance, when the great Linnaean herbarium was being photographed, specimen by specimen, in 1941 for record in case of destruction during the second World War, the photographer, Miss Gladys Brown, was stung on the arm, raising a blister, by one of the specimens of *Urtica* dried and mounted approaching two hundred years earlier.

urticifo′lius, -a, -um Nettle-leaved.

urticoi′des Resembling nettle.

usitatis′simus, -a, -um Most useful.

ustula′tus, -a, -um Burned, scorched.

u′tilis, -is, -e Useful.

Utricula′ria f. Bladderwort. L. *utriculus*, a small bottle; in allusion to insect-trapping bladders borne on the leaves and runners of these aquatic and terrestrial herbs.

utricula′tus, -a, -um; utriculo′sus, -a, -um Bladderlike.

Uva′ria f. From L. *uva*, bunch of grapes. In allusion to the somewhat grape-like fruits.

uva-ur′si Bear's grape. Specific epithet of *Arctostaphylos*.

uvariifo′lius, -a, -um With leaves like *Uvaria*.

u′vifer, uvi′fera, -um Bearing grapes.

Uvula′ria f. Bellwort. From the anatomical term *uvula*, the lobe hanging from the back of the soft palate in man, in reference to the hanging blossoms of this plant.

V

vaccinifo′lius, -a, -um With leaves like *Vaccinium*.

vaccinioi′des Resembling *Vaccinium*.

Vacci′nium n. Blueberry. Bilberry. Cranberry. A Latin name apparently derived from the same prehistoric Mediterranean language as the Gr. *hyakinthos* and transferred to these berry-bearing shrubs.

vacil'lans Variable.

va'gans Of wide distribution; wandering.

vagina'lis, -is, -e; vagina'tus, -a, -um Sheathed; having a sheath.

valdivia'nus, -a, -um From Valdivia, Chile.

valenti'nus, -a, -um From Valentia in Spain.

Valeria'na f. Valerian. The medieval Latin name for these perennial herbs, possibly derived from L. *valere*, to be healthy; in allusion to the plant's medicinal uses in nervousness and hysteria.

Valerianel'la f. Corn-salad. Lamb's lettuce. Diminutive of *Valeriana*, from the resemblance of these herbs to *Valeriana*.

va'lidus, -a, -um Strong; well-developed.

Valla'ris f. L. *vallus*, a stake in a palisade. These twining shrubs are said to be used in Java for fences.

Vallisne'ria f. Eelgrass. Named for Antonio Vallisnieri de Vallisnera (1661–1730), professor at Padua. These are fresh-water plants. Salt-water eelgrass so much appreciated by wild ducks is *Zostera marina*.

Vallo'ta f. Scarborough-lily. Bulbous plants named for Pierre Vallot (1594–1671), French physician and botanical author.

Vancouve'ria f. Perennial herbs named for Captain George Vancouver, Royal Navy (1758–1798), British explorer. As a young man he sailed twice with Captain Cook. His outstanding feat was his four-year voyage (1791–1795) during which he carried out the first detailed surveys of the Pacific coast of North America as far north as Cook Inlet. Vancouver Island is named for him. The botanical results were due to the surgeon Archibald Menzies (1754–1842).

Van'da f. From the Hindi name for one of these epiphytic orchids.

Vando'psis f. From *Vanda*, plus Gr. *opsis*, similar to (see *Vanda*).

vanhout'tei In honour of Louis Benoît Van Houtte (1810–1876), celebrated Belgian nurseryman, editor of *Flore des Serres*.

Vanil'la f. Vanilla bean. Sp. *vainilla*, a small pod; with reference to the shape of the fruit. *Vanilla* is the only genus of orchid from which a commercial product is derived but the value of the pods has been much reduced through the introduction of synthetic vanillas, some of which can be produced very cheaply in very large quantities.

varia'bilis, -is, -e; va'rians; varia'tus, -a, -um Variable; varying.

variega'tus, -a, -um Irregularly coloured; variegated.

va'rius, -a, -um Differing; diverse.

vase'yii In honour of George Vasey (1822–1893), medical man, later botanist of the U.S. Department of Agriculture.

vedrarien′sis, -is, -e From Verrières near Paris, nursery of Messrs. Vilmorin-Andrieux.

ve′getus, -a, -um Vigorous.

Veit′chia f. Palms named for James Veitch (1815–1869), and his son John Gould Veitch (1839–1870), of Exeter and Chelsea, England. The Veitches were the leading nurserymen of their day and their family was in business continuously for 106 years until the Chelsea firm was brought to an end in 1914 on the retirement of Sir Harry James Veitch (1840–1924). Meanwhile, many good plants were introduced by them, as shown by the number of times their name is commemorated in specific names (*veitchianus*, *veitchii*, *veitchiorum*). They sent their own collectors abroad and trained a legion of excellent gardeners. Dr. E. H. Wilson was among their collectors (see *Sinowilsonia*).

vela′ris, -is, -e Pertaining to a veil.

velle′reus, -a, -um Fleecy.

ve′lox Swift, quick-growing.

Velthei′mia f. Bulbous plants named for August Ferdinand von Veltheim (1741–1801) of Brunswick, German patron of botany.

veluti′nus, -a, -um Velvety.

vena′tor Hunter. In allusion to scarlet ('pink') coats of British foxhunters.

venena′tus, -a, -um Poisonous. Also, *venenosus*, very poisonous.

Veni′dium n. From L. *vena*, a vein; because of the appearance of the ribbed fruits of these ornamental herbs.

veno′sus, -a, -um Full of veins, prominently veined.

ventrico′sus, -a, -um Having a swelling on one side.

venus′tus, -a, -um Handsome; charming.

Vera′trum n. False hellebore. The Latin name.

verbascifo′lius, -a, -um With leaves like mullein.

Verbas′cum n. Mullein. The ancient Latin name for these mainly biennial herbs.

Verbe′na f. Vervain. The Latin name for leaves and shoots of laurel, myrtle, etc., used in religious ceremonies and also in medicine.

Verbesi′na f. From the resemblance of leaves of these herbs to *Verbena*.

verecun′dus, -a, -um Modest.

ve′ris Of the spring; spring-flowering. Specific epithet of the cowslip (*Primula veris*).

vermicula'ris, -is, -e; vermicula'tus, -a, -um Wormlike.

verna'lis, -is, -e Of spring; spring-flowering.

verni'cifer, vernici'fera, -um Producing varnish.

vernico'sus, -a, -um Varnished.

ver'nix Varnish.

Verno'nia f. Ironweed. North American perennial herbs named for William Vernon (d. *c.* 1711), English botanist who collected in Maryland in 1698.

ver'nus, -a, -um Of the spring.

Vero'nica f. Speedwell. Herbs, shrubs, and some trees, named for St. Veronica (see *Hebe*).

verruco'sus, -a, -um Warty.

verruculo'sus, -a, -um With small warts.

Verschaffel'tia f. A palm named for Ambrose Colletto Alexandre Verschaffelt (1825–1886) of Ghent, Belgian nurseryman and author of a book on camellias.

versi'color Variously coloured.

verticilla'ris, -is, -e; verticilla'tus, -a, -um Having whorls; forming a ring around an axis.

Verticor'dia f. Juniper-myrtle. The name of these shrubs is a compliment to Venus to whom myrtle was sacred (L. *verto*, I turn; *cor*, the heart).

ve'rus, -a, -um True to type; standard.

ves'cus, -a, -um Thin; feeble.

Vesica'ria f. L. *vesica*, a bladder; from the inflated pods of these herbs

vesica'rius, -a, -um Bladder-like.

vesiculo'sus, -a, -um Furnished with small bladders or vesicles.

vesperti'nus, -a, -um Of the evening; evening-blooming.

vesti'tus, -a, -um Covered; clothed, usually with hairs.

Vetive'ria f. From Southern Indian vernacular name *vettiveri*, derived like *Zingiber* from Dravidian *ver*, root.

vex'ans Annoying; wounding.

vexilla'ris, -is, -e Having a standard (flag). Also *vexillarius*, a standard-bearer.

via'rum Of the waysides.

viburnifo'lius, -a, -um With leaves like *Viburnum*.

Vibur'num n. Arrow-wood, wayfaring tree. Latin name of one species of this genus of evergreen and deciduous flowering shrubs.

Vicia

Vi'cia f. Vetch. The Latin name for these herbs.

viciifo'lius, -a, -um Vetch-leaved.

Victo'ria f. Royal water-lily. Queen Victoria water-lily. Named for Queen Victoria (1819–1901). The great water-lily, *V. amazonica* (*V. regia*), was found in Guiana by Schomburgk, having earlier been discovered by Haenke, but introduced into cultivation from Bolivia in 1846 by Thomas Bridges (1807–1865), who brought seeds to Kew. The first detailed account of it appeared in the *Bot. Mag.* t. 4275 (1847). A seedling was obtained from Kew by Joseph Paxton, head gardener to the Duke of Devonshire at Chatsworth, where a special house and pool were constructed to Paxton's design to contain it. The feat of bringing the plant into bloom caused a sensation and a special telegram was dispatched to announce the fact to Her Majesty. The leaves, which may be up to six feet across, are reputed to be able to support the weight of a small child—provided he does not fidget and his weight is distributed correctly.

victoria'lis, -is, -e Victorious. It is said that bulbs of *Allium victorialis* were once worn by German miners as a protection against evil spirits.

Vi'gna f. Twining and prostrate herbs named for Dominico Vigna (d. 1647), professor of botany at Pisa, author in 1625 of a commentary on Theophrastus.

Villare'sia f. Evergreen climbing shrubs or small trees named for Mattias Villares, a monk who had a botanic garden in the monastery of Santa Espina, Chile, *c.* 1790.

Villar'sia f. In honour of Dominique Villars (1745–1814), French botanist, chiefly noted for his work on the flora of Dauphiné. He also spelled his name as Villar, which is the name of his birth-place.

villo'sus, -a, -um Covered with soft hairs.

vilmorinia'nus, -a, -um; vilmorinii In honour of the firm of Vilmorin-Andrieux of Paris, celebrated French nurserymen and seedsmen.

vimina'lis, -is, -e; vimi'neus, -a, -um With long slender shoots, like osiers.

Vin'ca f. Periwinkle. Contracted from the Latin name *vinca pervinca* or *vincapervinca*, apparently from *vincio*, bind, wind around, in allusion to the use of the long flexible shoots in making wreaths.

Vinceto'xicum n. L. *vinco*, to conquer; *toxicum*, poison. This genus of herbs and sub-shrubs was supposed to contain an antidote to snake-bite.

vinea'lis, -is, -e Belonging to vines, growing in vineyards.

vi'nifer, vini'fera, -um Wine-bearing.

vino'sus, -a, -um Wine-red.

Vi'ola f. Pansy, heart's-ease, violet. The Latin name for various sweet-scented flowers, such as violets, stocks, wallflowers, and derived from the same source as the Gr. *ion* (which in its earlier form had an initial letter corresponding to *v* or *w*, the digamma, which was later lost); see p. 339.

viola'ceus, -a, -um Violet-coloured.

violas'cens Becoming violet-coloured.

virens Green. Also **virescens**, becoming green.

virga'tus, -a, -um Twiggy.

Virgi'lia f. A tree named for the Latin poet Virgil (70–19 B.C.).

virgina'lis, -is, -e; virgi'neus, -a, -um White; virginal.

virginia'nus, -a, -um; virgi'nicus, -a, -um; virginien'sis, -is, -e Virginian.

virgulto'rum Of thickets.

vi'ridis, -is, -e Green. Also in compound words *viridi.* Thus:
  virides'cens   becoming green
  viridiflo'rus   with green flowers
  viridifo'lius   green-leaved
  viridifus'cus   green-brown
  viridis'simus   very green
  viri'dulus   somewhat green

viscidifo'lius, -a, -um With sticky leaves.

vis'cidus, -a, -um; visco'sus, -a, -um Sticky; clammy.

Vis'cum n. Latin name for mistletoe.

vita'ceus, -a, -um Vinelike.

vitelli'nus, -a, -um Like the yolk of an egg.

Vi'tex f. Latin name for V. *agnus-castus* or chaste tree, one species of the aromatic deciduous flowering shrubs which comprise this genus.

vitifo'lius, -a, -um With leaves like the grapevine.

Vi'tis f. Grape. The Latin name for the grapevine.

    V. *vinifera* is the European grape. It has been used by man since prehistoric times but no one knows when the secret of wine-making was discovered. Long ago, grapes were grown and wine made in the southern counties of England, as instanced by many place names. Today, grape-growing for wine is hardly practised north of a line through Brittany, Liège in Belgium to Silesia. Whether this is due to climatic change or to more discriminating taste is not certain, perhaps a little of both.

From their first arrival in the New World, the American settlers made determined efforts to grow acceptable wine from imported stocks. As early as 1639, the Virginia Assembly made it mandatory for all growers of corn and tobacco to grow a certain number of vines. That measure being unsuccessful, a heavy premium of 10,000 pounds of tobacco was offered for each two tuns of wine from grapes grown in the Colony. In spite of these inducements, the industry failed and it took Americans some two hundred years to discover that imported stocks could not be made commercially productive. It was only when John Adlum in about 1820 introduced the Catawba grape that a truly satisfactory wine grape became available to American wine growers. Catawba is still the outstanding American wine grape.

It remained for Georg Engelmann (see *engelmannii*), the German-born St. Louis physician, to discover that American grape stocks were virtually proof against the minute plant louse, phylloxera. It had already devastated the vineyards of Madeira and was beginning serious attacks on the famous vineyards of Europe. Presumably they would have been lost had not Engelmann stepped in with his timely discovery.

Vittadi'nia f. Thick-stemmed herbs and sub-shrubs named for C. Vittadini (1800–1865), Italian botanical author.

vitta'tus, -a, -um Striped lengthwise. L. *vitta*, a band.

vit'tiger, vitti'gera, -um Marked with stripes.

volga'ricus, -a, -um Of the Volga River.

volu'bilis, -is, -e Twining.

volu'tus, -a, -um With rolled leaves.

vomito'rius, -a, -um Emetic.

Vrie'sia f. Tropical American plants with stiff leaves, named for Willem Hendrik de Vriese (1806–1862), Dutch botanist, professor of botany first at Amsterdam, then at Leyden.

vulca'nicus, -a, -um Growing on a volcano.

vulga'ris, -is, -e; vulgatus, a-, -um Common.

vulpi'nus, -a, -um Relating to foxes; used for species with an inferior sort of fruit.

W

Wachendor'fia f. In honour of Everardus Jacobus van Wachendorff (1702–1758), professor at Utrecht.

Wahlenber'gia f. Bellflower. Annual and perennial herbs named for

Georg Wahlenberg (1780–1851), Swedish botanist, professor at Uppsala, noted for his studies of European plant geography.

Waldstei'nia f. Creeping herbs like strawberries named for Count Franz Adam Waldstein-Wartenburg (1759–1823), Austrian botanist, part-author of a finely illustrated book on the plants of Hungary.

Wallich'ia f. Palms named for Nathaniel Wallich (1786–1854), originally Nathan Wolff, Danish botanist, who went out to India in 1807 as surgeon to the Danish settlement at Serampore near Calcutta, entered the service of the Hon. East India Company in 1813 and was superintendent of the Calcutta botanic garden from 1814 to 1841, during which period he made immense contributions to the knowledge of the Indian flora.

Wardaster m. See *wardii*.

war'dii Specific epithet of many plants from eastern Asiatic mountains and highlands where for many years Frank Kingdon-Ward (1885–1958) carried out intensive botanical exploration, bringing back many plants, some of which are now widely cultivated in temperate gardens.

warleyen'sis, -is, -e After Warley Place, Brentwood, Essex, England, the garden of Miss E. A. Wilmott.

Warscewiczel'la f. Epiphytic orchids named for Joseph Warsczewicz (1812–1866), who escaped from Poland after taking an active part in the 1830 rebellion. He travelled widely in South America collecting orchids for Messrs. van Houtte, Belgian nurserymen. Eventually he was able to return home to become Inspector of the botanic gardens in Krakow. (Name pronounced 'var-shev-i-chel-la'.)

Warzewiczia See *Warscewiczella*.

Washingto'nia Named for George Washington (1732–1799), first President of the United States. They are tall, not very handsome, fan-palms, native in California and the southern United States.

waso'nii In honour of Rear-Admiral Cathcart Romer Wason (1874–1941) of the Royal Navy.

Watso'nia f. Bugle lily. Bulbous herbs named for Sir William Watson (1715–1787), English scientist and physician.

watso'nii This epithet is likely to commemorate Sereno Watson (1826–1892) of Harvard University (see under *Serenoa*) if a North American plant, but it has also been used to commemorate Hewett Cottrell Watson (1804–1881), pioneer British plant-geographer, and William Watson (1858–1925), curator of the Royal Botanic Gardens, Kew.

wat'tii In honour of Sir George Watt (1851–1930), Scottish medical man and economic botanist, author of *Dictionary of Economic Products of India*.

Wedelia

Wede'lia f. Herbs and sub-shrubs named for Georg Wolfgang Wedel (1645–1721), professor of botany at Jena.

Wei'gela f. Deciduous flowering shrubs named for Christian Ehrenfried Weigel (1748–1831), German botanist, professor at Greifswald.

Weinman'nia f. Evergreen shrubs or trees with fernlike foliage named for Johann Wilhelm Weinmann (1737–1745) of Regensburg, German apothecary, who published between 1737 and 1745 a big illustrated work on plants entitled *Phytanthozaiconographia*.

Welde'nia f. A tuberous-rooted herb. In honour of Ludwig van Welden (1780–1853), master of the ordnance in the Austrian army.

werck'lei In honour of Karl (Carlos) Wercklé (1866–1924), investigator of flora of Costa Rica.

Westrin'gia f. Australian rosemary. Shrubs named for Johan Peter Westring (1753–1833), physician to the King of Sweden.

weyri'chii In honour of Heinrich Weyrich (1828–1868), Russian naval surgeon of German extraction who botanized in eastern Asia.

Wigan'dia f. Foliage plants named for Johannes Wigand (1523–1587), Bishop of Pomerania, who wrote on Prussian plants.

wi'ghtii In honour of Robert Wight (1796–1872), Superintendent of Madras Botanic Garden and author of important works on Indian plants.

Wilco'xia f. Cacti named for General Timothy E. Wilcox, U.S. Army, a keen student of plants, *c.* 1900.

wilkesia'nus, -a, -um In honour of Charles Wilkes (1798–1877), American naval officer, leader of the United States Exploring Expedition of 1838–1842 in the Pacific Ocean.

william'sii In honour of Williams, the family name of at least seven distinguished horticulturists or botanists.

wilmot'tiae, willmottia'nus, -a, -um In honour of Miss Ellen Ann Wilmott (1860–1934), celebrated English amateur gardener.

wilso'nii See *Sinowilsonia*.

Wistaria See *Wisteria*.

Wiste'ria f. Deciduous climbing flowering shrubs named for Caspar Wistar (1761–1818), professor of anatomy at the University of Pennsylvania and one of the early owners of what is now Vernon Park, Philadelphia. Despite the apparent anomaly Wisteria is the spelling that is correct according to the international rules of botanical nomenclature.

wolga'ricus, -a, -um Of the Volga River region.

Wood'sia f. Ferns named for Joseph Woods (1776–1864), English architect and botanical author, best known for his *Tourist's Flora* (1852).

Woodwar'dia f. Chain-fern. Named for Thomas Jenkinson Woodward (1745–1820), English botanist.

Wulfe'nia f. Herbaceous perennials named for Franz Xavier, Freiherr von Wulfen (1728–1805), Austrian botanical author.

Worsley'a f. Blue Amaryllis. In honour of Archington Worsley (1861–1943) English civil engineer and gardener.

Wye'thia f. Mule-ears. Named for Nathaniel J. Wyeth (1802–1856), American botanist who discovered these plants.

X

Xantheran'themum n. From Gr. *xanthos*, yellow; *Eranthemum* which is a related genus.

xan'thinus, -a, -um Yellow. In compound words *xanth-*. Thus:

xanthacan'thus	yellow-spined
xanthocar'pus	yellow-fruited
xanthochlo'rus	yellowish green
xanthoco'don	yellow bell
xantholeu'cus	yellowish white
xanthoner'vis	yellow-nerved
xanthophyl'lus	with yellow leaves
xanthorrhi'zus	yellow-rooted
xanthoxy'lon	with yellow heartwood

Xanthis'ma f. Gr. *xanthos*, yellow; alluding to the colour of the flowers of this daisylike annual.

Xantho'ceras n. From Gr. *xanthos*, yellow; *keras*, a horn; from the yellow hornlike growths between the petals of this deciduous flowering tree.

Xanthorhi'za f. From Gr. *xanthos*, yellow; *rhiza*, root.

Xanthorrho'ea f. Gr. *xanthos*, yellow; *rheo*, to flow. A yellow resinous gum is extracted from these perennials which are known as Botany Bay gums or blackboys.

Xanthoso'ma n. Gr. *xanthos*, yellow; *sōma*, a body; with reference to the yellow inner tissues of some species of these milky herbs.

xanthostepha'nus, -a, -um With a yellow crown or garland.

Xeran'themum n. Immortelle. From Gr. *xēros*, dry; *anthos*, a flower. The flower heads are dry and retain their form and colour for years.

Xerophyl'lum n. Turkey's beard. From Gr. *xeros*, dry; *phyllon*, a leaf. The leaves of this perennial sub-aquatic herb are dry and grass-like.

Xi'phium n. Greek name for a *Gladiolus* referring to the sword-like leaves, from *xiphŏs*, sword.

Xylo'bium Gr. *xylon*, wood; *bios*, life. These orchids grow on trees.

xylocan'thus, -a, -um With woody spines.

Xylos'teum n. From Gr. *xylon*, wood; *ŏstĕŏn*, bone.

Y

yakuinsula'ris, -is, -e, yakusima'nus, -a, -um From the island of Yakushioma, southern Japan.

yedoen'sis, -is, -e From Yedo (now Tokyo), Japan.

Yuc'ca f. From the Carib name for manihot or cassava (a genus belonging to the *Euphorbiaceae*), unfortunately applied to these liliaceous evergreen shrubs or small trees with rosettes of sword-shaped leaves. Adam's needle and Spanish bayonet are familiar names of some species.

yungningen'sis, -is, -e From Yungning, Szechwan, China.

yunnanen'sis, -is, -e; yunna'nicus, -a, -um From Yunnan, west China.

Z

zaleu'cus, -a, -um Very white.

za'lil Afghan vernacular name for a kind of Delphinium.

Zaluzian'skya f. Annual or perennial herbs named for Adam Zaluziansky von Zaluzian (1558–1613), botanist of Prague.

Za'mia f. Evergreen palmlike plants. Name derived from *zamiae*, a false rendering in some texts of Pliny for *azaniae*, referring to pinecones. The roots of the coontie (*Z. floridana*) were used by the Indians to make soap.

Zantedes'chia f. Calla-lily (U.S.), arum-lily (U.K.). Perennial herbs named for Francesco Zantedeschi (b. 1797), Italian botanist.

Zanthox'ylum n. Prickly ash, Hercules' club, tooth-ache tree. Gr. *xanthos*, yellow; *xylon*, wood; from the colour of the heartwood of some species of these ornamental shrubs.

Zauschne'ria f. In honour of Johann Baptist Zauschner (1737–1799), professor at Prague.

Ze'a f. Corn (U.S.), maize, Indian corn (U.K.). From the Greek name for another cereal. The plant is widely cultivated in the tropical and temperate zones from sea level to about 12,000 feet. It is the staple grain in Mexico, Central America, and southern Africa. In the United States, about 90 per cent of the crop is fed to animals and goes to market as meat, milk, and eggs. The leaves provide the wrappers of the 'whacking white cheroot' of Burma.

Zebri'na f. From Portuguese *zebra*; in allusion to the striped leaves of one species of these herbaceous trailing perennials.

zebri'nus, -a, -um Zebra-striped.

Zel'kova f. From the Caucasian name for these deciduous trees.

Zeno'bia f. An ornamental shrub named for Zenobia, Queen of Palmyra *c*. A.D. 266.

Zephyran'thes f. Zephyr-lily. Gr. *zĕphyrŏs*, the west wind; *anthos*, a flower. These bulbous herbs are natives of the Western hemisphere.

zeylan'icus, -a, -um Of Ceylon.

zibethi'nus, -a, -um Ill-smelling like a civet cat. From the Italian *zibetto* (derived from the Arabian *zabad* or *zubad*), a civet. Specific epithet of the durian (*Durio*) whose fruit contains a pith which can smell horribly. In it are embedded the fruit segments which have a delightful flavour.

Zigade'nus m. Gr. *zygŏs*, a yoke; *adēn*, a gland. The glands of these herbs are in pairs at the base of the perianth.

Zin'giber n. Ginger. From the Greek name which, in turn, is said to derive from an East Indian word cognate with Pali *singivera*. The root of *Z. officinale* is the source of commercial ginger; see p. 342.

Zin'nia f. Showy annuals named for Johann Gottfried Zinn (1727–1759), professor of botany, Göttingen.

Ziza'nia f. Wild rice, Canadian wild rice. Greek name for another wild grain, not this. *Z. aquatica* is eagerly eaten by wild ducks and gives their flesh an especially fine flavour. Similarly, it is especially good when eaten with wild duck.

zizanioi'des Resembling *Zizania*.

Zi'zyphus f. Jujube. From the Arabic name for this small tree.

zona'lis, -is, -e; zona'tus, -a, -um Banded or with a girdle usually of a distinct colour.

Zoy'sia f. Named for Karl von Zoys (1756–1800), Austrian botanist. There is a new vogue for this turf grass for sandy lawns in warmish climates.

Zygadenus

Zygade'nus See *Zigadenus*.

Zygocac'tus m. Christmas cactus. From Gr. *zygŏs*, a yoke, plus *Cactus*; from the way in which the stems are jointed.

Zygope'talum n. Gr. *zygŏs*, a yoke; *pĕtalŏn*, a petal; from the swelling at the base of the lip which seems to yoke together the lateral segments of the flowers of these epiphytic orchids.

VERNACULAR NAMES

An Introduction to Vernacular Names

WILLIAM T. STEARN

ALL the world over and from remote times onwards men have devised names for plants of special interest to them and in their naming have been motivated by much the same ways of thought because, as fishermen, hunters, primitive farmers and herdsmen, they have had much the same ways of life. Other plants were left nameless until in comparatively recent times, that is after A.D. 1500, the urge to describe and record in books the diversity of the whole natural world led to the development of botany as a special field of enquiry distinct from herbalism and medicine and thereby to the study of these hitherto unimportant plants. Hence the greater part of the world's plants lack common or vernacular names; they have been recorded only under botanical names, usually of Latin form, since 1753. Modern botanical nomenclature originated, however, in the vernacular names of plants used for economic, medicinal or decorative purposes by the Ancient Greeks and Romans, some of which had descended to them out of languages dead and gone long before the invention of writing. Thus the related Greek *rhŏdŏn* ('ροδον) and Latin *rosa*, Greek *lĕiriŏn* (λειριον) and Latin *lilium*, Greek *iŏn* ('ιον), earlier *wiŏn* (Fιον), and Latin *viola*, Greek *sukŏn* (συκον) and Latin *ficus*, Greek *ĕlaia* (έλαια), earlier *ĕlaiwa* (έλαιFα), and Latin *oliva*, Greek *kuparissŏs* (κυπαρισσος) and Latin *cupressus* were not derived the one from the other. They must have come independently from common sources, among them a supplanted Mediterranean language sometimes called Thraco-pelasgian spoken maybe four thousand years ago. Plant names like *Hyacinthus* and *Terebinthus* are also considered to be relics of a pre-Greek language. The origins of such English plant names as those of the rose, violet, fig, cypress, hyacinth, ginger, lily and crocus go back to very remote antiquity. As Walter de la Mare wrote:

> . . . No man knows
> Through what wild centuries
> Roves back the rose.

The connexion between the botanical and the vernacular nomenclature of plants is much closer, both as regards method and the names themselves, than is commonly noticed. Indeed during the past three hundred years many vernacular names out of many diverse languages have been taken into botanical nomenclature, with little or no alteration, for use as generic names (see p. 5). Even Linnaeus, despite his puristic ruling in his *Critica botanica* (1737) that 'generic names which have not a root derived from Greek and Latin are to be rejected', in fact accepted names having quite other origins. Thus the generic name *Trollius*, adopted by Linnaeus in 1753 for the globeflowers, originated as a contraction of the name *Trollius flos* used by Conrad Gessner in 1555 instead of the Swiss-German name *Trollblume* meaning a rounded flower. Vernacular names adopted as specific epithets are even more numerous; some examples illustrating their diversity are *Abies pinsapo, Actinidia kolomikta, Albizia lebbek, Brassica kaber, Cajanus cajan, Camellia sasanqua, Diospyros kaki, Gentiana kurroo, Krigia dandelion, Malus zumi, Myrica gale, Narcissus tazetta, Ophiopogon jaburan, Picea omorika, Pinus mugo, Pittosporum tobira, Prunus mume, Schinus molle, Silene schafta, Spondias mombin, Taraxacum kok-saghyz, Terminalia catappa, Veronica beccabunga, Zea mays*.

Clearly then vernacular names should not be despised and disregarded. When only a few plants need to be distinguished within a limited area, vernacular names can be just as useful, precise and stable for local use as their scientific equivalents intended for international use. It is important, however, to ascertain their application by examining the plants designated and so to equate them with standard scientific names, for the same vernacular name may be given to different plants in different places within the same country.

In 1892 Nathaniel Colgan of Dublin set out to ascertain the botanical identity of 'true shamrock'. That year and the next he collected and received about St. Patrick's Day, living specimens certified as 'true shamrock' from twenty Irish counties, grew them and identified them when in flower. Of these 19 were white clover (*Trifolium repens*), 12 were lesser yellow trefoil (*Trifolium dubium*, syn. *T. minus*), 2 red clover (*Trifolium pratense*) and 2 spotted medick (*Medicago arabica*); not one was wood-sorrel (*Oxalis acetosella*) sometimes called 'shamrock' in England. A summary of Colgan's investigation into the history of the shamrock (Irish *seamróg*, a diminutive of *seamar*, clover), originally published in the *Journal of the Royal Society of Antiquaries of Ireland* 26: 211–226, 349–361 (1896), will be found in Lloyd Praeger's *The Way that I went* (1937). Unfortunately all the early lives of St. Patrick are completely silent about shamrock, apparently an Irish famine-food down to about 1682, and Colgan traced the legend of it as exemplifying the Trinity no further back than 1727. This example will suffice to indicate the need for caution, indeed for painstaking enquiry, when accepting vernacular names. Their interpretation links botany, etymology, and folklore. Collections of such names, notably English ones assembled by Britten and Holland and commented upon by Prior and by Grigson, Danish ones by Lange, French ones by Rolland, German ones by Marzell, as well as those in the multilingual work of Gerth van Wijk, reveal how many variants may arise from a single basic concept or motive in the course of centuries of use as part of common speech over a wide area. They also illustrate, from the wealth of local names applied to a single species, how bewildering would be reliance for communication upon these alone. For example, the marsh-marigold or kingcup (*Caltha palustris*) has over 80 local names in Britain, about 60 in France and at least 140 in Germany, Austria and Switzerland, some of which may also be applied to other species.

English plant names can be divided into several groups according to their origin. Out of the linguistic heritage of the Teutonic peoples the Anglo-Saxons brought to England their names for trees or shrubs such as ash (Old English *aesc*, Old High German *ask*, Modern German *Esche*, Dutch *es*, Swedish *ask*), beech (Old English *bece*, Old High German *buohha*, Modern German *Buche*, Dutch *beuk*, Swedish *bok*), birch or birk (Old English *berc*, *birce*, Modern German *Birke*, Dutch *berk*, Swedish *björk*), hawthorn (Old English *haguthorn*, *hawethorn*, Modern German *Hagedorn*, Dutch *haagdoorn*, Swedish *hagtorn*), hazel (Old English *haesel*, Modern German *Hasel*, Dutch *hazelaar*, Swedish *hassel*), linden or lime (Old English *lind*, Modern German *Linde*, Dutch *linde*, Swedish *lind* with a Småland variant *linn*, whence the family name of Linnaeus), oak (Old English *ac*, Old High German *eih*, Modern German *Eiche*, Dutch *eik*, Swedish *ek*), and willow (Old English *welig*, *withig*, Dutch *wilg*, Modern German *Weide*). These were important woody plants in their forested Continental homeland. The name 'rowan' (Norwegian *rogn*, *raun*, Swedish *ronn*) for *Sorbus aucuparia* is a Scandinavian legacy to northern Britain.

Names of Latin or Greek origin attest indebtedness to southern Europe for such cultivated decorative or culinary plants as the lily (Latin *lilium*), fennel (*foeniculum*), peony (*paeonia*), rose (*rosa*), rue (*ruta*), sage (*salvia*) and thyme (*thymus*), which were probably medieval monastic introductions into Britain. Their short names have undergone little change during the past two thousand years. People receiving a plant from elsewhere usually retain or try to retain its original name if this is short, distinctive and memorable. Such names may accordingly indicate the origin and the transfer from people to people of cultivated plants. Thus the sweet-scented *Jasminum officinale* has the name 'jasmine' or 'jessamine' in England, *ye-si-min* in China and *dzasasimi* in Madagascar, all independently derived from the Persian *yasamin* or *yasmin*. The English word 'ginger', investigated in detail by Alan S. C. Ross, goes back through a complicated linguistic history to a plant name coined in remote antiquity somewhere in south-eastern Asia, probably southern India, and apparently alluding to the horny appearance of the rhizome of this ancient spice plant (*Zingiber officinale*);

one element is the Sanskrit *śrnga*, 'horn', another the Dravidian *vēr*, '*root*', giving the Pali *singivera*. Traders brought the pungent rhizomes from India to the Ancient Greeks, who rendered its name as *ziggibĕris*, and to the Romans, who rendered it as *zingiber*, a word giving rise later to Old French *gimibre*, Old High German *gingibero* and Middle English 'ginger', and so to the name 'wild ginger' for *Asarum*, and 'ginger-lily' for *Hedychium* or *Zingiberaceae* for its botanical family. Long names transferred into another language, with their original meaning no longer obvious, tend to be corrupted or changed into forms more easily memorized. Thus the Greek *glycorrhiza*, referring to the sweetish root of the liquorice plant, became *liquorita* in Late Latin, *licorys* in Middle English and thus 'licorice' or 'liquorice' in Modern English. The bitter strong-smelling leaves of tansy (*Tanacetum vulgare*) were formerly placed in coffins or rubbed on corpses, as mentioned by Linnaeus in his Swedish Flora (*Flora Suecica*, 2nd ed. 284; 1755); this was to preserve them against worms (cf. the Dutch name 'boerenwormkruid') and the plant accordingly acquired the Greek name *athanasia*, 'immortality' (from *a*, without, not; *thanatos* death) which became changed into Late Latin *tanasia*, Old French *tanesie* and English 'tansey' or 'tansy'.

Such names exemplify the tendency to simplify unfamiliar words. Associated with this, as Cassidy has noted in a study of Jamaican names and folklore, is the replacement of difficult or foreign words by easier native ones; even though the result may be nonsensical or comical, it may not seem more so than the original name to the uninitiated. Thus *Carpobrotus edulis*, now naturalized on the south-western coast of England and long included in the genus *Mesembryanthemum*, has become colloquially known as 'Sally-my-handsome'; *Galinsoga* has 'gallant soldier' as a vernacular equivalent, although there is nothing gallant or soldierly about this diffuse weed; *Antholyza*, in which *Curtonus* was formerly included, has been converted into 'Aunt Eliza'. Similarly, in Jamaica, *Delonix regia*, long included in *Poinciana*, has acquired, by way of the intermediate 'Panchilana', the name 'Fancy Anna' fitting its flamboyance. *Aloe*, at one time known as a *Sempervivum* or 'sempervive', is now sometimes called 'single bible' in Jamaica. The English name 'bullace' (earlier 'bolays') for *Prunus insititia* applied by English colonists to Jamaican species of *Bumelia* (*Dipholis*) with likewise small dark fruits had by 1657 been converted into 'bully' or 'bully tree', which a century or so later became 'bullet' or 'bullet tree', now the most commonly used name. The quaint Jamaican name 'Panchalum' for a species of *Cordia* would be inexplicable but for still used vernacular intermediates between this and the name 'Spanish Elm'. These are well-documented quite recent changes; if the original forms of such names were unknown, it would be difficult or impossible to deduce or reconstruct them: thus it should not be surprising that many plant names of great antiquity, which may have been similarly transformed, have no plausible meaning; possibly some names associated with mythical figures, e.g. *Hyacinthus* and *Narcissus*, were modified loan-words with originally no such association.

The value of such apparently facetious names as aids to memory, which has led to their retention and spread, used to be illustrated by Sir Herbert Maxwell (1845–1937) with tales about the *memoria technica* of Wyber, an old Scottish head gardener at Cardross, whose ingenuity and quiet sense of humour enabled him to triumph over the perplexities of Greek and Latin botanical names. Asked how his *memoria technica* worked, Wyber replied, 'Weel enough, Sir, weel enough. There's a tree now, the *Cryptomeria japonica*. When I'm teaching the lads I tell them when they look at that tree to say "Creep to the mear (mare) and jump on to her". Ane o' the lads was gaily surprised when I was giving him a lesson in the greenhouse ane day. I was showing him the *Selaginella stolonifera*, a kind o' moss, ye ken. If ye want to mind the name, say I, think of "Silly Jean and Nellie stole a neive-fu (a handful)". "Losh! Mr. Wyber", quo' he, "them's my twa sisters".' The Greek name *anemone* possibly originated in some such manner from a Semitic name connected with Na'aman.

Many English names for common plants, for example bindweed, daffodil, daisy, dandelion, henbane, hollyhock, honeysuckle, horehound, mugwort, poppy, primrose, wormwood and yarrow, originated during the Middle Ages. They first found their way into print in the anonymous *Herball* printed by Rycharde Banckes in 1525, the *Grete Herball* printed by Peter Treveris in 1526 and the more original works of the clergyman, physician and naturalist, William Turner (*c.* 1508–1568), whose little books *Libellus de Re herbaria* (1538) and *Names of Herbes* (1548) had become among the rarest of botanical publications until reprinted in 1877 and 1882 and again in 1965. In the first of these, Turner recorded the English names known to him as used in Northumberland and East Anglia. They include the curious name 'Alleluya' or 'Allelua' which, also in the form 'Hallelujah', still persists locally in Britain for wood sorrel (*Oxalis acetosella*). The earliest mention of this traced by Sprague is to be found in a thirteenth-century manuscript Latin herbal (first printed in 1946) by an Italian monk, Rufinus of Genoa, who stated that the herb-gatherers supplied it under that name to the druggists of Bologna for making a cooling syrup. It also occurs in a Latin–Flemish manuscript plant glossary of about 1350 (first printed by Vandewiele and Braekman in 1968 in *Scient. Historia* 10: 114–144). It was so called, according to Turner's *New Herbal* (1568), 'because it appeareth about Easter when Alleluya is sung again'. As stated by Sprague, 'the wood sorrel flowers in April and May, and a liturgical

343

A Gardener's Dictionary of Plant Names

hymn, each verse of which ends in Alleluia, is sung daily during the Office of Compline from Easter Eve to the Saturday in Whitsun week, that is during April and May. Hence the name Alleluia.' There are other plant names as indirectly connected through religious service with their time of flowering. A particularly notable one is 'St. John's wort' (*Hypericum perforatum*), called by Turner 'Saynte Johans grasse', a conspicuous plant with gland-dotted leaves which is in flower on St. John's Day (24 June). Apparently its magical associations make it particularly obnoxious to evil spirits and a young woman fearing rape by the Devil can effectively safeguard herself by sitting on it or keeping it under her dress. Thus frustrated long ago the Devil in revenge pricked the leaves all over with his needle; these needle marks can be seen by holding a leaf up against the light and provide evidence as convincing as other evidence for the inheritance of acquired characteristics.

The Devil's dealings with the plant world are commemorated in a diversity of names. The devil's bit scabious (*Succisa pratensis*) has a rootstock apparently bitten off at the end which formerly possessed medicinal properties so valuable that, according to the *Grete Herball*, 'the devyll had envy at the vertue thereof and bete the rute so for to have destroyed it'. The name 'devil's bit' was taken by early colonists to North America and there transferred to native plants with an abruptly terminated rootstock such as *Chamaelirium luteum*. The American botanist, Willard Clute, counting plant names in Europe and America associated with saints and demons, found that Europeans 'have more saints than devils in their flora while our situation is quite the reverse; for we have twice as many devils and witches as saints', a matter worthy of consideration by politicians and sociologists, for the European names connected with saints and saints' days originated before the Reformation, the American names connected with the Devil arose later in Protestant America. Plants associated with the Devil usually possess some noxious or unpleasant characteristic; thus 'Devil's guts' for dodder (*Cuscuta*), on account of its parasitic stems like a small mammal's intestines, 'Devil's milk' for *Euphorbia*, on account of its poisonous milky juice, or bindweed (*Calystegia* and *Convolvulus*) so difficult to eradicate on account of their long creeping roots, 'Devil's stinkpot' for the stinkhorn (*Phallus impudicus*), 'Devil's club' for the terribly prickly *Oplopanax horridus*. On the other hand there are also many old plant names associated with the Virgin Mary, among them 'Virgin Mary's milk-drops' for *Pulmonaria officinalis* and 'Lady's thistle' for *Silybum marianum*, both with spotted leaves.

A very interesting series of European plant names connected with keys show how names now meaningless were at one time quite apt. These have been the subject of a detailed and very scholarly paper in Norwegian by Rolf Nordhagen entitled 'Marilykjel, springstrå och jennurt'. In England among the many local names of the cowslip (*Primula veris*; Fig. *a*) are 'culverkeys' (also applied to the keys or fruits of ash), 'Herb Peter', 'Lady Keys', 'Peter', 'St. Peterwort'. Similar names for it have been used in Germany, such as *Schlüsselblume* and '*Peters-schlüssel*', in the Netherlands and Scandinavia, such as in Norway *marilykjel, himelsnykkel* and *nökleblom*. The German name *Himmelschlüsseli* has also been used for the spring gentian (*Gentiana verna*; Fig. *b*) which has solitary flowers, not bunched flowers as in

a: flower of cowslip ('Herb Peter', 'Schlüsselblume'; *Primula veris,*
b: flower of spring gentian ('Himmelschlüsseli'; *Gentiana verna*)
c: key of 'Chinese padlock' from Sweden (after Nordhagen, 1948)
(*Drawings by Roberta E. Young*)

Primula veris. The names for the cowslip have been explained as indicating a resemblance between its clustered flowers and a bunch of keys, which at first seems a far-fetched and implausible idea, for the individual flowers of the cowslip, as also of the gentian, in no way resemble keys now used. Nordhagen showed, however, in 1948, that these flowers with their long fluted calyx and smaller corolla-lobes somewhat resemble a type of key (Fig. *c*) belonging to a very old lock mechanism known in ancient Rome, probably introduced into Scandinavia during the Viking period, evidently once widespread in Europe, and probably of Asiatic invention, sometimes called a 'Chinese padlock'. This key has a slender handle corresponding to the pedicel, a straight broader shank corresponding to the calyx and a spreading lobed limb with holes in the middle corresponding to the limb of the corolla.

The chain of associations by which an originally sensible or at least understandable plant name may give rise to a nonsensical one can be illustrated by some European vernacular names for moonwort (*Botrychium lunaria*; Fig. *d*), a small fern with 4 to 7 pairs of fan-shaped leaflets (pinnae), crudely horseshoe-like in outline, along the midrib (rachis) of the leaf (frond). One of these is 'Unshoe-the-horse' as recorded by Culpepper in 1653: 'Moonwort is an herb which they say wil open locks and unshoo such horses as tread upon it . . . country people that I know, cal it *Unshoo the Horse*.' A variant of this is 'shoeless horse'. The superstition that moonwort possesses magical powers connected with locks and iron has been recorded in several European countries and gave rise to plant names such as German *Teufelsschlüssel* (latinized as *clavis diaboli*) and *Walpurgiskraut*, Swedish *låsort* (lockherb) and Norwegian *murulykjel* (i.e. mare's key), which connects with the Anglo-Saxon *mare*, Norse *mara*, for a nocturnal horse-riding female demon, preserved in 'nightmare'. They evidently derive from the unusual shape of the leaf, but the association seemed far-fetched until elucidated by Nordhagen. Searching for old locks in Norway in 1944 and 1945, he found three which had to be opened by keys having 3 to 5 pairs of symmetrical projections (Fig. *e*), thus resembling in plan the *Botrychium* leaf. This was evidently the magical herb used by a burglar tried for witchcraft at Valdres, Norway, in 1737–8. It is easy to see how, from being associated with iron keys and the opening of locks as well as with the moon, the moonwort, on account of its leaflet shape, ultimately became also associated with horseshoes and with loosening them. Vestiges of long-vanished beliefs founded on sympathetic magic and the doctrine of signatures may thus linger in local vernacular names for plants.

William Turner, mentioned above as the pioneer sixteenth-century recorder of English plant names, could find none for several plants: 'I have not hearde that it is named in englishe', 'it hathe no name in englishe that I knowe', 'howe that it is named in englishe, as yet I can not tel', he had several times to confess. He remedied this lack by introducing names of his own, mostly translated or modified from Greek, German or French. Thus of *Larix*, then unknown in Britain, he wrote 'the duch men [i.e. Germans] cal Laricem ein larchen baume. . . . It may be called in englishe a Larche tree', and 'larch' in English it has remained. Of *Osyris* (i.e. *Linaria vulgaris*, toadflax) he stated that it 'groweth plenteously in Englande, but I do not remember what name it hath. If it have no name it may be called in englishe Lynary or todes flax, for the Poticares cal it Linariam and the duch [Germans] cal it Krotenflaks.' Similarly 'Barba hirci named in greeke Tragopogon . . . may be called in englishe gotes bearde'. The enrichment of the English language by the deliberate coinage of new names or the adaptation of foreign names thus initiated by Turner has continued to the present time. The herbalist John Gerard (1545–1612) was an early coiner of

d: leaf of moonwort ('Teufelsschlüssel; *Botrychium lunaria*)
e: key of old Norwegian lock (after Nordhagen, 1947)
          (*Drawings by Roberta E. Young*)

such names, including 'traveller's joy' (*Clematis*). 'Shrimp plant' for *Beloperone* (or *Drejerella*) *guttata* (alias *Justicia brandegeana*) and 'Jade vine' for *Strongylodon macrobotrys*, for example, are among the welcome newcomers.

Since names are a means of communication and vernacular names belong to long oral tradition, passing from mouth to mouth without the intermediary of writing, they should be easy to hold in the mind and accordingly must have had originally some apt or reasonable association with the plant concerned, although this association may later be obscured or eliminated by transfer to some other plant or by corruption of the word itself as part of general linguistic change. Such associations may be very diverse. These include evident characters of the plant itself, e.g. 'arrowhead' or *Sagittaria* referring to leaf shape, 'reed-mace' or 'cat's-tail' (*Typha*), and 'mouse-tail' or *Myosurus* referring to inflorescence, 'tway-blade' (*Listera*) referring to paired leaves, 'woman's tongue' (*Albizia lebbek*) referring to long continually clattering pods, 'bindweed' or *Convolvulus* referring to habit of growth, 'rattle-box' or *Crotalaria* referring to rattling pods, 'buttercup' (*Ranunculus*) referring to flowers, 'ice-plant' (*Cryophytum*) from a frosted appearance, 'naked ladies' (*Colchicum*) referring to slender pink flowers bare of leaves; special properties including taste or smell, for example 'stinkhorn' (*Phallus*), 'soapwort' (*Saponaria*), 'skunk-cabbage' (*Symplocarpos*); 'bittersweet' (*Solanum dulcamara*), 'piss-a-bed' (*Taraxacum officinale*); place of growth, for example 'bog asphodel' (*Narthecium ossifragum*), 'cornflower' (*Centaurea cyanus*), 'wallflower' (*Cheiranthus cheiri*), 'mountain laurel' (*Kalmia*), 'marsh mallow' (*Althaea officinalis*), from the mucilaginous roots of which housewives formerly made their marsh-mallow sweetmeats; time of flowering, for example, pasque-flower (*Pulsatilla*). Others may refer to personages, real or mythical, for example 'herb-robert'; 'timothy', 'joe-pye-weed' and 'geiger tree'. Real or supposed medicinal virtues, often linked with the doctrine of signatures, for example lungwort or *Pulmonaria*, form the basis of some names. The prevalence of medicinal plant names in North America, where grow asthma-weed, bellyache-root, Indian physic, boneset, ague-grass, kill-wart, fever-twig, colic root, cramp bark, mouth-root, consumption weed, toothache tree and others, testify to the pioneer period of self-medication with wild herbs, some of course derived from Indian herb-lore. Names referring to places often indicate whence a given plant was obtained, for example 'alsike' (*Trifolium hybridum*) from Alsike in Sweden, 'peach' (Latin *persica*) from Iran (Persia), into which in pre-Roman times it had been introduced from China, sisal (*Agave sisalana*) from Sisal, Mexico, and 'Madagascar periwinkle' (*Catharanthus roseus*): they can, however, be quite misleading, for example 'Jerusalem artichoke', 'Jerusalem cowslip' and 'Mexican bamboo', or become almost unintelligible, for example the West Indian 'dasheen' supposed to be contracted from '*madere de Chine*'.

Many plant names refer to animals. These are often derogatory and attached to the inferior of two kinds, for example toadflax (*Linaria*) contrasted with flax itself (*Linum*), horsechestnut (*Aesculus*) contrasted with sweet chestnut (*Castanea*), dog violet (*Viola canina*) as contrasted with sweet violet (*Viola odorata*). They may also refer to some character recalling that of an animal, for example 'hound's tongue' or *Cynoglossum*, 'adder's tongue' or *Ophioglossum*, 'cat's-tail' (*Typha*), 'rattle-snake fern' (*Botrychium virginianum*), 'tickseed' (*Coreopsis*), or to a common habitat, for example 'frog lily' (*Nuphar advena*), 'buffalo-berry' (*Shepherdia argentea*). Yet others may simply indicate the liking of an animal for that kind of plant, for example 'hogweed' (*Heracleum spondylium*), 'catmint' (*Nepeta cataria*), 'partridge-berry' (*Mitchella repens*), 'butterfly-bush' (*Buddleja*), 'bee-balm' (*Melissa* and *Monarda*).

On the contrary some, such as 'lamb-kill' (*Kalmia*) and 'wolf's bane' or *Lycoctonum* (*Aconitum*), refer to their poisonous nature. 'Dogwood', however, has nothing to do with dogs and refers to its use as a skewer wood; it is cognate with 'dagger' from French *dague*.

Some names of course are fanciful or playful, such as 'love-in-a-mist' for *Nigella damascena*, 'love-in-idleness', 'jump-up-and-kiss-me', 'johnny-jump-up' and 'kiss-me-at-the-garden-gate' for *Viola tricolor*; 'welcome-home-husband-though-never-so-drunk' for *Euphorbia cyparissias* and *Sedum acre*, 'love-lies-bleeding' for *Amaranthus caudatus*.

Others may preserve obsolete or obsolescent words. Thus the word 'loblolly' formerly referred to a kind of thick gruel or porridge and was also applied to a loutish person who could be described as 'soft and useless'; hence in the West Indies a tree with rather soft and useless wood as *Cupania* received in the seventeenth and eighteenth centuries the uncomplimentary name 'loblolly tree' or 'loblolly wood' and in the Southern United States a few notable species often growing in soft damp places, such as the loblolly bay (*Gordonia lasianthus*) and the loblolly pine (*Pinus taeda*), both soft-wooded, as well as the loblolly magnolia (*Magnolia grandiflora*), were likewise designated.

Since it is usually easier to adopt or adapt existing names than to coin new ones, settlers in a new area often apply the names of plants of their homeland to others which in some way resemble them although possibly not closely related. Thus the American evening-primrose (*Oenothera*) has nothing but a yellow colour in common with the British primrose (*Primula*). The mountain laurels (*Kalmia* and *Rhododendron*) of North America agree with the European laurels only in their foliage. The Moreton Bay chestnut (*Castanospermum*) has large brown seeds in general appearance somewhat like those of chestnut (*Castanea*) or horsechestnut, but is not otherwise like either. The white beech (*Gmelina*) of eastern Australia is botanically far removed from the European beech (*Fagus*) but has a straight grey-barked trunk recalling that of beech. Only taste and smell connect the *Eucalyptus* species called 'peppermint' in Australia with the herbaceous peppermint (*Mentha*). Only in name does *Eucalyptus regnans*, called 'mountain ash' in Australia, resemble the mountain ash or rowan (*Sorbus aucuparia*) of Britain. Usually plants, such as species of *Eucalyptus*, having no counterparts elsewhere, receive new names or have their aboriginal names retained or modified; to the one class belong such names as 'stringybark', 'woollybutt', 'ghost gum', 'blackwood', 'bloodwood', 'blackbutt', 'scribbly gum', 'messmate', 'tallowwood'; to the other, 'marri', 'karri', 'bangaley', 'coolabah', 'jarrah', 'wandoo'.

A single name suffices when only one of a kind is known but has to be qualified by adding a distinguishing word or words when rather similar but distinct other ones come to notice. Thus the names 'ash' and *fraxinus* were adequate and precise when *Fraxinus excelsior* was the only recognized species and served for many centuries as combined generic and specific names (mononomials) for this species. The discovery of other species has led to *Fraxinus* becoming a scientific generic name covering these also and to the use of the name 'European ash' (*Fraxinus excelsior*) for this and of 'White ash' (*F. americana*), 'red ash' (*F. pennsylvanica*), 'water ash' (*F. caroliniana*) and 'black ash' (*F. nigra*) for other species. The significant fact about these vernacular specific names and their scientific equivalents is that they are two-word names or binomials consisting of a one-word generic name, 'ash' or *Fraxinus*, associated with a one-word specific epithet such as 'white' or *americana*, 'black' or *nigra*; moreover, they are simple designations, not mutually exclusive descriptive definitions. This binomial system of naming must go back to the very early stage of human speech and thought when men first found it necessary to distinguish several rather similar kinds of objects. It occurs, for example, in languages as unrelated as Greek, with species of poppy (*mēkōn*), for example, distinguished by Theophrastus as *mēkōn ē mĕlaina, mēkōn ē kĕratitis* and *mēkōn ē rhŏias*, and Malay, with species of *Eugenia* (*kelat*) distinguished as *kelat hitam, kelat jambu, kelat jangkang, kelat layu* and *kelat merah*; other languages likewise provide examples. Linnaeus coined Swedish binomial vernacular names for species of grasses in his *Flora Suecica* of 1745 before he applied binomial scientific names to them in his *Species Plantarum* of 1753. The two sets of names are more or less parallel, for example: 95. *Får-swingel* (that is sheep-fescue), *Festuca ovina*; 93. *Rōd-swingel* (red-fescue), *Festuca rubra*; 92. *Ång-swingel* (meadow fescue), *Festuca elatior*; 90. *Swin-swingel* (swine-fescue), *Festuca fluitans*; 84. *Rag-losta* (rye-brome), *Bromus secalinus*; 85. *Ren-losta* (fieldside brome), *Bromus arvensis*; 86. *Tak-losta* (roof-brome), *Bromus tectorum*; 88. *Läng-losta* (tall-brome), *Bromus giganteus*; 89. *Spärr-losta* (ratchet-brome), *Bromus pinnatus*.

Thus, as has been said elsewhere, Linnaeus by his introduction in 1753 of consistent two-word names of Latin form for species thereby brought the scientific nomenclature of plants back to the simplicity and brevity of the vernacular nomenclature out of which it had grown, back to the kinds of names used from time immemorial by peasants, woodmen, herdsmen, hunters and herb-gatherers in many lands. Scientific and vernacular names should be regarded as complementary; neither can supersede the other for all purposes.

Experience has everywhere made evident the need for vernacular names adapted to everyday speech and their association with botanical names serving as reference points. Such a need develops with increasing knowledge of more and more kinds of plants and the desire to communicate and perpetuate this; it reflects a relatively high stage of advancement in the popularisation of learning. This explains the independent development of parallel systems of plant nomenclature in Europe and Japan. Down to the middle of the eighteenth century, as Philipp Franz von Siebold noted in 1825 when resident in Japan, empirical botany had reached as high a level there as in Europe. Siebold found that the Japanese possessed a remarkably detailed knowledge of their wild and cultivated plants, had produced much herbalistic and botanical literature about them and had given them precise and stable names in accordance with a dual system of nomenclature. They had coined vernacular Japanese names for everyday use comparable to European vernacular names and associated with these names they had learned names given a so-called Chinese reading and written in Chinese characters which were comparable in their use and status to the Latin-based botanical names of Europe. Siebold recorded, for example, the Japanese name of the *Hydrangea* now called *H. macrophylla* as *azisai* (now rendered *ajisai*) and its Chinese equivalent as *zu-hats-sen* (now *shū hassen*) with Chinese idiographs now romanized as *chü pa hsien*. The Japanese thus found it easy to correlate their own vernacular names with a third set of names, that is the European botanical names made known to them by Thunberg, Siebold, Maximowicz, Savatier and Franchet, since they could substitute these for their Chinese learned names as international standards of reference and link them with excellent woodcut illustrations in such works as the *kwa-wi*, *Honzo dzufu* and *Somoku dzusetsu*, while retaining and indeed adding to their own Japanese vernacular names. In consequence they made more rapid progress in the study of their native plants and the adoption of Western systems of classification and nomenclature than any other Oriental nation. Romanized versions of these Japanese names are given along with botanical names throughout Ohwi's *Flora of Japan* (1965).

Because the diversity and abundance of local names and their varied applications necessarily restrict their use for general purposes, it has now become usual to select from among these one or two vernacular names as standard equivalents for the scientific names of conspicuous, interesting or economically important species. Such names are those adopted in Bailey's *Manual of Cultivated Plants*, 2nd ed. (1949), Fernald's *Gray's Manual of Botany*, 8th ed. (1950), Clapham, Tutin & Warburg's *Flora of the British Isles*, 2nd ed. (1962), Correll & Johnston's *Manual of the Vascular Plants of Texas* (1970) and like comprehensive works. Their acceptance, which is undoubtedly desirable for convenience and precision, must lead sooner or later to the disappearance of the others; since these may be just as rich in linguistic and folklore associations and apt for the plants themselves, they should not be regarded with contempt and allowed to pass away unrecorded and unexplained.

W. T. S.

February 1971

Some Sources of Further Information

Bartlett, H. H. 1940. History of the generic concept in botany. *Bull. Torrey Bot. Club* 67: 349–362.

Bartlett, H. H. 1953. English names of some East Indian plants and plant products. *Asa Gray Bull.* 2: 149–176.

Britten, J. & Holland, R. 1878–86. *A Dictionary of English Plant-Names.* 2 vols. London.

Cameron, J. 1883. *Gaelic Names of Plants, Scottish and Irish.* Edinburgh.

Cassidy, F. G. 1953. Language and folklore. *Caribbean Quarterly* 3: 4–12.

Cassidy, F. G. & Le Page, R. B. 1967. *Dictionary of Jamaican English.* Cambridge (Cambridge University Press).

Clute, W. N. 1923. *A Dictionary of American Plant Names.* Joliet, Illinois.

Clute, W. N. 1939. *A Second Book of Plant Names and their Meanings.* Indianapolis.

Clute, W. N. 1942. *The Common Names of Plants and their Meanings.* 2nd ed. Indianapolis.

Comber, T. 1876. The etymology of plant names. *Trans. Hist. Soc. Lancash. & Cheshire* 28: 15–71 (1876), 29: 43–83 (1877), 30: 1–38 (1878).

Dawkins, R. M. 1936. The semantics of Greek names for plants. *Journal of Hellenic Studies* 56: 1–11.

Earle, J. 1880. *English Plant Names from the tenth to the fifteenth Centuries.* Oxford.

Fernald, M. L. 1942. Some historical aspects of plant taxonomy. *Rhodora* 44: 21–43.

Fisher, R. 1932–4. *The English Names of our commonest wild Flowers.* Arbroath.

Friend, H. 1884. *Flowers and Flower Lore.* 2nd ed. London.

Egli, M. 1930. *Benennungsmotive bei Pflanzen an schweizerdeutschen Pflanzennamen untersucht.* Bulach.

Gerth van Wijk, H. L. 1911. *Dictionary of Plant-Names.* 2 vols. The Hague. Reprinted 1971.

Grigson, G. 1958. *The Englishman's Flora.* London (Phoenix House).

Guyot, L. & Gibassier, P. 1960. *Les Noms des Plantes (Que sais-je 856)* Paris (Presses Universitaires de France).

Guyot, L. & Gibassier, P. 1960. *Les Noms des Arbres (Que sais-je 861).* Paris (Presses Universitaires de France).

Lange, J. 1959–61. *Ordbog over Danmarks Plantenavne.* 3 vols. Copenhagen (Munksgaard).

Some Sources of Further Information

Laurell, F. 1904. *Svenska Växtnamn och binä Nomenklatur.* Uppsala.

Leighton, A. (Mrs. I. L. L. Smith). 1970. *Early English Gardens in New England.* London (Cassell).

McClintock, D. 1966. *Companion to Flowers.* London (G. Bell).

Marzell, H. 1937 et seq. *Wörterbuch der Deutschen Pflanzennamen.* Leipzig (Hirzel).

Maxwell, H. 1923. *Flowers, a Garden Notebook.* Glasgow.

Nordhagen, R. 1948. Studien over gamle plantenavn: marilykjel, springstrå og jennurt. *Bergens Mus. Årb.* 1946–47, Hist. Antikv. no. 3.

Prior, R. C. A. 1879. *On the Popular Names of British Plants.* 3rd ed. London.

Rolland, E. 1896–1914. *Flore populaire.* 11 vols. Paris.

Ross, A. S. C. 1952. *Ginger; a loan-word Study.* London (Philological Society).

Ross, A. S. C. 1958. *Etymology with especial Reference to English.* London (André Deutsch; 1965, Methuen).

Siebold, Ph. F. von. 1828. Einige Worte über den Zustand der Botanik auf Japan. *Nova Acta Acad. Nat. Curios. Caes-Leopold-Carol.* 14: 671–693.

Smith, P. G. & Welch, J. E. 1964. Nomenclature of vegetables and condiment herbs grown in the United States. *Proc. Amer. Soc. Hort. Sci.* 84: 535–548.

Smith, T. A. 1966. Common Names of South African Plants. *Republic S. Afr. Bot. Survey Mem.* 35. Pretoria.

Sprague, T. A. 1957. English plant-names. *Proc. Cotteswold Nat. Field Club* 32, ii: 22–27.

Stromberg, R. 1940. *Griechische Pflanzennamen (Göteborgs Högsk. Årsskr.* 46). Göteborg.

Turner, W. 1965. *Libellus de Re herbaria, 1538: The Names of Herbes, 1548 Facsimiles with introductory matter by J. Britten, B. D. Jackson and W. T. Stearn.* London (Ray Society).

A

absinthium	*Artemisia absinthium*
acacia	*Acacia*
acacia, false	*Robinia pseudacacia*
acacia, rose	*Robinia hispida*
aconite	*Aconitum*
aconite, winter	*Eranthis*
Adam's needle	*Yucca smalliana*
adder's mouth	*Pogonia*
adder's tongue	*Erythronium*
adder's tongue fern	*Ophioglossum*
African daisy	*Arctotis; Lonas*
African corn-lily	*Ixia*
African cypress	*Widdringtonia*
African lily	*Agapanthus*
African marigold	*Tagetes*
African peach	*Nauclea latifolia*
African ragwort	*Othonna*
African tuliptree	*Spathodea campanulata*
African valerian	*Fedia*
African violet	*Saintpaulia*
agrimony	*Agrimonia*
ague-weed	*Eupatorium*
air-plant	*Aerides; Bryophyllum pinnatum*
akee tree	*Blighia sapida*
alder	*Alnus*
alderbuckthorn	*Frangula alnus*
alder, white	*Clethra*
alexanders	*Smyrnium olusatrum*
Alexandrian laurel	*Danaë racemosa*
alfalfa	*Medicago sativa*
Alice, sweet	*Pimpinella anisum*
alkanet	*Alkanna tinctoria*
Allegheny vine	*Adlumia fungosa*
alligator pear	*Persea*
allspice	*Calycanthus floridus; Pimenta dioica*
allspice tree	*Pimenta diocia*
almond	*Prunus dulcis (Amygdalus communis)*
almond, dwarf	*Prunus tenella*
alsike clover	*Trifolium hybridum*
alum-root	*Heuchera*
alyssum, sweet	*Lobularia maritima*
amaranth, Chinese	*Amaranthus tricolor*
amaranth, globe	*Gomphrena globosa*
amaryllis	*Amaryllis; Hippeastrum*
amaryllis, blue	*Worsleya procera*
Amazon lily	*Eucharis grandiflora*
American cowslip	*Dodecatheon*
American cranberry	*Vaccinium macrocarpon*
American cress	*Barbarea*
American feverfew	*Chrysanthemum parthenium*
American pennyroyal	*Hedeoma*
American wayfaring-tree	*Viburnum alnifolium*
anchor-plant	*Colletia cruciata*
Andromeda	*Pieris*
anemone	*Anemone*
anemone, Japanese	*Anemone hybrida*
anemone, poppy	*Anemone coronaria*
anemone, snowdrop	*Anemone sylvestris*
anemone, wood	*Anemone nemorosa*
angel's trumpet	*Datura suaveolens*
angelica	*Angelica anchangelica*
angelica-tree	*Aralia spinosa*
angelica-tree, Chinese	*Aralia chinenis*
angelica-tree, Japanese	*Aralia japonica*
angelica, wild	*Angelica sylvestris*
anise	*Pimpinella anisum*
anise, sweet	*Foeniculum vulgare*
aniseed tree	*Illicium*
annatto	*Bixa orellana*
Antarctic forget-me-not	*Myosotidium hortensia*
apple	*Malus pumila*
apple berry	*Billardiera*
apple-of-Peru	*Nicandra physalodes*
apple, Otaheite	*Spondias dulcis*

355

apricot	*Prunus amygdalus*	autumn crocus	*Colchicum*
		avens	*Geum urbanum*
apricot, Japanese	*Prunus mume*	avens, mountain	*Dryas octopetala*
aralia, false	*Dizygotheca*	avocado	*Persea americana*
arbor-vitae	*Thuja*	azalea	*Rhododendron*
arbutus, trailing	*Epigaea repens*		(deciduous
arrow-arum	*Peltandra*		species)
arrowhead	*Sagittaria*		
arrowroot	*Maranta arundinacea*		

B

arrow-wood	*Viburnum dentatum*	baboon-root	*Babiana*
artichoke, Chinese	*Stachys affinis*	baby blue-eyes	*Nemophila menziesii*
artichoke, globe	*Cynara scolymus*		
artichoke, Japanese	*Stachys affinis*	baby's breath	*Gypsophila paniculata*
artichoke, Jerusalem	*Helianthus tuberosus*	baby's tears	*Soleirolia soleirolii*
artillery plant	*Pilea microphylla*	bachelor's button	Name applied to
arum-lily	*Zantedeschia*		double-flowered
ash	*Fraxinus*		forms of Ranun-
asoka tree	*Saraca indica*		culus and Ach-
asparagus	*Asparagus officinalis*		illea, also to Centaurea and
asparagus fern	*Asparagus plumosus*		Craspedia
		bael fruit	*Aegle marmelos*
asparagus pea	*Tetragonolobus purpureus*	bald cypress	*Taxodium distichum*
aspen	*Populus tremula*	baldmoney	*Meum athamanticum*
asphodel	*Asphodelus*		
asphodel, yellow	*Asphodeline*	balloon flower	*Platycodon*
aster, annual ⎫	*Callistephus*	balloon vine	*Cardiospermum halicacabum*
aster, China ⎭	*chinensis*		
aster, Stokes'	*Stokesia laevis*	balm	*Melissa officinalis*; *Ocimum basilicum*
aster, tree	*Olearia*		
astilbe	*Astilbe*	balm, bastard	*Melittis melissophyllum*
Aunt Eliza	*Curtonus* (formerly *Antholyza*)		
		balm, bee	*Melissa officinalis*
Australian currant	*Leucopogon*	balm, lemon	*Melissa officinalis*
		balm-of-Gilead	*Cedronella tri-phylla*; *Populus gileadensis*
Australian Desert kumquat	*Eremocitrus*		
Australian honeysuckle	*Banksia*	balsam	*Impatiens*
		balsam-apple	*Momordica balsamina*
Australian laurel	*Pittosporum*		
Australian rosemary	*Westringia*	balsam-pear	*Momordica charantia*
Australian sarsaparilla	*Hardenbergia violacea*	balsam poplar	*Populus balsamifera*
Australian sword lily	*Anigozanthos*	bamboo	*Bambusa, Sasa, Thamnocalamus,* etc.
Austrian briar	*Rosa foetida*		
Austrian copper briar	*Rosa foetida bicolor*	bamboo, heavenly	*Nandina domestica*
		banana	*Musa*

baneberry	*Actaea*	beard grass	*Andropogon*
banyan	*Ficus benghalensis*	beard-tongue	*Penstemon*
baobab tree	*Adansonia digitata*	bear's breech	*Acanthus*
Barbados-cherry	*Malpighia*	beauty-berry	*Callicarpa*
Barbados-gooseberry	*Pereskia aculeata*	beauty bush	*Kolkwitzia amabilis*
Barbados-pride	*Caesalpinia pulcherrima*	bedstraw	*Galium*
		bee-balm	*Monarda*
barberry	*Berberis*	beech	*Fagus*
bark, paper	*Melaleuca*	beech fern	*Thelypteris phegopteris*
barley	*Hordeum*		
barna, sacred	*Crateva nurvala*	beech, southern	*Nothofagus*
barren strawberry	*Waldsteinia*	beefsteak plant	*Perilla*
barrenwort	*Epimedium*	beefwood	*Casuarina*
basket willow	*Salix viminalis*	beet	*Beta vulgaris*
basil	*Ocimum basilicum*	beggar's ticks	*Bidens*
bass-wood	*Tilia*	bell, Canterbury	*Campanula medium*
bastard balm	*Melittis melissophyllum*	belladonna	*Atropa bella-donna*
bastard jasmine	*Cestrum*	bellbine	*Calystegia*
bastard pennyroyal	*Trichostema*	bellflower	*Campanula; Wahlenbergia*
bats-in-the-belfry	*Campanula trachelium*	bellflower, Chilean	*Nolana*
bay	*Laurus*	bellflower, Chinese	*Platycodon*
bay, sweet	*Magnolia virginiana*	bellflower, giant	*Ostrowskia magnifica*
bayberry	*Myrica pensylvanica; Pimenta racemosa*	bellflower, gland	*Adenophora*
		bellflower, pendulous	*Symphyandra*
bay-tree	*Laurelia; Laurus*	bell-heather	*Erica cinerea*
bay willow	*Salix pentandra*	bellwort	*Uvularia*
beach grass	*Ammophila*	belvedere	*Kochia*
beach heather	*Hudsonia*	Benjamin bush	*Lindera benzoin*
beach pea	*Lathyrus*	bent grass	*Agrostis*
beach plum	*Prunus maritima*	benzoin	*Lindera benzoin*
bead-plant	*Nertera granadensis*	bergamot	*Monarda didyma*
bead-tree	*Melia azedarach*	Bermuda buttercup	*Oxalis pes-caprae*
bean, asparagus	*Vigna sinensis sesquipedalis*	Bermuda grass	*Cynodon dactylon*
bean, broad	*Vicia faba*	betelnut	*Areca catechu*
bean, Jack	*Canavalia ensiformis*	Bethlehem sage	*Pulmonaria*
bean, kidney	*Phaseolus vulgaris*	betonica	*Stachys officinalis*
bean, Lima	*Phaseolus limensis*	betony	
		betony, wood	*Pedicularis*
bean, runner	*Phaseolus coccineus*	bhendi tree	*Thespesia populnea*
bean, soya	*Glycine max*	bidgee-widgee	*Acaena anserinifolia*
bean, yam	*Pachyrhizus*		
bearberry	*Arctostaphylos*	bididy-bid	*Acaena*
bear-grass	*Camassia; Nolina*	big tree	*Sequoiadendron giganteum*
bear-tongue	*Clintonia*		

357

bilberry	*Vaccinium myr-tillus* (Britain) *Vaccinium uligin-osum* (U.S.A.)	black wattle	*Acacia decurrens mollis*
		black willow	*Salix nigra*
		bladder campion	*Silene vulgaris*
bilimbi	*Averrhoa bilimbi*	bladder-cherry	*Physalis*
		bladder fern	*Cystopteris*
bindweed	*Convolvulus Calystegia*	bladdernut	*Staphylea*
		bladder senna	*Colutea*
birch	*Betula*	bladderwort	*Utricularia*
birch, canoe	*Betula papyrifera*	blanket-flower	*Gaillardia*
birch, cherry	*Betula lenta*	blazing star	*Chamaelirium luteum*; *Liatris*; *Tritonia*
birch, dwarf	*Betula nana*		
birch, grey	*Betula populifolia*		
birch, paper	*Betula papyrifera*	bleeding-heart	*Dicentra*
birch, silver	*Betula pendula*	blessed thistle	*Cnicus benedictus*; *Silybum marianum*
birch, yellow	*Betula lutea*		
bird cherry	*Prunus padus*		
bird-of-paradise flower	*Strelitzia*	blistercress	*Erysimum*
		blood-leaf	*Iresine*
bird, pepper	*Capsicum annuum*	blood-lily	*Haemanthus*
bird's foot	*Ornithopus perpusillus*	bloodroot	*Sanguinaria canadensis*
bird's foot trefoil	*Lotus corniculata*	bloodwood tree	*Haematoxylum campechianum*
birthroot	*Trillium*		
birthwort	*Aristolochia*	bluebead-lily	*Clintonia borealis*
bishopsweed	*Aegopodium podagraria*	bluebell	*Campanula* (Scotland); *Mertensia* (Virginia); *Endymion* (England); *Wahlenbergia* (Australia)
bistort	*Polygonom*		
bitterbush	*Picramnia pentandra*		
bitternut	*Carya cordiformis*		
bittersweet	*Celastrus scandens*; *Solanum dulcam-ara*		
		bluebell, California	*Phacelia*
bitterwort	*Lewisia*	blueberry	*Vaccinium*
blackberry	*Rubus*	bluebonnet, Texas	*Lupinus subcarnosus*
blackberry lily	*Belamcanda chinensis*		
		bluebush	*Kochia*
blackboy	*Xanthorrhoea*	blue cohosh	*Caulophyllum thalictroides*
black bryony	*Tamus communis*		
black cohosh	*Cimicifuga*	bluecurls	*Trichostema*
black-eyed Susan	*Rudbeckia hirta*; *Rudbeckia sero-tina*; *Thunbergia alata*	blue daisy	*Felicia amelloides*
		blue-eyed grass	*Sisyrinchium*
		blue-eyed Mary	*Collinsia*
		blue gum	*Eucalyptus*
black horehound	*Ballota*	blue lace-flower	*Trachymene caerulea*
black lily	*Fritillaria camschatcensis*		
		blue oxalis	*Parochetus communis*
black medick	*Medicago*		
black mulberry	*Morus nigra lupulina*	blue palmetto	*Rhapidophyllum*
		blue poppy	*Meconopsis*
black oak	*Quercus velutina*	blue sowthistle	*Cicerbita*
black poplar	*Populus nigra*	blue spiraea	*Caryopteris*

358

blue wattle	*Acacia cyanophylla*	broom, sweet	*Scoparia dulcis*
bluet	*Houstonia*	bryony	*Bryonia*
bogbean	*Menyanthes trifoliata*	bryony, black	*Tamus communis*
bog heather	*Erica tetralix*	bryony, white	*Bryonia dioica*
bog myrtle	*Myrica*	buckbean	*Menyanthes trifoliata*
bog-orchid	*Arethusa bulbosa; Calypso bulbosa*	bucket orchid	*Coryanthes*
bog rosemary	*Andromeda glaucophylla*	buckeye	*Aesculus*
		buckthorn	*Rhamnus*
boneset	*Eupatorium perfoliatum*	buckthorn, alder	*Frangula alnus*
		buckthorn, sea	*Hippophae rhamnoides*
borage	*Borago officinalis*	buckwheat	*Fagopyrum esculentum*
Boston ivy	*Parthenocissus tricuspidata*	buffalo-berry	*Shepherdia argentea*
Botany Bay gum	*Xanthorrhoea*	bugbane	*Cimicifuga*
bottlebrush tree	*Callistemon; Melaleuca*	bugle	*Ajuga*
		bugle lily	*Watsonia*
bottle-gourd	*Lagenaria*	bugleweed	*Ajuga*
bottle-tree	*Brachychiton*	bugloss	*Anchusa*
bouncing Bet	*Saponaria officinalis*	bugloss, viper's	*Echium*
		bullace	*Prunus spinosa insititia*
bowstring hemp	*Sansevieria*		
box	*Buxus; Eucalyptus (Australia)*	bulbous chervil	*Chaerophyllum*
		bullock's heart	*Annona reticulata*
boxberry	*Gaultheria procumbens*	bulrush	*Scirpus; Typha (Britain)*
box-elder	*Acer negundo*		
box, sweet	*Sarcococca*	bunchberry	*Cornus*
box-thorn	*Lycium*	bunchflower	*Melanthium*
bracken; brake	*Pteridium aquilinum; Pteris*	bur, New Zealand	*Acaena*
		bur-cucumber	*Sicyos angulatus*
bramble	*Rubus*	burdock	*Arctium*
brasiletto	*Caesalpinia vesicaria*	bur-marigold	*Bidens*
		burnet	*Poterium; Sanguisorba minor*
Brazil-nut	*Bertholletia*		
breadfruit	*Artocarpus altilis*	burnet rose	*Rosa spinosissima*
bread-nut	*Brosimum alicastrum*	burning bush	*Dictamnus albus; Euonymus atropurpureas*
bridal wreath	*Francoa; Spiraea*		
bristlecone pine	*Pinus aristata*	burstwort	*Herniaria*
broad bean	*Vicia faba*	bush-clover	*Lespedeza*
brome grass	*Bromus*	bush-honeysuckle	*Diervilla*
Brompton stock	*Matthiola incana*	bush-pea	*Pultenaea*
brooklime	*Veronica beccabunga*	butcher's broom	*Ruscus aculeatus*
		butter-and-eggs	*Linaria vulgaris*
brookweed	*Samolus valerandi*	butter-bur	*Petasites*
broom	*Cytisus; Genista; Spartium*	buttercup	*Ranunculus*
		buttercup, Bermuda	*Oxalis pes-caprae*
broom, butcher's	*Ruscus aculeatus*		
broom, hedgehog	*Erinacea anthyllis*	butterfly bush	*Buddleja davidii*
broom, Spanish	*Spartium junceum*		

359

butterfly flower	*Bauhinia monandra* (tropics) *Schizanthus*	canary creeper	*Tropaeolum peregrinum*
butterfly iris	*Moraea*	canary-grass	*Phalaris canariensis*
butterfly pea	*Clitoria ternatea*; *Centrosema*	candle-plant	*Kleinia articulata*
butterfly-weed	*Asclepias tuberosa*	candle-tree	*Parmentiera cereifera*
butternut	*Juglans cinerea*	candlewick	*Verbascum thapsus*
butterwort	*Pinguicula*		
button-bush	*Cephalanthus*	candlewood	*Fouquiera*
buttonwood	*Platanus occidentalis*	candy carrot	*Athamanta*
		candy mustard	*Aethionema*
Buxbaum's speedwell	*Veronica persica*	candytuft	*Iberis*
		cane, dumb	*Dieffenbachia seguine*
		cane palm	*Calamus*
		cane, southern	*Arundinaria*
C		cannon-ball tree	*Couroupita guianensis*
cabbage	*Brassica oleracea capitata*	canoe birch	*Betula papyrifera*
cabbage rose	*Rosa centifolia*	cantaloupe	*Cucumis melo*
cabbage tree	*Andira*; *Cordyline*	Canterbury-bell	*Campanula medium*
Cacao	*Theobroma cacao*		
calaba tree	*Calophyllum*	Cape chestnut	*Calodendrum capense*
calabash gourd	*Lagenaria ciceraria*	Cape cowslip	*Lachenalia*
calabash tree	*Crescentia cujete*	Cape figwort	*Phygelius capensis*
calamint	*Calamintha*	Cape gooseberry	*Physalis*
calico-bush	*Kalmia*	Cape honeysuckle	*Tecomaria capensis*
California bluebell	*Phacelia*	Cape jasmine	*Gardenia jasminoides*
California golden bells	*Emmenanthe*	Cape marigold	*Dimorphotheca*
		Cape pondweed	*Aponogeton*
California laurel	*Umbellularia californica*	Cape primrose	*Streptocarpus*
		Cape stock	*Heliophila*
California poppy	*Eschscholzia*	caper	*Capparis spinosa*
		caper spurge	*Euphorbia lathyris*
calla-lily	*Zantedeschia*	carambola	*Averrhoa carambola*
calvary-clover	*Medicago echinus*		
camas	*Camassia*	caraway	*Carum carvi*
camomile (see chamomile)		cardamon	*Elettaria cardamomum*
campeachy-wood	*Haematoxylum campechianum*	cardinal-flower	*Lobelia cardinalis*
		cardoon	*Cynara cardunculus*
campion	*Silene*		
campion, bladder	*Silene vulgaris*	carline thistle	*Carlina*
campion, evening	*Lychnis*	carnanba palm	*Copernica cerifera*
campion, moss	*Silene acaulis*	carnation	*Dianthus caryophyllus*
campion, red	*Silene dioica*		
campion, rose	*Lychnis coronaria*	carob	*Ceratonia siliqua*
campion, sea	*Silene maritima*	Carolina allspice	*Calycanthus floridus*
campion, white	*Silene alba*		
Canada Ted	*Gaultheria*	carrion-flower	*Smilax*; *Stapelia*
Canadian wild rice	*Zizania aquatica*	carrot	*Daucus carota*

cashew nut	*Anacardium occidentale*	chamomile, wild	*Matricaria recutita*
cassava, bitter	*Manihot esculenta*	chard, Swiss	*Beta vulgaris cicla*
cassava, sweet	*Manihot dulcis*	chaste tree	*Vitex*
castor bean; castor-oil plant	*Ricinus communis*	chayote	*Sechium edule*
catbrier	*Smilax*	checker-berry	*Gaultheria procumbens*
catchfly	*Lychnis; Silene*	cherimoya	*Annona cherimola*
catchfly, Nottingham	*Silene nutans*	Cherokee rose	*Rosa laevigata*
catjang	*Vigna sinensis cylindrica*	cherry	*Prunus*
		cherry birch	*Betula lenta*
catmint; catnip	*Nepeta cataria*	cherry, bird	*Prunus padus*
cat's claw	*Doxantha unguis-cati*	cherry-bladder	*Physalis*
		cherry, ground	*Physalis pruinosa*
cat's ear	*Hypochoeris radicata*	cherry, Jerusalem	*Solanum pseudo-capsicum*
cat-tail	*Phleum; Typha*	cherry-laurel	*Prunus laurocerasus*
cauliflower	*Brassica oleracea botrytis*	cherry-pie	*Heliotropium*
		cherry, wild	*Prunus avium*
cedar	*Cedrus*	chervil	*Anthriscus*
cedar, Atlas	*Cedrus atlantica*	chervil, bulbous	*Chaerophyllum bulbosum*
cedar, incense	*Libocedrus*		
cedar, Japanese	*Cryptomeria japonica*	chervil, salad	*Anthriscus cerefolium*
cedar of Lebanon	*Cedrus libani*	chervil, wild	*Cryptotaenia*
cedar, northern white	*Thuja occidentalis*	chestnut	*Castanea*
		chestnut, horse	*Aesculus*
cedar, red	*Juniperus virginiana*	chestnut, Spanish	*Castanea sativa*
		chestnut-oak	*Lithocarpus*
cedar, stinking	*Torreya taxifolia*	chick pea	*Cicer arietinum*
cedar, wattle	*Acacia elata*	chickweed	*Stellaria*
cedar, white	*Chamaecyparis thyoides*	chickweed, mouse-ear	*Cerastium*
celandine, greater	*Chelidonium majus*	chicory	*Cichorium intybus*
celandine, lesser	*Ranunculus ficaria*	Chilean bellflower	*Nolana*
celandine poppy	*Stylophorum diphyllum*	Chilean boldo tree	*Peumus boldus*
celery	*Apium graveolens dulce*	chili pepper	*Capsicum annuum*
		China aster	*Callistephus chinensis*
centaury	*Centaurium*	chinaberry	*Melia azedarach*
centaury, American	*Sabatia*	China-fir	*Cunninghamia lanceolata*
century plant	*Agave*	China-grass	*Boehmeria nivea*
cereus, night-blooming	*Selenicereus grandiflorus*	Chinese gooseberry	*Actinidia chinensis*
ceriman	*Monstera deliciosa*	Chinese hat	*Holmskioldia sanguinea*
chain-fern	*Woodwardia*		
chamomile	*Anthemis nobilis*	Chinese-lantern lily	*Sandersonia aurantiaca*
chamomile, corn	*Antnemis arvensis*		
chamomile, false	*Boltinta*	Chinese moonseed	*Sinomenium acutum*
chamomile, yellow	*Anthemis tinctoria*		

Chinese sacred lily	*Narcissus tazetta orientalis*	cockscomb	*Celosia argentea cristata*
chinkerinchee	*Ornithogalum thyrsoides*	cock's-foot grass	*Dactylus glomerata*
chives	*Allium schoenoprasum*	cockspur	*Echinochloa crus-galli*
chives, Chinese	*Allium tuberosum*	cockspur thorn	*Crataegus crus-galli*
choctaw root	*Apocynum cannabinum*	cocoa	*Theobroma cacao*
chokeberry	*Aronia*	coco-de-mer	*Lodoicea maldavica*
cholla	*Opuntia*	coconut	*Cocos nucifera*
Christmas-berry	*Heteromeles*; *Schinus terebinthifolius*	coconut, double	*Lodoicea maldavica*
Christmas-cactus	*Schlumbergera buckleyi*	coco-plum	*Chrysobalanus icaco*
Christmas fern	*Polystichum acrostichoides*	codlins-and-cream	*Epilobium hirsutum*
Christmas-rose	*Helleborus niger*	coffee	*Coffea*
Christ-thorn	*Paliurus spina-christi*	coffee, wild	*Psychotria*
chufa	*Cyperus esculentus*	cohosh	*Actaea*
cicely, sweet	*Myrrhis odorata*	cohosh, black	*Cimicifuga*
cigar-plant	*Cuphea ignea*	cohosh, blue	*Caulophyllum thalictroides*
cinnamon fern	*Osmunda cinnamomea*	coltsfoot	*Tussilago*
cinnamon vine	*Apios americana*	coltsfoot, alpine	*Homogyne alpina*
cinnamon, wild	*Canella*	columbine	*Aquilegia*
cinquefoil	*Potentilla*	Colville's glory	*Colvillea racemosa*
citron	*Citrus medica*; *Citrullus lanatus citroides*	comb fern	*Schizaea*
		comfrey	*Symphytum*
		comfrey, Russian	*Symphytum uplandicum*
citronella	*Collinsonia*	compass-plant	*Silphium laciniatum*
clary	*Salvia sclarea*		
cleavers	*Galium aparine*	cone-flower	*Echinacea*; *Lepachys*; *Rudbeckia*
cliff-brake fern	*Pellaea*		
climbing cactus	*Acanthocereus*	cone-plant	*Conophytum*
climbing-fern	*Lygodium*	Cootamunda wattle	*Acacia baileyana*
climbing palas	*Butea superba*	copper leaf	*Acalypha*
cloak fern	*Notholaena*	copper-pod	*Peltophorum pterocarpum*
clove tree	*Syzygium aromaticum*	coral tree	*Erythrina*
clover, alsike	*Trifolium hybridum*	coral vine	*Antigonum leptopos*
clover, bush	*Lespedeza*		
clover, hop	*Trifolium agrarium*	coriander	*Coriandrum sativum*
clover, Mexican	*Richardia scabra*		
clover, sweet	*Melilotus*	cork oak	*Quercus suber*
club of Hercules	*Aralia spinosa*; *Xanthoxylum clava-herculis*	cork-tree	*Phellodendron*
		corn	*Triticum* (Britain); *Zea mays* (U.S.A.);
club-moss	*Lycopodium*; *Selaginella*	corncockle	*Agrostemma githago*
cob-nut	*Corylus avellana*		

cornel	*Cornus*
cornel, dwarf	*Chamaeperi-clymem*
cornelian cherry	*Cornus mas*
cornflower	*Centaurea cyanus*
corn-marigold	*Chrysanthemum segetum*
corn mint	*Mentha arrensis*
corn poppy	*Papaver rhoeas*
corn-salad	*Valerianella olitoria*
Cornish heath	*Erica vagans*
costmary	*Chrysanthemum balsamita*
cottage pink	*Dianthus caryophyllus*
cotton	*Gossypium*
cotton, lavender	*Santolina chamaecyparissus*
cottongrass	*Eriophorum*
cotton-gum	*Nyssa aquatica*
cotton tree	*Bombax*
cotton-weed	*Diotis candidissima*
cottonwood	*Populus*
cowbane	*Cicuta virosa*
cow-itch	*Urera*
cow-parsley	*Anthriscus sylvestris*
cow-parsnip	*Heracleum*
cowpea	*Vigna sinensis*
cowslip	*Caltha palustris* (U.S.A.); *Mertensia* (Virginia); *Primula veris* (Britain)
cowslip, American	*Dodecatheon*
cow tree	*Brosimum*
crab-apple	*Malus*
crab-apple, Siberian	*Malus adstringens*; *Malus baccata*; *Malus prunifolia*
crack willow	*Salix fragilis*
cranberry	*Vaccinium oxycoccus*; *Vaccinium macrocarpon*
cranberry-bush	*Viburnum trilobum*
cranesbill	*Geranium*
crape-myrtle	*Lagerstroemia*
cream cups	*Platystemon californicus*
creeper, Rangoon	*Quisqualis indica*
creeping-Charlie	*Lysimachia nummularia*; *Pilea nummularifolia*
creeping-Jenny	*Lysimachia nummularia*
creeping willow	*Salix repens*
creosote bush	*Larrea*
cress, American	*Barbarea*
cress, bitter	*Cardamine*
cress, garden	*Lepidium sativum*
cress, Indian	*Tropaeolum majus*
cress, rock	*Arabis*
cress, spring	*Cardamine*
cress, stone	*Aethionema*
cress, upland	*Barbarea verna*
cress, water	*Rorippa*
cress, winter	*Barbarea*
crested poppy	*Argemone*
Cretan dittany	*Amaracus dictamnus*
cricket-bat, willow	*salix alba caerulen*
crocus	*Crocus*
cross-leaved heath	*Erica tetralix*
crocus, saffron	*Crocus sativus*
crocus, Autumn	*Colchicum*
cross-vine	*Bignonia capreolata*
crosswort	*Cruccia laevipes*
croton	*Codiaeum variegatum*
crowberry	*Empetrum nigrum*
crowfoot	*Ranunculus*
crow garlic	*Allium vineale*
crown-imperial	*Fritillaria imperialis*
crown-vetch	*Coronilla varia*
crystal wort	*Riccia*
cuckoo-flower	*Cardamine pratensis*
cuckoo-pint	*Arum maculatum*
cucumber	*Cucumis sativus*
cucumber, prickly	*Echinocystis lobata*
cucumber, squirting	*Ecballium elaterium*
cucuzzi	*Lagenaria siceraria*
cudweed	*Gnaphalium*
cup-flower	*Nierembergia rivularis*
Cupid's dart	*Catananche caerulea*

cup-and-saucer-flower	*Cobaea scandens*
curly-grass fern	*Schizaea*
currant	*Ribes*
currant, black	*Ribes nigrum*
currant, buffalo	*Ribes odoratum*
currant, flowering	*Ribes sanguineum*
currant, golden	*Ribes aureum*
currant, red	*Ribes sylvestris*
curry-leaf	*Murraya paniculata*
custard-apple	*Annona*
cypress	*Cupressus*
cypress, bald	*Taxodium distichum*
cypress, false	*Chamaecyparis*
cypress, Italian	*Cupressus sempervirens*
Cypress, Monterey	*Cupressus macrocarpa*
cypress, pond	*Taxodium ascendens*
cypress, summer	*Kochia scoparia*
cypress, swamp	*Taxodium*
cypress-pine	*Callitris*

D

daffodil	*Narcissus*
dahlia	*Dahlia*
dahlia, climbing	*Hidalgoa wercklei*
dahoon	*Ilex cassine*
daikon	*Raphanus sativus*
daisy	*Bellis perennis*
daisy, African	*Arctotis; Lonas*
daisy, Barberton	*Gerbera jamesonii*
daisy, blue	*Felicia*
daisy, bush	*Olearia*
daisy, Michaelmas	*Aster*
daisy, New Holland	*Vittadenia triloba*
daisy, New Zealand	*Celmisia*
daisy, ox-eye	*Chrysanthemum leucanthemum*
daisy, Shasta	*Chrysanthemum maximum*
daisy, Swan River	*Brachycome iberidifolia*
daisy, Transvaal	*Gerbera jamesonii*
daisy, white	*Chrysanthemum leucanthemum*

dame's violet	*Hesperis matronalis*
dammar pine	*Agathis*
dandelion	*Taraxacum*
dandelion, dwarf	*Krigia*
Darling River pea	*Swainsona greyana*
dasheen	*Colocasia esculenta*
date palm	*Phoenix dactylifera*
date plum	*Diospyros*
date-yucca	*Samuela*
dawn redwood	*Metasequoia glyptostroboides*
day-flower	*Commelina*
day-lily	*Hemerocallis*
dead-nettle	*Lamium*
dead-nettle, spotted	*Lamium maculatum*
dead-nettle, white	*Lamium album*
deadly nightshade	*Atropa bella-donna*
deerberry	*Vaccinium stamineum*
deer-foot	*Achlys*
deer-grass	*Rhexia*
desert-candle	*Eremurus*
desert rod	*Enemostachys*
devil's bit	*Chamaelirium luteum* (U.S.A.); *Succisa pratensis*
devil's paintbrush	*Hieracium*
dewberry	*Rubus caesius*
diamond-flower	*Ionopsidium acaule*
dildo	*Cereus*
dill	*Anethum graveolens*
dish-cloth gourd	*Luffa cylindrica*
distaff thistle	*Carthamus tinctorius*
ditchmoss	*Elodea*
dittany	*Cunila origanoides; Dictamnus albus*
dittany, Cretan	*Amaracus dictamnus*
divi-divi	*Caesalpinia coriaria*
dock	*Rumex*
dock, spinach	*Rumex patientia*
dockmackie	*Viburnum acerifolium*
dodder	*Cuscuta*
dogbane	*Apocynum*
dog rose	*Rosa canina*
dog's-tooth violet	*Erythronium*

dogwood	*Cornus*	elm, Cornish	*Ulmus angustifolia*
Dorset heath	*Erica ciliaris*	elm, Dutch	*Ulmus hollandica*
Douglas fir	*Pseudotsuga*	elm, English	*Ulmus procera*
	menziesii	elm, Huntingdon	*Ulmus hollandica*
dove orchid	*Peristeria*		*vegeta*
dove tree	*Davidia*	elm, Siberian	*Ulmus pumila*
dragon arum	*Arisaema;*	elm, slippery	*Ulmus fulva*
	Dracunculus	elm, smooth-	*Ulmus carpinifolia*
	vulgaris	leaved	
dragon's mouth	*Helicodiceros*	elm, Spanish	*Cordia*
	muscivorus	elm, wych	*Ulmus glabra*
dragonhead	*Dracocephalum*	empress of Brazil	*Worsleya procera*
dragonhead, false	*Physostegia*	enchanter's	*Circaea lutetiana*
dragon-root	*Arisaema*	nightshade	
dropwort	*Filipendula*	endive	*Cichorium endiva*
	vulgaris	English oak	*Quercus robur*
dropwort, water	*Oenanthe fistula*	eryngo	*Eryngium*
duckweed	*Lemna*	evening flower	*Hesperantha*
dumb cane	*Dieffenbachia*	evening primrose	*Oenothera*
durian	*Durio zibethinus*	evening-star	*Cooperia*
durmast oak	*Quercus petraea*	evergreen oak	*Quercus ilex*
dusty miller	*Centaurea;*	everlasting	*Antennaria;*
	Artemisia		*Helichrysum;*
	stellerana		*Helipterum;*
			Xeranthemum
Dutchman's	*Dicentra*	everlasting,	*Helipterum*
breeches	*cucullaria*	Australian	
Dutchman's pipe	*Aristolochia durior*	everlasting, pearly	*Anaphalis*
dwale	*Atropa bella-donna*		*margaritacea*
dwarf alder	*Fothergilla*	everlasting,	*Ammobium alatum*
dwarf almond	*Prunus tenella*	winged	
dwarf birch	*Betula nana*	everlasting, pea	*Lathyrus latifolius*
dwarf dandelion	*Krigia*	ever-ready onion	*Allium cepa*
dyer's greenweed	*Genista*		*perutile*

E

earthnut	*Arachnis hypogaea*
ebony	*Diospyros*
edelweiss	*Leontopodium*
	alpinum
eelgrass	*Zostera;*
	Vallisneria
eggplant	*Solanum*
	melongena
eglantine	*Rosa eglanteria*
elder	*Sambucus*
elder, box-	*Acer negundo*
elecampane	*Inula helenium*
elephant's-ear	*Colocasia*
elephant-ear fern	*Elaphoglossum*
elm	*Ulmus*
elm, American	*Ulmus americana*
elm, Chinese	*Ulmus parvifolia*

F

fairy moss	*Azolla*
fairybells	*Disporum*
false acacia	*Robinia*
	pseudacacia
false aralia	*Dizygotheca*
false chamomile	*Boltonia*
false cypress	*Chamaecyparis*
false dragon-head	*Physostegia*
false goatsbeard	*Astilbe*
false gromwell	*Onosmodium*
false heath	*Fabiana imbricata*
false hellebore	*Veratrum*
false indigo	*Amorpha*
false mallow	*Malvastrum;*
	Sidalcea
false mitrewort	*Tiarella*

false plantain	*Heliconia*	fern, ostrich	*Matteucia*
false saffron	*Carthamus*		*struthiopteris*
false Solomon's	*Smilacina*	fern, polypody	*Polypodium*
seal; false		fern, rattlesnake	*Botrychium*
spikenard		fern, rock brake	*Cryptogramma*
false spirea	*Sorbaria*	fern, royal	*Osmunda*
false tamarisk	*Myricaraia*	fern, rush	*Schizaea*
fancy anna	*Delonix regia*	fern, rusty-back	*Ceterach*
fanwort	*Cabomba*		*officinarum*
	caroliniana	fern, sensitive	*Onoclea sensibilis*
farkleberry	*Vaccinium*	fern, shield	*Dryopteris;*
	arboreum		*Polystichum*
feather-grass	*Stipa*	fern, silver	*Dryopteris*
feather hyacinth	*Muscari comosum*	fern, spleenwort	*Asplenium*
	monstrosum	fern, staghorn	*Platycerium*
fennel	*Foeniculum vulgare*	fern, swamp	*Athyrium*
fennel-flower	*Nigella*	spleenwort	
fenugreek	*Trigonella foenum-*	fern, sweet	*Comptonia*
	graecum	fern, sword	*Nephrolepis*
fern, adder's	*Ophioglossum*	fern, tree	*Cyathea*
tongue		fern, walking	*Camptosorus*
fern, asparagus	*Asparagus*		*rhizophyllus*
	plumosus	fern, wall	*Polypodium*
fern, bladder	*Cystopteris*	fern, water	*Salvinia*
fern, bracken or	*Pteridium*	fern, wood	*Dryopteris*
brake	*aquilinum*	fescue	*Festuca*
fern, chain	*Woodwardia*	feverbush	*Lindera*
fern, Christmas	*Polystichum*	feverfew	*Chrysanthemum*
fern, cinnamon	*Osmunda*		*parthenium*
	cinnamomea	fiddle-wood	*Citharexylum*
fern, cliffbrake	*Pellaea*	fig	*Ficus*
fern, climbing	*Lygodium*	fig-marigold	*Mesembryan-*
fern, cloak	*Notholaena*		*themum*
fern, comb	*Schizaea*	fig-wort	*Scrophularia*
fern, curly-grass	*Schizaea*	filbert	*Corylus maxima*
fern, elephant-ear	*Elaphoglossum*	finger-grass	*Digitaria*
fern, flowering	*Osmunda*	finger-lime	*Microcitrus*
fern, gold	*Pityrogramma*	finocchio	*Foeniculum*
fern, grape	*Botrychium*		*vulgare dulce*
fern, hare's foot	*Davallia*	fir	*Abies*
fern, hart's	*Phyllitis*	fir, Douglas	*Pseudotsuga*
tongue	*scolopendrium*		*menziesii*
fern, hayscented	*Dennstaedtia*	firebush, Mexican	*Kochia scoparia*
fern, holly	*Polystichum*		*culta*
fern, interrupted	*Osmunda*	fire-thorn	*Pyracantha*
fern, Japanese	*Cyclophorus*	fireweed	*Epilobium*
fern, lady	*Athyrium filix-*	fish-grass	*Cabomba*
	femina		*caroliniana*
fern, lip	*Cheilanthes*	flag	*Iris*
fern, maidenhair	*Adiantum*	flag, sweet	*Acorus calamus*
fern, male	*Dryopteris filix-*	flamboyant tree	*Delonix regia*
	mas	flame-lily	*Pyrolirion*
fern, moonwort	*Botrychium*	flame-of-the-	*Butea monosperma*
		forest	

flamingo flower	*Anthurium*	funkia	*Hosta*
flannel-bush	*Fremontia*	furze	*Ulex*
	(Fremonto-		
	dendron)		
flannel-flower	*Actinotus helianthi*	**G**	
flax	*Linum*	gale	*Myrica gale*
flax, New	*Phormium*	galingale	*Cyperus*
Zealand		gallant soldier	*Galinsoga*
flax-lily	*Dianella*	gall-of-the-earth	*Prenanthes*
fleabane	*Erigeron*	garbanzo	*Cicer arietinum*
fleur-de-lis	*Iris*	garden cress	*Lepidium*
floral firecracker	*Brevoortia*	gardener's garters	*phalaris*
floss silk tree	*Chorisia*		*arundinacea*
flower fern	*Aneimia*		*picta*
flowering fern	*Osmunda*	gardenia	*Gardenia*
flowering moss	*Pyxidanthera*	garget	*Phytolacca*
flowering rush	*Butomus*	garland flower	*Hedychium*
	umbellatus		*coronarium*
flowering spurge	*Euphorbia*	garlic	*Allium sativum*
	corollata	garlic, crow	*Allium vineale*
foamflower	*Tiarella cordifolia*	garlic mustard	*Alliaria petiolata*
fool's parsley	*Aethusa cynapium*	gas plant	*Dictamnus albus*
forget-me-not	*Myosotis*	gay-wings	*Polygala*
forget-me-not,	*Myosotidium*	gean	*Prunus avium*
Antarctic	*hortensia*	geiger tree	*Cordia sebestana*
four-o'clock	*Mirabilis jalapa*	genip	*Melicocca bijuja*
foxglove	*Digitalis*	genista, florist's	*Cytisus canariensis*
foxglove, false	*Gerardia*	gentian	*Gentiana*
fox grape	*Vitis labrusca*	gentian, horse	*Triosteum*
foxtail grass	*Alopecurus*	gentian, willow	*Gentiana*
foxtail lily	*Eremurus*		*asclepiadea*
flamingo flower	*Anthurium*	geranium	*Pelargonium*
frangipani	*Plumeria*	germander	*Teucrium*
fraxinella	*Dictamnus albus*	germander	*Veronica*
French marigold	*Tagetes patula*	speedwell	*Chamaedrys*
French mulberry	*Callicarpa*	ghostflower	*Monotropa uniflora*
fringe flower	*Schizanthus*	giant bamboo	*Dendrocalamus*
fringe-tree	*Chionanthus*	giant bellflower	*Ostrowskia*
fringed orchis	*Habenaria*		*magnifica*
fritillary	*Fritillaria*	giant fennel	*Ferula*
frogbit	*Hydrocharis*	giant scabious	*Cephalaria*
	morsus-ranae	giant summer	*Galtonia*
frost flower	*Aster*	hyacinth	*candicans*
fruiting myrtle	*Eugenia*	giant water	*Lagarosiphon*
fumitory	*Corydalis;*	thyme	
	Fumaria	gill-over-the-	*Glechoma*
fumitory,	*Adlumia fungosa*	ground	*hederacea*
climbing	(U.S.A.);	ginger	*Zingiber officinale*
	Corydalis clavi-	ginger, wild	*Asarum*
	culata (Britain)	ginger-lily	*Hedychium*
fuchsia	*Fuchsia*	gingseng	*Panax*
fuchsia,	*Correa*	girasole	*Helianthus*
Australian			*tuberosus*
		glasswort	*Salicornia*

globe daisy	*Globularia*	gorse	*Ulex*
globe flower	*Trollius*	gourd	*Cucurbita;*
globe-mallow	*Sphaeralcea*		*Lagenaria*
globe thistle	*Echinops*	gourd, calabash	*Lagenaria siceraria*
glory-bush	*Tibouchina*	gourd, dish-cloth	*Luffa cylindrica*
glory flower	*Eccremocarpus*	gourd, ivy	*Coc cinea*
	scaber		*cordifolia*
glory-lily	*Gloriosa*	gourd, snake	*Trichosanthes*
glory-of-the-snow	*Chionodoxa*		*anguina*
glory-of-the-sun	*Leucocoryne*	gourd, wax	*Benincasa hispida*
	ixioides	goutweed	*Aegopodium*
glory-pea	*Clianthus*		*podagraria*
gloxinia, florist's	*Sinningia*	grape	*Vitis vinifera*
goat's-beard	*Aruncus; Spiraea;*	grape, fox	*Vitis labrusca*
	Tragopogon	grape, Oregon	*Mahonia*
goatsrue	*Galega officinalis*		*aquifolium*
goat willow	*Salix caprea*	grape, seaside	*Coccoloba*
gobo	*Arctium lappa*	grape fern	*Botrychium*
	anagyroides	grapefruit	*Citrus paradisi*
gold fern	*Pityrogramma*	grape-hyacinth	*Muscari*
gold thread	*Coptis*	grass, brome	*Bromus*
golden aster	*Chrysopsis*	grass, canary	*Phalaris*
golden-bell	*Forsythia*		*canariensis*
golden chain	*Laburnum*	grass, fish	*Cabomba*
golden club	*Orontium*		*caroliniana*
golden drop	*Onosma*	grass, Guinea	*Panicum maximum*
golden larch	*Pseudolarix*	grass, Kentucky	*Poa pratensis*
	amabilis	grass, Manila-blue	*Zoisia matrella*
golden	*Anthemis tinctoria*	grass, Pampas	*Cortaderia*
marguerite		grass, pink	*Calopogon*
goldenrod	*Solidago*	grass, quaking	*Briza*
goldenrod,	*Bigelovia*	grass-of-	*Parnassia*
plumed		Parnassus	*palustris*
golden saxifrage	*Chrysosplenium*	great burnet	*Poterium*
golden seal	*Hydrastis*	Greek valerian	*Polemonium*
golden shower	*Cassia fistula*	green hellebore	*Helleborus viridis*
	canadensis	green wattle	*Acacia decurrens*
golden spider-	*Lycoris aurea*	greenbrier	*Smilax*
lily		grey birch	*Betula populifolia*
golden star	*Chrysogonum*	grey poplar	*Populus canescens*
	virginianum	gromwell	*Lithospermum*
golden wattle	*Acacia pycnantha*		*officinale*
goldilocks	*Ranunculus*	ground-cherry	*Physalis pruinosa*
	auricomus;	ground-ivy	*Glechoma*
	Aster linosyris		*hederacea*
good-King-	*Chenopodium*	ground-laurel	*Epigaea repens*
Henry	*bonus henricus*	groundnut	*Arachis hypogaea*
gombo	*Abelmoschus*	ground-oak	*Chamaedrys*
	esculentus	ground pine	*Lycopodium*
gooseberry	*Ribes uva-crispa*	ground rattan	*Rhapis*
gooseberry,	*Pereskia aculeata*	groundsel	*Senecio vulgaris*
Barbados		groundsel-tree	*Baccharis*
goosefoot	*Chenopodium*	gru gru	*Acrocomia aculeata*
	album	guava	*Psidium guavjava*

guava, strawberry	*Psidium cattleianum*
guelder-rose	*Viburnum*
Guernsey lily	*Nerine sarniensis*
guinea flower	*Hibbertia*
gul mohur	*Delonix regia*
gum, black	*Nyssa sylvatica*
gum myrtle	*Angophora*
gum tree	*Eucalyptus*
gumbo	*Abelmoschus esculentus*
gumplant	*Grindelia*

H

hackberry	*Celtis*
hair grass	*Aira*
handkerchief tree	*Davidia*
harbinger-of-spring	*Erigenia bulbosa*
hardhack	*Spiraea tomentosa*
hardheads	*Centaurea nigra*
harebell	*Campanula rotundifolia*
hare's ear	*Bupleurum*
hare's-foot fern	*Davallia*
hare's-tail grass	*Lagurus ovatus*
hart's-tongue fern	*Phyllitis scolopendrium*
hatchet vetch	*Securigera securidaca*
hawk's-beard	*Crepis*
hawkbit	*Leontodon*
hawkweed	*Hieracium*
hawkweed, yellow	*Tolpis*
hawthorn	*Crataegus*
hay-scented fern	*Dennstaedtia*
hazel	*Corylus*
heal-all	*Prunella vulgaris*
heartnut	*Juglans ailantifolia cordiformis*
heart's-ease	*Viola tricolor*
heath	*Erica; Epacris* (Australia); *Styphelia* (Australia)
heath, Cornish	*Erica vagans*
heath, cross-leaved	*Erica tetralix*
heath, Dorset	*Erica ciliaris*
heath, fake	*Fabiana imbricata*
heath, Irish	*Erica erigena*
heath, sea	*Frankenia laevis*

heath, St. Dabeoc's	*Daboëcia cantabrica*
heath, spike	*Bruckenthalia spiculifolia*
heather	*Calluna vulgaris*
heather, beach	*Hudsonia*
heather, bell-	*Erica cinerea*
heather, bog	*Erica tetralix*
hedge-bindweed	*Calystegio; Convolvulus*
hedgehog broom	*Erinacea antryllis*
hedge-hyssop	*Gratiola*
hedge-mustard	*Erysimum*
hedge-nettle	*Stachys*
hedge-parsley	*Torilis*
hedge wattle	*Acacia armata*
heliotrope	*Heliotropium*
heliotrope, garden	*Valeriana*
heliotrope, winter	*Petasites fragrans*
hellebore	*Helleborus*
hellebore, false	*Veratrum*
hellebore, green	*Helleborus viridis*
hellebore, stinking	*Helleborus foetidus*
helmet, policeman's	*Impatiens glandulifera*
helmet flower	*Coryanthes*
hemlock	*Tsuga*
hemlock, poison	*Conium maculatum*
hemlock, water	*Cicuta virosa*
hemp	*Cannabis sativa*
hemp-agrimony	*Eupatorium cannabinum*
hemp, Indian-	*Apocynum cannabinum*
hempweed, climbing	*Mikania scandens*
hen-and-chickens houseleek	*Sempervivum soboliferum*
henbane	*Hyoscyamus niger*
henna	*Lawsonia inermis*
herald's trumpet	*Beaumontia*
herb-bennet	*Geum urbanum*
herb-Christopher	*Actaea spicata*
herb-grace	*Ruta graveolens*
herb-Mercury	*Mercurialis*
herb-Paris	*Paris quadrifolia*
herb patience	*Rumex patientia*
herb-Robert	*Geranium robertianum*

herb-of-St. Barbara	*Barbarea vulgaris*	horse-chestnut	*Aesculus*
Hercules' club	*Aralia spinosa;*	horse-chestnut, red	*Aesculus rubra*
	Zanthoxylum clava-herculis	horse-chestnut, white	*Aesculus hippocastanum*
heron's bill	*Erodium*	horse-gentian	*Triosteum*
herniary	*Herniaria*	horse-mint	*Mentha longifolia*
hickory	*Carya*		*Monarda*
hickory, Shagbark	*Carya ovata*	horse-radish	*Amoracia*
Himalaya-berry	*Rubus procerus*		*rusticana*
Himalayan poppy	*Meconopsis*	horse-radish, Japanese	*Eutrema wasabi*
hoarhound	*Marrubium*	horse-radish tree	*Moringa*
hoarhound, water	*Lycopus*	horseshoe vetch	*Hippocrepis*
hobblebush	*Viburnum alnifolium*		*comosa*
hogweed	*Heracleum sphondylium*	horsetail	*Equisetum*
		horse-weed	*Collinsonia*
holly	*Ilex*	Hottentot fig	*Carpobrotus edulis*
holly, American mountain	*Nemopanthus mucronatus*	hound's tongue	*Cynoglossum*
		house-leek	*Sempervivum*
hollyhock	*Althaea*	huckleberry	*Gaylussacia*
holm oak	*Quercus ilex*	huckleberry, garden	*Solanum melanocerasum*
holy grass	*Hierochloë*		
hondapara	*Dillenia indica*	hyacinth	*Hyacinthus orientalis*
honesty	*Lunaria annua*		
honey-bell	*Mahernia verticillata*	hyacinth, feather	*Muscari comosum monstrosum*
honey-berry	*Melicocca bijuga*	hyacinth, grape	*Muscari*
honey-bush	*Melianthus*	hyacinth, summer	*Galtonia candicans*
honey-locust	*Gleditsia*	hyacinth, tassel	*Muscari comosum*
honeysuckle	*Lonicera*	hyacinth, water	*Eichhornia crassipes*
honeysuckle, Australian	*Banksia*	hyacinth, wild	*Endymion*
honeysuckle, bush-	*Diervilla*	hyssop	*Hyssopus officinalis*
honeysuckle, Cape	*Tecomaria capensis*	hyssop, giant	*Agastache*
honeysuckle, Japanese	*Lonicera japonica*		
honeywort	*Cerinthe*	**I**	
hop	*Humulus*	iceplant	*Cryophytum crystallinum;*
hop-hornbeam	*Ostrya carpinifolia*		*Sedum spectabile*
horehound	*Marrubium vulgare*	immortelle	*Helichrysum;*
horehound, black	*Ballota*		*Helipterum;*
horehound, water	*Lycopus*		*Xeranthemum*
		incense cedar	*Libocedrus*
hornbeam	*Carpinus*	incense plant	*Calomeria amaranthoides (Humea elegans)*
horned poppy	*Glaucium*		
horn-of-plenty	*Fedia*		
hornwort	*Ceratophyllum demersum*	India hawthorn	*Raphiolepis*
		India lotus	*Nymphaea*
horse-balm	*Collinsonia*	India-rubber plant	*Ficus elastica*
horse bean	*Vicia faba*		

Indian almond	*Terminalia catappa*	ivy	*Hedera*
Indian cress	*Tropaeolum*	ivy arum	*Scindapsus*
Indian crocus	*Pleione*	ivy, Boston	*Parthenocissus tricuspidata*
Indian cucumber-root	*Medeola virginiana*	ivy gourd	*Coccinia cordifolia*
Indian fig	*Opuntia*	ivy, ground	*Glechoma hederacea*
Indian grass	*Arundo*	ivy, poison	*Rhus toxicodendron*
Indian-hemp	*Apocynum cannabinum*		
Indian mulberry	*Morinda*		
Indian paint-brush	*Castilleja*		
		J	
Indian physic	*Porteranthus trifoliatus*	jaboticaba	*Myrciaria cauliflora*
Indian pink	*Spigelia marilandica*	jackfruit	*Artocarpus heterophyllus*
Indian pipe	*Monotropa uniflora*	Jack-go-to-bed-at-noon	*Tragopogon pratensis*
Indian poppy	*Meconopsis*	Jack-in-the-pulpit	*Arisaema*
Indian shot	*Canna*		
Indian strawberry	*Duchesnea indica*	Jacobea	*Senecio jacobaea*
Indian tobacco	*Lobelia inflata*	Jacobean lily	*Sprekelia formosissima*
Indian turnip	*Arisaema*		
indigo	*Indigofera*	Jacob's ladder	*Polemonium*
indigo, false	*Amorpha*	Jacob's rod	*Asphodeline lutea*
indigo, wild	*Baptisia*	jade vine	*Strongylodon macrobotrys*
inkberry	*Ilex glabra*; *Phytolacca americana*	jak	*Artocarpus heterophyllus*
innocence	*Houstonia caerulea*		
interrupted fern	*Osmunda claytoniana*	Jamaica pepper	*Pimenta dioica*
		Jamaica sorrel	*Hibiscus sabdariffa*
Iris	*Iris*	Jamestown weed	*Datura stramonium*
iris, English	*Iris xiphioides*	Japanese bellflower	*Platycodon*
iris, German	*Iris germanica*		
iris, mourning	*Iris susiana*	Japanese bitter orange	*Poncirus trifoliata*
iris, scarlet-seeded	*Iris foetidissima*	Japanese bunching onion	*Allium fistulosum*
iris, Spanish	*Iris xiphium*		
iris, stinking	*Iris foetidissima*	Japanese cedar	*Cryptomeria japonica*
iris, widow	*Hermodactylus tuberosus*	Japanese honeysuckle	*Lonicera japonica*
Irish heath	*Erica erigena*	Japanese horse-radish	*Eutrema wasabi*
ironbark	*Eucalyptus*		
iron tree	*Metrosideros*	Japanese persimmon	*Diospyros kaki*
ironweed	*Veronia*		
ironwood	*Carpinus*; *Ostrya virginiana*	Japanese quince	*Chaenomeles*
		Japanese raisin tree	*Hovenia dulcis*
ironwort	*Sideritis*		
Italian cypress	*Cupressus sempervirens*	Japanese toad-lily	*Tricyrtis*
ivory-nut palm	*Phytelephas macrocarpa*	Japanese yew	*Cephalotaxus harringtoniana*

Japonica	*Chaenomeles speciosa*; *Camellia japonica*; *Kerria japonica*	kenaf	*Hibiscus cannabinus*
		kidney bean	*Phaseolus vulgaris*
		kidney vetch	*Anthyllis vulneraria*
jasmine	*Jasminum*		
jasmine, bastard	*Cestrum*	king-cup	*Caltha palustris*
jasmine, Madagascar	*Stephanotis floribunda*	kiwifruit	*Actinidia chinensis*
		knapweed	*Centaurea*
Jerusalem artichoke	*Helianthus tuberosus*	knotweed	*Polygonum*
		kohlrabi	*Brassica caulorapa*
Jerusalem cherry	*Solanum pseudo-capsicum*	kudzu-vine	*Pueraria thunbergiana*
Jerusalem cowslip	*Pulmonaria*	kumquat	*Fortunella*
Jerusalem oak	*Chenopodium botrys*	kumquat, Marumi	*Fortunella japonica*
Jerusalem sage	*Phlomis fruticosa*	kumquat, Nagami	*Fortunella margarita*
Jerusalem thorn	*Parkinsonia aculeata*; *Paliurus spina-christi*		
			L
jewel-weed	*Impatiens*		
Jew's mallow	*Corehorus olitorios*	Labrador tea	*Ledum groenlandicum*
Job's tears	*Coix lacryma-jobi*	laburnum	*Laburnum*
Joe-Pye-weed	*Eupatorium*	laburnum, Indian	*Cassia fistula*
jonquil	*Narcissus jonquilla*	lace-leaf	*Aponogeton fenestralis*
Joshua tree	*Yucca brevifolia*		
joy-weed	*Alternanthera*	ladies' fingers	*Abelmoschus esculentus*; *Anthyllis vulneraria*
Judas tree	*Cercis siliquastrum*		
June berry	*Amelanchier*		
juniper	*Juniperus*	ladies' tresses	*Spiranthes*
juniper-myrtle	*Verticordia*	lad's love	*Artemisia*
Jupiter's beard	*Anthyllis*; *Centranthus ruber*	Lady Doorley	*Ipomoea horsfalliae*
		lady fern	*Athyrium*
jute	*Corchorus olitorius*	lady-of-the-night	*Brunfelsia americana*
		lady's finger	*Abelmoschus esculentus*
	K	lady's mantle	*Alchemilla*
Kaffir lily	*Clivia*; *Schizostylis*	lady's slipper	*Cypripedium*
kaki	*Diospyros kaki*	lady's smock	*Cardamine pratensis*
kale	*Brassica oleracea acephala*	lady's thistle	*Silybum marianum*
kale, sea	*Crambe maritima*	lady's thumb	*Polygonum*
Kamila tree	*Mallotus*	lamb-kill	*Kalmia*
kangaroo apple	*Solanum laciniatum*	lamb's lettuce	*Valerianella*
		lamb's-quarters	*Chenopodium*
kangaroo paw	*Anigozanthus*	lamb's tail grass	*Alopecurus*
kangaroo vine	*Cissus antarctica*	lantana, wild	*Abronia*
Kansas gay-feather	*Liatris*	larch	*Larix*
		larch, golden	*Pseudolarix amabilis*
Kauri pine	*Agathis*		

larkspur	*Consolida;*	lily, Amazon	*Eucharis grandi-*
	Delphinium		*flora*
lattice-leaf	*Aponogeton*	lily, blackberry	*Belamcanda*
	fenestralis		*chinensis*
laurel	*Laurus nobilis*	lily, blood	*Haemanthus*
laurel,	*Danae racemosa*	lily, day	*Hemerocallis*
Alexandrian		lily, golden spider	*Lycoris aurea*
laurel, bay	*Laurus nobilis*	lily, Guernsey	*Nerine sarniensis*
laurel, cherry	*Prunus*	lily, Jacobean	*Sprekelia*
	laurocerasus		*formosissima*
laurel, mountain	*Kalmia*	lily, Kaffir	*Clivia;*
laurel, New	*Corynocarpus*		*Schizostylis*
Zealand		lily, leopard	*Sansevieria*
laurel, Portugal	*Prunus lusitanica*	lily, Madonna	*Lilium candidum*
laurel, sheep;	*Kalmia*	lily, Mariposa	*Calochortus*
laurel, swamp		lily,	*Pancratium*
laurestinus	*Viburnum tinus*	Mediterranean	*maritimum*
lavender	*Lavandula*	lily, nankeen	*Lilium testaceum*
lavender, sea	*Limonium*	lily, Peruvian	*Alstroemeria*
lavender-cotton	*Santolina*	lily, regal	*Lilium regale*
	chamaecyparissus	lily, Scarborough	*Vallota speciosa*
leadwort	*Plumbago*	lily, St.	*Anthericum liliago*
leather fern	*Acrostichum*	Bernard's	
	danaeifolium	lily, St. Bruno's	*Paradisea*
leatherwood	*Dirca palustris;*		*liliastrum*
	Cyrilla racemi-	lily, spider	*Hymenocallis*
	flora	lily, tiger	*Lilium lancifolium*
leek	*Allium ampelo-*		*(L. tigrinum)*
	prasum porrum	lily, zephyr	*Zephyranthes*
lemon	*Citrus limon*	lily-of-the-Nile	*Agapanthus*
lemon balm	*Melissa officinalis*	lily-of-the-valley	*Convallaria*
lemon verbena	*Aloysia triphylla*	lily-of-the-valley,	*Maianthemum*
lentil	*Lens culinaris*	wild	
leopard flower	*Belamcanda*	lily-thorn	*Catesbaea spinosa*
	chinensis	lime	*Citrus aurantifolia*
leopard lily	*Sansevieria*		(tropics); *Tilia*
leopard plant	*Ligularia*		(Britain)
	kaempferi	lime, finger	*Microcitrus*
	aureomaculata	limeberry	*Triphasia trifolia*
leopard's bane	*Doronicum*	Lincolnshire	*Chenopodium*
lettuce	*Lactuca sativa*	asparagus	
licorice	*Glycyrrhiza*	linden, or lime	*Tilia*
	glabra	ling, ling-heather	*Calluna vulgaris*
licorice, wild	*Abrus precatorius*	lion's ear	*Leonotis leonurus*
life plant	*Bryophyllum*	lion's foot	*Leontopodium*
	pinnatum		*alpinum*
lignum	*Muehlenbeckia*	lion's-tail	*Leonurus*
lignum vitae	*Guiacum officinale;*	lip fern	*Cheilanthes*
	Guiacum	liquorice	*Glycyrrhiza*
	sanctum		*glabra*
lilac	*Syringa*	litchi	*Litchi chinensis*
lilly-pilly	*Acmena smithii*	little Mary	*Bellium minutum*
lily	*Lilium*	liverleaf,	*Hepatica*
lily, African	*Agapanthus*	liverwort	

373

lizard orchid	*Himantoglossum hircinum* (Britain); *Burnettia cuneata* (Australia)
lizard's-tail	*Saururus cernuus*
loblolly bay	*Gordonia lasianthus*
loblolly magnolia	*Magnolia grandiflora*
loblolly pine	*Pinus taeda*
locust	*Robinia*
locust, honey	*Gleditsia*
loganberry	*Rubus loganobaccus*
logwood	*Haematoxylum campechianum*
Lombardy poplar	*Populus nigra italica*
London plane	*Platanus hybrida*
London pride	*Saxifraga urbium*
loosestrife, purple	*Lythrum salicaria*
loosestrife, swamp	*Decodon verticillatus*
loosestrife, yellow	*Lysimachia vulgaris*
lords-and-ladies	*Arum maculatum*
lotus, blue	*Nymphaea stellata*
lotus, sacred	*Nelumbo nucifera*
lotus, white	*Nymphaea lotus*
louse-wort	*Pedicularis*
lovage	*Levisticum officinale*; *Ligusticum scoticum*
love-apple	*Lycopersicon esculentum*
love-grass	*Eragrostis*
love-in-a-mist	*Nigella damascena*
love-lies-bleeding	*Amaranthus caudatus*
lucerne	*Medicago sativa*
Lucombe oak	*Quercus hispanica lucombeana*
lungwort	*Pulmonaria*; *Mertensia*
lupin, lupine	*Lupinus*

M

Macartney rose	*Rosa bracteata*
Madagascar jasmine	*Stephanotis floribunda*
Madagascar palm	*Chrysalidocarpus lutescens*
Madagascar periwinkle	*Catharanthus roseus*
madder	*Rubia tinctoria*
Madeira vine	*Boussingaultia gracilis pseudo-basseloides*
madwort	*Alyssum*
mahoe	*Thespesia populnea*
mahogany tree	*Swietenia mahogani*
maidenhair fern	*Adiantum*
maidenhair tree	*Gingko biloba*
maize	*Zea mays*
Malabar nightshade	*Basella alba*
male-fern	*Aspidium*; *Dryopteris*
mallee	*Eucalyptus*
mallee, white	*Eucalyptus gracilis*
mallow	*Malva*
mallow, false	*Malvastrum*
mallow, globe	*Sphaeralcea*
mallow, Jew's	*Corchorus olitorius*
mallow, marsh	*Althaea officinalis*
mallow, musk	*Malva moschata*
mallow, rose	*Hibiscus*
mallow, tree	*Lavatera arborea*
Maltese cross	*Lychnis chalcedonica*
mammee-apple	*Mammea americana*
manchineel	*Hippomane mancinella*
mandrake	*Mandragora* (Europe); *Podophyllum peltatum* (U.S.A.)
mango	*Mangifera indica*
mangosteen	*Garcinia mangostana*
manna grass	*Glyceria*
maple	*Acer*
maple, field	*Acer campestre*
maple, flowering	*Abutilon*
maple, hedge	*Acer campestre*
maple, Japanese	*Acer palmatum*
maple, Norway	*Acer platanoides*
maple, sugar	*Acer saccharum*
mare's tail	*Ephedra*; *Hippuris vulgaris*

marguerite	*Chrysanthemum leucanthemum*; *Chrysanthemum frutescens*	meadow rue	*Thalictrum*
		meadow saffron	*Colchicum*
		meadow-sweet	*Filipendula*
		medick	*Medicago*
marguerite, golden	*Anthemis tinctoria*	medick, black	*Medicago lupulina*
		medick, spotted	*Medicago arabica*
marigold	*Calendula*	Mediterranean lily	*Pancratium maritimum*
marigold, African	*Tagetes erecta*		
marigold, fig	*Mesembryanthemum*	medlar	*Mespilus germanica*
		melic grass	*Melica*
marigold French	*Tagetes patula*	melilot	*Melilotus*
		melon	*Cucumis melo*
marigold, marsh	*Caltha palustris*	melon, bitter	*Momordica charantia*
marigold, pot	*Calendula officinalis*		
		melon, water	*Citrullus lanatus*
Mariposa lily	*Calochortus*	mercury	*Chenopodium bonus-henricus* *Mercurialis*
marjoram	*Origanum vulgare*		
marjoram, pot	*Marjorana onites*		
marjoram, sweet	*Majorana hortensis*	mercury, dog's	*Mercurialis perennis*
marjoram, wild	*Origanum vulgare*		
marking nut	*Anacardium*	mermaid weed	*Proserpinaca*
marlberry	*Ardisia*	mesquite	*Prosopis*
marram grass	*Ammophila arenaria*	Mexican bamboo	*Polygonum cuspidatum*
marrow, vegetable	*Cucurbita pepo medullosa*	Mexican buckeye	*Ungnadia speciosa*
		Mexican clover	*Richardia scabra*
marsh felwort	*Swertia*	Mexican firebush	*Kochia scoparia culta*
marsh-marigold	*Caltha palustris*		
marsh rosemary	*Limonium*	Mexican lily	*Hippeastrum*
marsh samphire	*Salicornia*	Mexican orange	*Choisya ternata*
marvel-of-Peru	*Mirabilis jalapa*	Mexican poppy	*Argemone mexicana*
mask-flower	*Alonsoa warscewiczii*		
		Michaelmas daisy	*Aster*
masterwort	*Astrantia*	mignonette	*Reseda odorata*
matrimony vine	*Lycium barbarum*	mignonette-vine	*Boussingaultia gracilis pseudo-basseloides*
may	*Crataegus monogyna*		
May-apple	*Podophyllum peltatum*	mint-bush	*Prostanthera*
		milfoil	*Achillea millefolium*
Mayflower	*Epigaea repens*		
mayweed	*Anthemis*; *Matricaria* *Tripleurospermum*	milk thistle	*Silybum marianum*
		milk vetch	*Astragalus*
mayweed, rayless	*Matricaria matricarioides*	milkweed	*Asclepias*; *Gomphocarpus*
mayweed, scentless	*Tripleurospermum maritimum*	milkwort	*Polygala*;
		millet	*Eleusine*; *Panicum*; *Paspalum*
mayweed, stinking	*Anthemis cotula*		
meadow-beauty	*Rhexia*	mimosa, florist's	*Acacia*
meadow foam	*Limnanthes douglasii*	mind-your-own-business	*Soleirolia soleirolii*
meadow foxtail	*Alopecurus pratensis*	mint, apple	*Mentha rotundi-folia*

mint-bush	*Prostanthera*
mint, cat-	*Nepeta cataria*
mint, corn	*Mentha arvensis*
mint, horse	*Mentha longifolia*
mint, lemon	*Monarda punctata*
	Monarda citriodora
mint, mountain	*Pycnanthemum*
mint, pepper	*Eucalyptus*
	(Australia);
	Mentha piperita
mint, spear-	*Mentha spicata*
mint, water	*Mentha aquatica*
missey-moosey	*Sorbus americana*
mistletoe	*Amyena*, etc.
	(Australia);
	Phoradendron
	Viscum (Britain)
mitrewort	*Mitella*
mitrewort, false	*Tiarella*
moccasin flower	*Cypripedium*
mock-orange	*Philadelphus*
mock strawberry	*Duchesnea indica*
mombin, red	*Spondias purpurea*
mombin, yellow	*Spondias lutea*
monarch-of-the-	*Sauromatum*
East	*venosum*
moneywort	*Lysimachia*
	nummularia
monkey-bread	*Adansonia*
tree	*digitata*
monkey-	*Hura*
dinnerbell	
monkey-flower	*Mimulus*
monkeynut	*Arachis hypogaea*
monkey-pod tree	*Samanea saman*
monkey-puzzle	*Araucaria*
tree	
monkshood	*Aconitum*
Monterey cypress	*Cupressus*
	macrocarpa
moon-daisy	*Chrysanthemum*
	leucanthemum
moonflower	*Ipomoea alba*
moonseed	*Cocculus*;
	Menispermum
moonwort	*Botrychium*
moonwort, blue	*Soldanella*
moosewood	*Dirca australe*
morning-glory	*Ipomoea*
moss campion	*Silene acaulis*
moth orchid	*Phalaenopsis*

mother of	*Cymbalaria*
thousands	*muralis*; *Saxi-*
	fraga stolonifera
mother-of-thyme	*Thymus serpyllum*
motherwort	*Leonurus*
Mount Atlas	*Anacylcus*
daisy	
mountain ash	*Sorbus aucuparia*
mountain avens	*Dryas*
mountain-fringe	*Adlumia fungosa*
mountain holly	*Nemopanthus*
	mucronatus
mountain-laurel	*Kalmia*
mountain mint	*Pycnanthemum*
mountain spider-	*Lloydia*
wort	
mountain spurge	*Pachysandra*
mountain-tea	*Gaultheria*
mourning bride	*Scabiosa*
mouse-ear	*Cerastium*
chickweed	
moutan	*Paeonia*
	suffruticosa
mouthroot	*Coptis*
mud plantain	*Alisma*;
	Heteranthera
mudgee wattle	*Acacia spectabilis*
mugga	*Eucalyptus*
	sideroxylon
mugwort	*Artemisia vulgaris*
mulberry, black	*Morus nigra*
mulberry, Indian	*Morinda*
mulberry, paper-	*Broussonetia*
	papyrifera
mulberry, red	*Morus rubra*
mulberry, white	*Morus alba*
mullein	*Verbascum*
mullein pink	*Lychnis*
mung	*Phaseolus aureus*
muscadine	*Vitis rotundifolia*
musk hyacinth	*Muscari*
	muscarimi
musk orchis	*Herminium*
	monorchis
muskmallow	*Malva moschata*
muskmelon	*Cucumis melo*
musk rose	*Rosa moschata*
mustard, black	*Brassica nigra*
mustard, brown	*Brassica juncea*
mustard, garlic	*Alliaria petiolata*
mustard, white	*Brassica hirta*
	Sinapis alba
myrobalan plum	*Prunus cerasifera*

myrtle	*Myrtus communis*	northern white	*Thuja*
myrtle, running	*Vinca*	cedar	*occidentalis*
		nut, cob-	*Corylus avellana*
		nut, marking	*Anacardium*

N

nailwort	*Draba; Paronychia*
naseberry	*Manilkara zapota*
nasturtium,	*Tropaeolum*
garden	
navelwort	*Omphalodes;*
	Umbilicus
nectarine	*Prunus persica*
	nectarina
nemu tree	*Albizia*
netfern	*Gleichenia*
nettle	*Urtica*
nettle, tree	*Dendrocnide*
New Holland	*Vittadenia triloba*
daisy	
New Jersey tea	*Ceanothus*
New Zealand bur	*Acaena*
New Zealand	*Celmisia*
daisy	
New Zealand	*Leucogenes*
edelweiss	
New Zealand flax	*Phormium*
New Zealand	*Corynocarpus*
laurel	
New Zealand	*Tetragonia*
spinach	*tetragonioides*
New Zealand	*Aristotelia*
wineberry	
nicker-bean, grey	*Caesalpinia*
	bonducella
nicker-bean,	*Caesalpina bonduc*
yellow	
night-blooming	*Selenicereus*
cereus	*grandiflorus*
nightshade	*Solanum*
nightshade,	*Atropa bella-donna*
deadly	
nightshade,	*Basella alba*
Malabar	
nightshade,	*Solanum*
woody	*dulcamara*
nine bark	*Physocarpus*
	opulifolius
none-so-pretty	*Saxifraga urbium*
nopal	*Opuntia*
	megacantha
Norfolk Island	*Araucaria*
pine	*heterophylla*

O

oak	*Quercus*
oak, black	*Quercus velutina*
oak, cork	*Quercus suber*
oak, durmast	*Quercus petraea*
	(Q. sessiliflora)
oak, English	*Quercus robur*
	(Q. pedunculata)
oak, evergreen	*Quercus ilex*
oak-fern	*Thelypteris*
	dryopteris
oak, holm	*Quercus ilex*
oak, Lucombe	*Quercus hispanica*
	lucombeana
oak, poison	*Toxicodendron*
oak, red	*Quercus borealis*
oak, scarlet	*Quercus coccinea*
oak, she-	*Casuarina*
oak, Turkey	*Quercus cerris*
oak, water	*Quercus nigra*
oak, white	*Quercus alba*
oak-of-Jerusalem	*Chenopodium*
	botrys
oats	*Avena*
obedient-plant	*Physostegia*
okra	*Abelmoschus*
	esculentus
old man	*Artemisia*
	abrotanum
old-man cactus	*Cephalocereus*
old man's beard	*Clematis vitalba*
old woman	*Artemisia*
	stellerana
oleander	*Nerium oleander*
olive	*Olea europaea*
olive, Russian	*Elaeagnus*
olive, spurge	*Cneorum*
olive, sweet	*Osmanthus*
	fragrans
onion	*Allium cepa*
onion, ever-ready	*Allium cepa*
	perutile
onion, Japanese	*Allium*
bunching	*fistulosum*
onion, potato	*Allium cepa*
	aggregatum

377

onion, tree	*Allium cepa proliferum*	palm, queen	*Arecastrum romanz offianum*
onion, Welsh	*Allium fistulosum*	palm, sago-	*Cyas revoluta*
opopanax	*Acacia farnesiana*	palm, toddy	*Caryota urens*
orach	*Atriplex hortensis*	palm, thatch	*Thrinax*
orange	*Citrus*	palmetto	*Sabal*
orange, hardy	*Poncirus trifoliata*	palmetto, saw	*Serenoa repens*
orange, mandarin	*Citrus reticulata*	pampas grass	*Cortaderria*
orange, Seville	*Citrus aurantium*	Panama candle-tree	*Parmentiera cereifera*
orange, sour	*Citrus aurantium*	panamigo	*Pilea involucrata*
orange, sweet	*Citrus sinensis*	pansy, garden	*Viola wittrockiana*
orange jessamine	*Murraya*	papaya	*Carica papaya*
orange root	*Hydrastis canadensis*	paper bark	*Melaleuca*
orchid	*Orchis* or members of *Orchidaceae* in general	paper mulberry	*Broussonetia*
		papyrus	*Cyperus papyrus*
		parsley	*Petroselinum crispum*
Oregon grape	*Mahonia aquifolium*	parsley, fool's	*Aethusa cynapium*
orpine	*Sedum telephium*	parsley, hedge-	*Torilis*
osage orange	*Maclura pomifera*	parsnip	*Pastinaca sativa*
osoberry	*Osmaronia cerasiformis*	parsnip, cow	*Heracleum sphondylium*
ostrich fern	*Matteucia struthiopteris*	partridgeberry	*Mitchella repens*
		partridge-pea	*Cassia fasciculata*
Oswego-tea	*Monarda*	pasque-flower	*Pulsatilla*
Otaheite apple	*Spondias dulcis*	passion-flower	*Passiflora*
oxeye, yellow	*Buphthalmum*	Paterson's curse	*Echium plantagineum*
ox-eye daisy	*Chrysanthemum leucanthemum*	patience	*Rumex patientia*
oxlip	*Primula elatior*	paw-paw	*Asimina tribola; Carica papaya*
oyster-plant	*Tragopogon porrifolius*	pea	*Pisum sativum*
oyster-plant, Spanish	*Scolymus hispanicus*	pea, asparagus	*Psophocarpus tetragonolobus*
		pea, bush	*Pultenaea*
		pea, chick	*Cicer orietinum*
		pea, Darling River	*Swainsona greyana*
P		pea, everlasting	*Lathyrus latifolius*
paeony	*Paeonia*	pea, partridge	*Cassia faciculata*
pagoda tree	*Plumeria rubra*	pea, shamrock	*Parochetus communis*
paigle	*Primula veris*		
painted-cup	*Castilleja*	pea, swainson	*Swainsona*
palas, climbing	*Butea superba*	pea, sweet	*Lathyrus odoratus*
palm, cabbage	*Roystonea oleracea*	pea, tree	*Caragana*
palm, date	*Phoenix dactylifera*	pea, wedge-	*Comphalobium*
		peach	*Prunus persica*
palm, fishtail	*Caryota*	peanut	*Arachis hypogaea*
palm, king	*Archontophoenix*	pear	*Pyrus communis*
palm, Madagascar	*Chrysalidocarpus lutescens*	pear, alligator	*Persea americana*
palm, oil	*Elaeis guineensis*	pear, balsam	*Momordica charantia*
palm, royal	*Roystonea*		
palm, pindo	*Butia*	pear, prickly-	*Opuntia*

pear, wooden	*Xylomelum pyriforme*	pine	*Pinus*
pearl-bush	*Exochorda*	pine, Austrian	*Pinus nigra*
pearlwort	*Sagina*	pine, bristlecone	*Pinus aristata*
pearly everlasting	*Anaphalis margaritacea*	pine, jack	*Pinus banksiana*
		pine, loblolly	*Pinus taeda*
pebble plants	*Lithops*	pine, Scots	*Pinus sylvestris*
pecan	*Carya illinoensis*	pine, screw-	*Pandanus*
pelican flower	*Aristolochia grandiflora*	pine, stone	*Pinus cembra*; *Pinus pinea*
pennywort	*Umbilicus rupestris*	pineapple	*Ananas comosus*
		pineapple weed	*Matricaria matricarioides*
peony	*Paeonia*		
pepino	*Solanum muricatum*	pinguin	*Bromelia pinguin*
		pink	*Dianthus*
pepper	*Piper*	pink, cottage	*Dianthus caryophyllus*
pepper-and-salt	*Erigenia bulbosa*	pink, ground	*Phlox*
pepper, bird	*Capsicum annuum*	pink, Indian	*Spigelia marilandica*
pepper, Tabasco	*Capsicum frutescens*	pink, moss	*Phlox*
pepperbush, sweet	*Clethra*	pink, mullein	*Lychnis*
		pink, rose	*Sabatia*
peppergrass	*Lepidium sativum*	pink, sea	*Armeria*
pepperidge	*Nyssa*	pink, swamp	*Calopogon, Helnias*
peppermint	*Mentha piperita*		
pepper-root	*Dentaria*	pink-root	*Spigelia marilandica*
pepper-tree	*Schinus molle*		
pepper-vine	*Ampelopsis*	pink siris	*Albizia*
pepperwort	*Lepidium sativum*	pinxterflower	*Rhododendron nudiflorum*
periwinkle	*Vinca*		
periwinkle, Madagascar	*Catharanthus roseus*	pipsissewa	*Chimaphila*
persicaria	*Polygonum*	pistachio	*Pistacia*
persimmon	*Diospyros*	pitcherplant	*Nepenthes*; *Sarracenia*
Peruvian lily	*Alstroemeria*		
pheasant's-eye	*Adonis annua*	plane tree	*Platanus*
physic-nut	*Jatropha multifida*	plantain	*Plantago*
pickaback plant	*Tolmiea menziesii*	plantain banana	*Musa*
pickerel-weed	*Pontederia cordata*	plantain, false	*Heliconia*
picotee	*Dianthus caryophyllus*	plaintain, poor Robin's	*Hieracium venosum*
pigeonberry	*Duranta repens*; *Phytolacca americana*	plantain, rattlesnake	*Goodyera*
		plantain, water	*Alisma*
		plantain lily	*Hosta*
pigeon pea	*Cajanus cajan*	pleurisy-root	*Asclepias tuberosa*
pignut	*Carya; Conopodium majus*; *Omphalea triandra* (Jamaica)	plum	*Prunus domestica*
		plum, beach	*Prunus maritima*
		plum, hog-	*Spondias mombin*
		plum, myrobalan	*Prunus cerasifera*
		plum-yew	*Cephalotaxus*
pignut palm	*Hyophorbe*	plumbago	*Plumbago*
pimpernel, water	*Samolus*	plume-grass	*Erianthus*
pincushion flower	*Scabiosa*	plume poppy	*Bocconia*; *Macleaya*
pindo palm	*Butia*		

poison hemlock	*Conium maculatum*	poppy-mallow	*Callirhoe*
poison-ivy, -oak, -sumach	*Rhus toxicodendron*	portia-tree	*Thespesia populnea*
pokeberry; pokeweed	*Phytolacca americana*	Portugal laurel	*Prunus lusitanica*
polyanthus	*Primula polyantha*	postman, running	*Kennedia prostrata*
polypody	*Polypodium*	pot, marigold	*Calendula officinalis*
pomegranate	*Punica granatum*		
pond cypress	*Taxodium ascendens*	potato	*Solanum tuberosom*
pond-lily, yellow	*Nuphar*	potato-bean	*Apios americana*
pondweed	*Potamogeton*	potato, Irish	*Solanum tuberosum*
pondweed, Cape	*Aponogeton*	potato onion	*Allium cepa aggregatum*
poor man's weatherglass	*Anagallis arvensis*	potato, sweet	*Ipomoea batatas*
poor Robin's plantain	*Hieracium venosum*	powder-puff bush	*Calliandra haematocephala*
poplar; popple	*Populus*	prairie-button	*Liatris*
poplar, balm-of-Gilead	*Populus gileadensis*	prairie-clover	*Petalostemon purpureum*
poplar, balsam	*Populus balsamifera*	prairie-gentian	*Eustoma*
		prairie-lily	*Cooperia; Mentzelia*
poplar, black	*Populus nigra*		
poplar, grey	*Populus canescens*	prickly ash	*Zanthoxylum*
poplar, Lombardy	*Populus nigra italica*	prickly cucumber	*Echinocystis lobata*
poplar, white	*Populus alba*	prickly date-palm	*Acanthophoenix*
poppy	*Papaver*		
poppy, blue	*Meconopsis*	prickly pear	*Opuntia*
poppy, bush-	*Dendromecon rigidum*	prickly-poppy	*Argemone*
		prickly thrift	*Acantholimon*
poppy, California	*Eschscholzia*	primrose	*Primula*
poppy, celandine-	*Stylophorum diphyllum*	primrose, Cape-	*Streptocarpus*
		primrose, evening	*Oenothera*
poppy, corn	*Papaver rhoeas*	primrose willow	*Ludwigia*
poppy, crested	*Argemone*	Prince Albert's yew	*Saxegothaea conspicua*
poppy, Himalayan	*Meconopsis*	prince's feather	*Amaranthus hybridus hypo-chondriacus; Polygonum orientale*
poppy, horned	*Glaucium*		
poppy, Iceland	*Papaver nudicaule*		
poppy, Indian	*Meconopsis*		
poppy, opium	*Papaver somniferum*	prince's pine	*Chimaphila*
poppy, oriental	*Papaver orientale*	privet	*Ligustrum*
poppy, plume	*Bocconia; Macleaya cordata*	prophet-flower	*Arnebia; Aipyanthus echioides*
poppy, prickly	*Argemone*	puccoon	*Lithospermum*
poppy, sea	*Glaucium*	pulse	*Lathyrus*
poppy, snow	*Eomecon chionantha*	pummelo	*Citrus grandis*
		pumpkin	*Cucurbita maxima*
poppy, water	*Hydrocleys nymphoides*	purple loosestrife	*Lythrum salicaria*
poppy, Welsh	*Meconopsis cambrica*	purple sage	*Salvia officinalis purpurea*

purple wreath	*Petrea*	rampion, horned	*Phyteuma*
purslane	*Portulaca oleracea*	ramsons	*Allium ursinum*
purslane, tree	*Atriplex halimus*	Rangoon creeper	*Quisqualis indica*
pussy's toes	*Antennaria*	raspberry	*Rubus idaeus*
puttyroot	*Aplectrum*	raspberry,	*Rubus*
	hyemale	strawberry	*illecebrosus*
pyrethrum,	*Chrysanthemum*	rat-tail cactus	*Aporocactus*
common	*coccineum*		*flagelliformis*
pyrethrum,	*Chrysanthemum*	rattan	*Calamus*
Dalmatian	*cinerariifolium*	rattan-vine	*Berchemia*
pyrola, one-	*Moneses uniflora*	rattle-box	*Crotalaria*
flowered		rattlesnake-fern	*Botrychium*
pyxie	*Pyxidanthera*		*virginianum*
	barbulata	rattlesnake-	*Agave virginica;*
		master	*Eryngium yucci-*
			folium; E.
			aquaticum
Q		rattlesnake-orchid	*Pholidota*
Quaker ladies	*Houstonia*	rattlesnake-	*Goodyera*
quaking grass	*Briza*	plantain	
Queen Anne's	*Anthriscus sylves-*	rattlesnake-root	*Chamaelirium*
lace	*tris* (England);		*luteum;*
	Daucus carota		*Prenanthes*
	(U.S.A.)	rattlesnake-weed	*Hieracium*
queen-of-the-	*Filipendula rubra*		*venosum*
prairie		redbud	*Cercis canadensis*
queen palm	*Arecastrum*	red Cape tulip	*Haemanthus*
	romanzoffianum	red cedar	*Juniperus*
Queensland	*Acacia*	red-hot poker	*Kniphofia*
wattle	*podalyriifolia*	red mulberry	*Morus rubra*
Queen Victoria	*Victoria amazonica*	red oak	*Quercus borealis*
waterlily		redroot	*Ceanothus*
quillwort	*Isoetes*	red valerian	*Centranthus ruber*
quince	*Cydonia oblonga*	redwood	*Sequoia*
quince, Japanese	*Chaenomeles*		*sempervirens*
		reed, common	*Phragmites*
			australis
R		reed-mace	*Typha*
radish	*Raphanus sativus*	rein orchis	*Habenaria*
radish, horse-	*Armoracia*	rest-harrow	*Ononis*
	rusticana	resurrection plant	*Anastatica*
radish, wild	*Raphanus*		*hierochuntica*
	raphanistrum	rheumatism root	*Jeffersonia diphylla*
raffia	*Raphia*	rhododendron	*Rhododendron*
ragged Robin	*Lychnis flos-cuculi*	rhodora	*Rhododendron*
ragweed	*Ambrosia*		*canadense*
ragwort	*Senecio jacobaea*	rhubarb	*Rheum*
ragwort, African	*Othonna*	ribbon-grass	*Phalaris*
ragwort, Oxford	*Senecio squalidus*	ribbon wood	*Plagianthus*
rakkyo	*Allium chinense*	rice	*Oryza*
rain-tree	*Samanea saman*	rice, Canadian	*Zizania aquatica*
ramie	*Boehmeria nivea*	wild	
rampion	*Campanula*	rice-flower	*Pimelea*
	rapunculus	rice paper tree	*Tetrapanax*

rock-cress	*Arabis*	rubber tree	*Hevea*
rock-cress, purple	*Aubrieta*	rubber vine	*Cryptostegia*
rocket, salad	*Eruca sativa*	rue	*Ruta graveolens*
rocket, sweet	*Hesperis*	rue-anemone	*Anemonella*
rocket, yellow	*Barbarea*		*thalictroides*
rockfoil	*Saxifraga*	running postman	*Kennedia prostrata*
rock jasmine	*Androsace*	rupturewort	*Herniaria*
rock-rose	*Cistus;*	rush	*Juncus*
	Helianthemum	rush fern	*Schizaea*
rogation flower	*Polygala*	rush, pool	*Scirpus*
rose	*Rosa*	Russian comfrey	*Symphytum*
rose acacia	*Robinia hispida*		*uplandicum*
rose, Austrian	*Rosa foetida bicolor*	Russian olive	*Elaeagnus*
copper		Russian sage	*Perovskia*
rose, burnet	*Rosa spinosissima*		*atriplicifolia*
rose, cabbage	*Rosa centifolia*	Russian thistle	*Salsola*
rose, Cherokee	*Rosa laevigata*	Russian vine	*Polygonum*
rose, cinnamon	*Rosa cinnamomea*		*baldschuanicum*
rose, confederate	*Hibiscus mutabilis*	Russian worm-	*Artemisia*
rosebay	*Nerium oleander;*	wood	*sacrorum*
	Rhododendron	rutabaga	*Brassica*
	maximum		*napobrassica*
rosebay	*Epilobium*	rye, wild	*Elymus*
willowherb	*angustifolium*	rye-grass	*Lolium; Secale*
rose, damask	*Rosa damascena*		
rose, dog	*Rosa canina*		
rose, guelder	*Viburnum*		
rose, Macartney	*Rosa bracteata*		**S**
rose, musk	*Rosa moschata*		
rose, Scotch	*Rosa spinosissima*	safflower	*Carthamus*
rose-campion	*Lychnis coronaria*		*tinctorius*
roselle	*Hibiscus sabdariffa*	saffron	*Crocus sativus*
rose-mallow	*Hibiscus*	saffron, false	*Carthamus*
	moscheutos		*tinctorius*
rosemary	*Rosmarinus*	saffron, meadow	*Colchicum*
	officinalis	sage	*Salvia*
rosemary, marsh-	*Limonium*	sage, Bethlehem	*Pulmonaria*
rosemary, wild	*Ledum*	sage, Jerusalem	*Phlomis fruticosa*
rose-of-China	*Hibiscus rosa-*	sago palm	*Cycas revoluta*
	sinensis	saguaro	*Carnegiea gigantea*
rose-of-Sharon	*Hibiscus rosa-*	sainfoin, saintfoin	*Onobrychis*
	sinensis; Hyperi-		*viciifolia*
	cum calycinum	Saint Agnes'	*Leucojum*
	(Britain)	flower	
rose-pink	*Sabatia*	Saint Andrew's	*Ascyrum*
rose-root	*Sedum rosea*	cross	
rosinweed	*Silphium*	Saint Barbara's	*Barbarea*
rouge plant	*Rivina humilis*	herb	
rowan	*Sorbus aucuparia*	Saint Barnaby's	*Centaurea*
royal fern	*Osmunda regalis*	thistle	*solstitialis*
royal poinciana	*Delonix regia*	Saint Bernard's	*Anthericum*
royal waterlily	*Victoria*	lily	*liliago*
	amazonica	Saint Bruno's lily	*Paradisea*
rubber plant	*Ficus elastica*		*liliastrum*

Saint Dabeoc's heath	*Daboëcia cantabrica*	satinwood tree	*Murraya*
Saint George's herb	*Valeriana*	sausage-tree	*Kigelia pinnata*
		savin	*Juniperus*
Saint John's bread	*Ceratonia siliqua*	savory	*Satureja*
		savory, summer	*Satureja hortensis*
Saint John's wort	*Hypericum*	savory, winter	*Satureja montana*
Saint Martin's flower	*Alstroemeria*	savoy	*Brassica olermacea sabauda*
Saint Mary's wood	*Calophyllum*	saw cabbage palm	*Acoelorraphe*
		saw-palmetto	*Serenoa repens*
Saint Patrick's cabbage	*Saxifraga umbrosa*	saw-wort	*Serratula*
		saxifrage	*Saxifraga*
Saint Peter's wort	*Ascyrum*; *Hypericum*; *Primula veris*	scabious	*Scabiosa*
		scallion	*Allium cepa*
		scammony	*Convolvulus scammonia*
salad chervil	*Anthriscus*	Scarborough-lily	*Vallota speciosa*
sallow, common	*Salix cinerea*	scarlet oak	*Quercus coccinea*
sallow, great	*Salix caprea*	scarlet runner	*Phaseolus coccineus*
salsify	*Tragopogon porrifolius*	sclarea	*Salvia*
salsify, black	*Scorzonera hispanica*	scolymus	*Scolymus hispanicus*
salt-bush	*Atriplex*; *Rhagodia* (Australia); *Enchylaena* (Australia)	Scotch rose	*Rosa spinosissima*
		Scotch thistle	*Onopordum acanthium*
salt tree	*Halimodendron*	Scots pine	*Pinus sylvestris*
saltwort	*Salsola*	screw pine	*Pandanus*
samphire, marsh	*Salicornia*	scurfy pea	*Psoralea*
samphire, rock	*Crithmum maritimum*	scurvy grass	*Cochlearia*
		scythian lamb	*Cibotium*
sandalwood	*Santalum*	sea-buckthorn	*Hippophaë rhamnoides*
sandalwood, red	*Adenanthera pavonina*	sea campion	*Silene maritima*
sand-box tree	*Hura crepitans*	sea daffodil	*Pancratium maritimum*
sand lily	*Leucocrinum*		
sand-myrtle	*Leiophyllum*	sea-heath	*Frankenia*
sand-verbena	*Abronia*	sea-holly	*Eryngium maritimum*
sandwort	*Arenaria*		
sanicle, bear's-ear	*Cortusa matthioli*	seakale	*Crambe maritima*
sapodilla	*Manilkara zapota* (*Achras zapota*)	sea-lavender	*Limonium*
		sea-pink	*Armeria*
sapote	*Pouteria sapota* (*Calocarpum sapota*)	sea poppy	*Glaucium*
		sea-purslane	*Atriplex*
		sea-ragwort	*Cineraria*
sarsaparilla	*Aralia*	sea rocket	*Cakile*
sarsaparilla plant	*Smilax*	seaside grape	*Coccoloba*
sassafras	*Sassafras albidum*	sea-urchin cactus	*Echinopsis*
sassafras, southern	*Atherosperma moschatum* (Australia)	sedge	*Carex*
		sedge, umbrella-	*Cyperus*
		seersucker plant	*Geogenanthus*
satin-flower	*Lunaria*; *Sisyrinchium*	self-heal	*Prunella vulgaris*
		senna, bladder	*Colutea*

383

senna, wild	*Cassia marilandica*	silver birch	*Betula pendula*
sensitive fern	*Onoclea sensibilis*	silver fern	*Pityrogramma*
sensitive-plant	*Mimosa pudica*	silver morning-	*Argyreia speciosa*
service-berry	*Amelanchier*	glory	
service tree	*Sorbus domestica*	silver-rod	*Solidago;*
sesame	*Sesamum indicum*		*Asphodeline*
setwall	*Valeriana*	silver tree	*Leucadendron*
shadblow;	*Amelanchier*	silverweed	*Argyreia speciosa*
shadbush			(tropics);
shaddock	*Citrus grandis*		*Potentilla*
shagbark hickory	*Carya ovata*		*anserina*
shallot	*Allium*	simpler's joy	*Verbena*
	cepa aggregatum	sisal	*Agave sisalana*
shamrock	*Oxalis acetosella*	skirret, skirwort	*Sium sisarum*
	(England);	skullcap	*Scutellaria*
	Trifolium dubium	skunk-cabbage	*Symplocarpus*
	(Ireland);		*foetidus*
	Trifolium repens	slipper-flower	*Pedilanthus*
	(Ireland, U.S.A.)	slipper-wort	*Calceolaria*
sheep-laurel	*Kalmia*	slippery elm	*Ulmus fulva*
sheep's-bit	*Jasione*	sloe-tree	*Prunus spinosa*
scibious		smallage	*Apium graveolens*
shell-flower	*Molucella*	smartweed	*Polygonum*
she-oak	*Casuarina*	smilax, florist's	*Asparagus*
	equisetifolia		*asparagoides*
shepherd's beard	*Andropogon*	smoke bush	*Comesperma*
shepherd's club	*Verbascum*	smoke-tree	*Cotinus coggygria*
shepherd's purse	*Capsella bursa-*	snail-flower	*Phaseolus*
	pastoris		*caracalla*
shield fern	*Dryopteris;*	snail-seed	*Cocculus carolinus*
	Polystichum	snakegourd	*Trichosanthes*
shield-wort	*Peltaria*		*anguina*
shingle-plant	*Monstera delicosa*	snake plant	*Sansevieria*
shinleaf	*Pyrola*	snakeroot	*Aristolochia;*
shoe-black	*Hibiscus rosa-*		*Cimicifuga;*
	sinensis		*Eupatorium;*
shooting star	*Dodecatheon*		*Liatris*
shower, golden	*Cassia fistula*	snake's head	*Fritillaria*
shrimp plant	*Drejerella guttata*	fritillary	*meleagris*
	(*Justicia*	snake's-head iris	*Hermodactylus*
	brandegeana)		*tuberosus*
side-saddle flower	*Sarracenia*	snake's mouth	*Pogonia*
silk-cotton tree	*Bombax ceiba*	orchid	
	(*Salmalia*	snake's-tongue	*Ophioglossum*
	malabarica)	snapdragon	*Antirrhinum*
silk-cotton tree,	*Cochlospermum*	snapweed	*Impatiens*
yellow	*religiosum*	sneezeweed	*Helenium*
silk-grass	*Yucca filamentosa*	sneezewort	*Achillea*
silk-tassel bush	*Garrya elliptica*	snowball tree	*Viburnum macro-*
silk-tree	*Albizia*		*cephalum sterile;*
silk vine	*Periploca graeca*		*Viburnum plica-*
silver bell tree	*Halesia*		*tum plicatum*
silver-berry	*Elaeagnus*		
	commutata	snowbell	*Styrax*

snowberry	*Chiococca; Symphoricarpos*	Spanish oyster- plant	*Scolymus hispanicus*
snowberry, creeping	*Chiogenes hispidula*	spatter-dock	*Nuphar*
snowdrop	*Galanthus*	spearflower	*Ardisia*
snowdrop-tree	*Halesia*	spear-grass	*Stipa*
snowflake	*Leucojum*	spear lily	*Doryanthes*
snow-glory	*Chionodoxa*	spearmint	*Mentha spicata*
snow-in-summer	*Cerastium tomentosum*	spearwort, great	*Ranunculus lingua*
snow-on-the mountain	*Euphorbia marginata*	spearwort, lesser	*Ranunculus flammula*
snow poppy	*Eomecon chionantha*	speedwell	*Veronica*
		speedwell, Buxbaum's	*Veronica persica*
snow wreath	*Neviusia alabamensis*	speedwell, Derwent	*Veronica derwentiana*
snowy mespilus	*Amelanchier*	speedwell, germander	*Veronica chamaedrys*
soap-bark tree	*Quillaja saponaria*		
soap-berry	*Sapindus saponaria; Shepherdia canadensis*	spice-bush	*Lindera benzoin*
		spider-flower	*Cleome*
		spider-flower, Brazilian	*Tibouchina*
soap-bush	*Noltea africana*	spider lily	*Hymenocallis*
soapwort	*Saponaria officinalis*	spider lily, golden	*Lycoris aurea*
		spiderwort	*Tradescantia*
Solomon's seal	*Polygonatum*	spike heath	*Bruckenthalia spiculifolia*
Solomon's seal, false	*Smilacina*		
		spike rush	*Eleocharis*
sorghum	*Sorghum*	spikemoss	*Selaginella*
sorrel, garden	*Rumex acetosa*	spikenard, American	*Aralia racemosa*
sorrel, Jamaica	*Hibiscus sabdariffa*		
sorrel, sheep's	*Rumex acetosella*	spikenard, false	*Smilacina*
sorrel, wood	*Oxalis acetosella*	spinach	*Spinacia oleracea*
sorrel-tree	*Oxydendrum arboreum*	spinach dock	*Rumex patientia*
		spinach, New Zealand	*Tetragonia tetragonioides*
sour gum	*Nyssa*	spindle-tree	*Euonymus*
soursop	*Annona muricata*	spine-areca	*Acanthophoenix*
sour-wood	*Oxydendrum arboreum*	spiral-flag	*Costus*
		spire lily	*Galtonia candicans*
southern beech	*Nothofagus*	spirea	*Spiraea*
southern cane	*Arundinaria*	spirea, blue	*Caryopteris*
southernwood	*Artemisia abrotanum*	spirea, false	*Sorbaria*
		spleenwort	*Asplenium*
sowbread	*Cyclamen*	spleenwort, swamp	*Athyrium*
sowthistle	*Sonchus*		
soybean	*Glycine max*	spoonwood	*Kalmia latifolia*
Spanish bayonet	*Yucca*	spring-beauty	*Claytonia*
Spanish bluebell	*Endymion hispanicus*	spruce	*Picea*
		spruce, black	*Picea mariana*
Spanish broom	*Spartium junceum*	spruce, Norway	*Picea abies*
Spanish elm	*Cordia*	spruce, red	*Picea rubens*
Spanish iris	*Iris xiphioides*	spruce, Sitka	*Picea sitchensis*
Spanish moss	*Tillandsia usneoides*	spruce, white	*Picea glauca*
		spurge	*Euphorbia*

spurge, caper	*Euphorbia lathyris*
spurge, flowering	*Euphorbia corollata*
spurge, wood	*Euphorbia amygdaloides*
spurge-laurel	*Daphne laureola*
spurge-olive	*Cneorum*
squash	*Cucurbita*
squaw-berry	*Mitchella repens*
squaw-weed	*Senecio*
squill	*Scilla; Urginea*
squill, striped	*Pushkinia scilloides*
squirrel-corn	*Dicentra canadensis*
squirrel-tail grass	*Hordeum*
squirting cucumber	*Ecballium elaterium*
stagbush	*Viburnum prunifolium*
stagger-bush	*Lyonia mariana*
staghorn fern	*Platycerium*
staghorn sumach	*Rhus typhima*
star, golden	*Chrysogonum virginianum*
star-apple	*Chrysophyllum cainito*
star-flower	*Trientalis*
star-grass	*Aletris; Hypoxis*
star-of-Bethlehem	*Ornithogalum*
star thistle	*Centaurea calcitrapa*
star wattle	*Acacia verticillata*
starwort	*Aster; Stellaria*
statice	*Limonium*
steeple-bush	*Spiraea*
stick-tight	*Bidens*
stinging nettle	*Urtica*
stinkweed	*Pluchea*
stitchwort	*Stellaria*
stock	*Matthiola*
stock, Brompton	*Matthiola incana maritima*
stock, night-scented	*Matthiola longi-petala bicornis*
stock, Virginia	*Malcomia maritima*
Stokes' aster	*Stokesia laevis*
stone cress	*Aethionema*
stonecrop	*Sedum*

stonecrop, biting	*Sedum acre*
stone pine	*Pinus cembra* Pinus pinea*
stone-root	*Collinsonia*
storax-tree	*Styrax*
stork's-bill	*Pelargonium*
straw flower	*Helichrysum*
strawberry	*Fragaria*
stawberry, barren	*Waldsteinia*
strawberry, Indian or mock	*Duchesnea indica*
strawberry tree	*Arbutus*
striped squill	*Puschkinia scilloides*
stud flower	*Helonias bullata*
succory	*Cichorium intybus*
sugar-apple	*Annona squamosa*
sugarberry	*Celtis laevigata*
sugarcane	*Saccharum officinarum*
sugar maple	*Acer saccharum*
sumac	*Rhus*
sumac, poison	*Toxicodendron*
sumac, staghorn	*Rhus typhina*
summer-cypress	*Kochia scoparia culta*
summer savory	*Satureja hortensis*
sundew	*Drosera*
sundrops	*Oenothera*
sunflower	*Helianthus*
sunray	*Helipterum*
sunrose	*Helianthemum*
sunshine wattle	*Acacia botrycephala*
supple-Jack	*Paullinia*
Swainson pea	*Swainsona*
swamp bay	*Magnolia virginiana*
swamp cypress	*Taxodium*
swamp loosestrife	*Decodon verticillatus*
swamp pink	*Arethusa bulbosa; Calopogon; Helonias bullata*
swamp wattle	*Acacia elongata*
swampweed	*Selliera radicans*
Swan River daisy	*Brachycome iberidifolia*
swede turnip	*Brassica napobrassica*
sweet Alice	*Pimpinella anisum*
sweet alyssum	*Lobularia maritima*
sweet basil	*Ocimum basilicum*

sweet bay	*Laurus nobilis*; *Magnolia virginiana* (U.S.A.)	tapegrass	*Vallisneria*
		tapioca	*Manihot*
		tarata	*Pittosporum eugenioides*
sweet-box	*Sarcococca*		
sweet briar	*Rosa eglanteria*	tare	*Vicia*
sweet broom	*Scoparia dulcis*	taro-root	*Colocasia esculenta*
sweet cicely	*Myrrhis odorata* (Britain); *Osmorrhiza* (U.S.A.)		
		tarragon	*Artemisia dracunculus*
sweet clover	*Melilotus*	tarweed	*Madia*
sweet corn	*Zea mays*	tassel flower	*Emilia sagittata*
sweet fern	*Comptonia peregrina*	tawny-berry	*Ilex krugiana*
		tea	*Camellia sinensis*
sweet flag	*Acorus calamus*	tea-berry	*Gaultheria procumbens*
sweet gale	*Myrica gale*		
sweet gum	*Liquidambar styraciflua*	tea-tree	*Leptospermum* (Australia); *Lycium barbarum* (Britain)
sweet marjoram	*Majorana hortensis*		
sweet olive	*Osmanthus*	teak tree	*Tectona grandis*
sweet pea	*Lathyrus odoratus*	tear-thumb	*Polygonum arifolium*; *Polygonum sagittatum*
sweet pepperbush	*Clethra alnifolia*		
		teasel	*Dipsacus fullonum*
sweet potato	*Ipomoea batatas*	teonsinte	*Euchlaena mexicana*
sweet rocket	*Hesperis matronalis*		
sweetsop	*Annona squamosa*	thatch palm	*Thrinax*
sweet sultan	*Centaurea moschata*	thimbleweed	*Anemone*
		thistle	*Carduus*; *Cirsium*; *Cnicus*
sweet vernal grass	*Anthoxanthum odoratum*		
		thistle, blessed	*Cnicus benedictus*; *Silybum marianum*
sweet William	*Dianthus barbatus*		
sweet wormwood	*Artemisia annua*		
sword fern	*Nephrolepis*	thistle, cotton	*Onopordum acanthium*
sword-lily	*Gladiolus*		
sycamore	*Acer*; *Platanus*	thistle, creeping	*Cirsium arvense*
sycamore-maple	*Acer pseudoplatanus*	thistle, distaff	*Carthamus*
		thistle, globe	*Echinops*
Sydney golden wattle	*Acacia longifolia*	thistle, golden	*Scolymus*
		thistle, hedgehog	*Echinocactus*
		thistle, plumed	*Cirsium*
		thistle, plumeless	*Carduus*
	T	thistle, Russian	*Salsola*
		thistle, saffron	*Carthamus*
tacamahac	*Populus balsamifera*	thistle, Scotch	*Onopordum acanthium*
tamarack	*Larix laricina*	thistle, South African	*Berkheya*
tamarind	*Tamarindus indica*		
tamarisk	*Tamarix*	thistle, sow	*Sonchus*
tamarisk, false	*Myricaria*	thistle, star	*Centaurea calcitrapa*
tanbank	*Lithocarpus*		
tangerine	*Citrus reticulata*	thorn, Jerusalem	*Paliurus spina-christi*; *Parkinsonia aculeata*
tansy	*Tanacetum vulgare*		

387

thorn, kangaroo	*Acacia armata*	tree-mallow	*Lavatera arborea*
thorn, lily-	*Catesbaea spinosa*	tree-of-heaven	*Ailanthus*
thorn-apple	*Datura*		*altissima*
thoroughwort	*Eupatorium*	tree onion	*Allium cepa*
thunder-plant	*Sempervivum*		*proliferum*
thrift	*Armeria*	tree-poppy	*Dendromecon*
throatwort	*Trachelium*		*rigidum*
	caeruleum	tree-tomato	*Cyphomandra*
thyme	*Thymus*		*betacea*
thyme, common	*Thymus vulgaris*	trefoil	*Trifolium*
thyme, mother-	*Thymus drucei*;	trefoil, bird's foot	*Lotus corniculatus*
of-	*Thymus*	trout-lily	*Erythronium*
	serpyllum	true-love	*Paris quadrifolia*
tickseed	*Bidens; Coreopsis*	trumpet creeper	*Bignonia; Campsis*
tick-trefoil	*Desmodium*	trumpet flower	*Beaumontia;*
tidy-tips	*Layia elegans*		*Clytostoma;*
tiger-flower	*Tigridia*		*Datura*
tiger-lily	*Lilium lancifolium*	trumpet, herald's	*Beaumontia*
	(L. tigrinum)	trumpet tree	*Cecropia peltata*
tiger's chaps	*Faucaria*	tuberose	*Polianthes*
timothy	*Phleum pratense*		*tuberosa*
toadflax	*Linaria*	tulip	*Tulipa*
toadflax, ivy-	*Cymbalaria*	tulip, lady	*Tulipa clusiana*
leaved	*muralis*	tulip tree	*Liriodendron*
toad-lily	*Tricyrtis*		*tulipifera*
tobacco	*Nicotiana*	tupelo	*Nyssa*
toddy palm	*Caryota urens*	turkey's beard	*Xerophyllum*
tomatillo	*Physalis*	turmeric	*Curcuma*
	philadelphica	turnip	*Brassica rapa*
tomato	*Lycopersicon*	turnip, Indian	*Arisaema*
	esculentum	turnip, swede	*Brassica napo-*
tomato, cherry	*Lycopersicon*		*brassica*
	esculentum	turtlehead	*Chelone*
	ceraisiforme	turtle-vine	*Bauhinia*
tomato, currant	*Lycopersicon*		*cumanensis*
	pimpinellifolium	tutsan	*Hypericum*
tomato, tree-	*Cyphomandra*		*androsaemum*
	betacea	twayblade	*Liparis; Listera*
tooth-ache-tree	*Zanthoxylum*	twinflower	*Linnaea borealis*
	americanum	twin-leaf	*Jeffersonia diphylla*
toothwort	*Dentaria*		*(U.S.A.);*
torch-lily	*Kniphofia*		*Zygophyllum*
touch-me-not	*Impatiens*		*(Australia)*
	noli-tangere	twist-arum	*Helicodiceros*
toyon	*Heteromeles*		*muscivorus*
	arbutifolia	twisted stalk	*Streptopus*
trailing arbutus	*Epigaea repens*		
trailing cactus	*Acanthocereus*		
Transvaal daisy	*Gerbera*		**U**
traveller's joy	*Clematis*		
traveller's tree	*Ravenala*	udo	*Aralia cordata*
	madagascariensis	umbrella fern	*Gleichenia*
tree-aster	*Olearia*	umbrella palm	*Hedyscepe*
tree-fern	*Cyathea etc.*		*canterburyana*

umbrella pine	*Sciadopitys verticillata*
umbrella plant	*Cyperus alternifolius; Peltiphyllum peltatum*
umbrella tree	*Magnolia tripetala; Melia azedarach; Thespesia populnea*
unicorn plant	*Proboscidea louisianica*
upas tree	*Antiaris toxicaria*

V

valerian	*Valeriana*
valerian, African	*Fedia*
valerian, red	*Centranthus ruber*
vanilla	*Vanilla*
vanilla, Carolina	*Trilisa*
vanilla-grass	*Hierochloe odorata*
vanilla-leaf	*Achlys*
varnish tree	*Aleurites moluccana*
vegetable-oyster	*Tragopogon porrifolius*
vegetable-sponge	*Luffa*
velvet-bean	*Stizolobium*
Venus flytrap	*Dionaea muscipula*
Venus' hair	*Adiantum capillus-veneris*
Venus' looking-glass	*Specularia*
Venus' navelwort	*Omphalodes*
verbena, lemon	*Aloysia triphylla*
verbena, sand	*Abronia*
vervain	*Verbena officinalis*
vetch	*Vicia*
vetch, crown	*Coronilla varia*
vetch, kidney	*Anthyllis*
vetch, milk	*Astragalus*
vine	*Vitis*
vine, kangaroo	*Cissus antarctica*
vine, kudzu	*Pueraria thunbergiana*
vine, mignonette	*Boussingaultia gracilis*
vine, rattan-	*Berchemia*
vine, Russian	*Polygonum baldschuanicum*
vine, wonga	*Pandorea pandorana*
violet	*Viola*

violet, African	*Saintpaulia*
violet cress	*Ionopsidium acaule*
viper's bugloss	*Echium*
Virginia cowslip	*Mertensia virginica*
Virginia creeper	*Ampelopsis; Parthenocissus*
Virginia-stock	*Malcomia*
Virginian poke	*Phytolacca americana*
virgin's bower	*Clematis*
viscaria	*Lychnis*

W

wahoo	*Euonymus atropurpurea*
wake-robin	*Trillium*
wallcress	*Arabis*
wall fern	*Polypodium*
wallflower	*Cheiranthus cheiri*
wall-pennywort	*Umbilicus rupestris*
wall-pepper	*Sedum acre*
wall-rue	*Asplenium ruta-muraria*
walnut	*Juglans*
walnut, black	*Juglans nigra*
walnut, English	*Juglans regia*
walnut, white	*Juglans cinerea*
wandering-Jew	*Tradescantia; Zebrina*
wandflower	*Dierama; Sparaxis*
wand plant	*Galax aphylla*
warratah	*Telopea*
Washington plant	*Cabomba caroliniana*
water-arum	*Calla palustris*
water carpet	*Chrysosplenium*
water-chestnut	*Trapa natans*
water convolvulus	*Ipomoea aquatica*
water-cress	*Nasturtium; Rorippa*
water-elm	*Planera aquatica*
water-flag	*Iris pseudacorus*
water hemlock	*Cicuta virosa*
water hoarhound, horehound	*Lycopus*
water-hyacinth	*Eichhornia crassipes*
water-leaf	*Hydrophyllum*
water-lettuce	*Pistia stratiotes*
water-lily	*Nuphar; Nymphaea*

389

water-lily, yellow	*Nuphar*	weed, pineapple	*Matricaria*
watermelon	*Citrullus lanatus*		*matricarioides*
water-milfoil	*Myriophyllum*	weeping willow	*Salix babylonica*
water moss	*Fontinalis*	weld	*Reseda luteola*
water oak	*Quercus nigra*	wellingtonia	*Sequoiadendron*
water-pennywort	*Hydrocotyle*		*giganteum*
water-pimpernel	*Samolus*	Welsh onion	*Allium fistulosum*
water plantain	*Alisma*	Welsh poppy	*Meconopsis*
water poppy	*Hydrocleys*		*cambrica*
	nymphoides	West Indian kale	*Colocasia*
water-shield	*Brasenia schreberi*	wheat	*Triticum*
water-soldier	*Stratiotes aloides*	whin	*Genista; Ulex*
water starwort	*Callitriche*	white-alder	*Clethra*
water thyme	*Anacharis*	white-beam	*Sorbus aria*
water trumpet	*Cryptocoryne*	white byrony	*Bryonia dioica*
water violet	*Hottonia*	white campion	*Silene alba*
waterweed	*Anacharis*	white cedar,	*Thuja*
	canadensis	northern	*occidentalis*
waterwillow	*Justicia*	white dead-nettle	*Lamium album*
	americana	white mulberry	*Morus alba*
wattle	*Acacia*	white oak	*Quercus alba*
wattle, black	*Acacia decurrens*	white poplar	*Populus alba*
	mollis	white willow	*Salix alba*
wattle, blue	*Acacia cyanophylla*	whitlow-grass	*Draba*
wattle, cedar	*Acacia elata*	whitlow-wort	*Paronychia*
wattle,	*Acacia*	whorl-flower	*Morina*
Cootamunda	*baileyana*	willow	*Salix*
wattle, golden	*Acacia pycnantha*	willow, basket	*Salix viminalis*
wattle, green	*Acacia decurrens*	willow, bay	*Salix pentandra*
wattle, hedge	*Acacia armata*	willow, black	*Salix nigra*
wattle, mudgee	*Acacia spectabilis*	willow, crack	*Salix fragilis*
wattle,	*Acacia*	willow, creeping	*Salix repens*
Queensland	*podalyriifolia*	willow, cricket-	*Salix alba*
wattle, star	*Acacia verticillata*	bat	*caerulea*
wattle, sunshine	*Acacia*	willow, desert	*Chilopsis linearis*
	botrycephala	willow, goat	*Salix caprea*
wattle, swamp	*Acacia elongata*	willow moss	*Fontinalis*
wattle, Sydney	*Acacia longifolia*	willow, weeping	*Salix babylonica*
golden		willow, white	*Salix alba*
wattle, winged	*Acacia alata*	willow-herb	*Epilobium*
wax-flower	*Hoya carnosa;*	widow tears	*Tradescantia*
	Stephanotis	wild cinnamon	*Canella*
	floribunda	wild rice	*Zizania aquatica*
wax gourd	*Benincasa hispida*	wind-flower	*Anemone*
wax-myrtle	*Myrica cerifera*	wing-nut	*Pterocarya*
wax-palm	*Ceroxylon*	winged	*Ammobium alatum*
wax-tree	*Rhus succedanea*	everlasting	
waxweed	*Cuphea*	winged wattle	*Acacia alata*
waxwork	*Celastrus scandens*	winter aconite	*Eranthis*
wayfaring tree	*Viburnum*	winterberry	*Ilex verticillata*
weavers' broom	*Spartium junceum*	winter daffodil	*Sternbergia lutea*
wedding bush	*Ricinocarpos*	wintergreen	*Gaultheria*
wedge-pea	*Gompholobium*		*procumbens;*
			Pyrola

winter heliotrope	*Petasites fragrans*
wintersweet	*Chimonanthus praecox; Acokanthera spectabilis* (tropics)
wire-plant	*Muehlenbeckia complexa*
witch-grass	*Agropyron repens*
witch-hazel	*Hamamelis*
witch-hobble	*Viburnum alnifolium*
withe-rod	*Viburnum cassinoides*
woad, dyer's	*Isatis tinctoria*
woad, waxen	*Genista*
woad, wild	*Reseda luteola*
wolfberry	*Symphoricarpus*
wolfchop	*Faucaria*
wolf's bane	*Aconitum lycoctonum*
wolf's-milk	*Euphorbia*
woman's-tongue-tree	*Albizia lebbek*
wonga vine	*Pandorea pandorana*
wood-apple	*Feronia limonia*
wood-betony	*Pedicularis*
woodbine	*Lonicera; Ampelopsis; Parthenocissus*
wood fern	*Dryopteris*
wood-rose, wooden rose	*Merremia tuberosa*
woodruff	*Asperula odorata (Galium odoratum)*
wood-sorrel	*Oxalis acetosella*
wood spurge	*Euphorbia amygdaloides*

woody nightshade	*Solanum dulcamara*
wormwood	*Artemisia*
wormwood, common	*Artemisia absinthium*
wormwood, Russian	*Artemisia sacrorum*
wormwood, sweet	*Artemisia annua*

Y

yam	*Dioscorea*
yam bean	*Pachyrhizus erosus*
yarrow	*Achillea millefolium*
yellow birch	*Betula lutea*
yellow ox-eye	*Buphthalmum*
yellow-wood, Kentucky	*Cladastris lutea*
yellowwort	*Blackstonia perfoliata*
yerna buena	*Micromeria*
yew	*Taxus*
yew, plum	*Cephalotaxus*
Yorkshire fog	*Holcus lanatus*
Youth-and-old-age	*Zinnia elegans*
youth-on-age	*Tolmiea menziesii*
yulan	*Magnolia denudata*

Z

zephyr-lily	*Zephyranthes*